TEXTBOOKS in MATHEM

T0077272

Exploring Geometry

Second Edition

Michael Hvidsten

Gustavus Adolphus College
St. Peter, Minnesota, USA

 **CRC Press**
Taylor & Francis Group
Boca Raton London New York

CRC Press is an imprint of the
Taylor & Francis Group, an **informa** business
A CHAPMAN & HALL BOOK

TEXTBOOKS in MATHEMATICS

Series Editors: Al Boggess and Ken Rosen

PUBLISHED TITLES CONTINUED

EXPLORING LINEAR ALGEBRA: LABS AND PROJECTS WITH MATHEMATICA®
Crista Arangala

GRAPHS & DIGRAPHS, SIXTH EDITION
Gary Chartrand, Linda Lesniak, and Ping Zhang

INTRODUCTION TO ABSTRACT ALGEBRA, SECOND EDITION
Jonathan D. H. Smith

INTRODUCTION TO MATHEMATICAL PROOFS: A TRANSITION TO ADVANCED MATHEMATICS, SECOND EDITION
Charles E. Roberts, Jr.

INTRODUCTION TO NUMBER THEORY, SECOND EDITION
Marty Erickson, Anthony Vazzana, and David Garth

LINEAR ALGEBRA, GEOMETRY AND TRANSFORMATION
Bruce Solomon

MATHEMATICAL MODELLING WITH CASE STUDIES: USING MAPLE™ AND MATLAB®, THIRD EDITION
B. Barnes and G. R. Fulford

MATHEMATICS IN GAMES, SPORTS, AND GAMBLING—THE GAMES PEOPLE PLAY, SECOND EDITION
Ronald J. Gould

THE MATHEMATICS OF GAMES: AN INTRODUCTION TO PROBABILITY
David G. Taylor

A MATLAB® COMPANION TO COMPLEX VARIABLES
A. David Wunsch

MEASURE THEORY AND FINE PROPERTIES OF FUNCTIONS, REVISED EDITION
Lawrence C. Evans and Ronald F. Gariepy

NUMERICAL ANALYSIS FOR ENGINEERS: METHODS AND APPLICATIONS, SECOND EDITION
Bilal Ayyub and Richard H. McCuen

ORDINARY DIFFERENTIAL EQUATIONS: AN INTRODUCTION TO THE FUNDAMENTALS
Kenneth B. Howell

RISK ANALYSIS IN ENGINEERING AND ECONOMICS, SECOND EDITION
Bilal M. Ayyub

SPORTS MATH: AN INTRODUCTORY COURSE IN THE MATHEMATICS OF SPORTS SCIENCE AND SPORTS ANALYTICS
Roland B. Minton

TRANSFORMATIONAL PLANE GEOMETRY
Ronald N. Umble and Zhigang Han

CRC Press
Taylor & Francis Group
6000 Broken Sound Parkway NW, Suite 300
Boca Raton, FL 33487-2742

First issued in paperback 2022

© 2017 by Taylor & Francis Group, LLC
CRC Press is an imprint of Taylor & Francis Group, an Informa business

No claim to original U.S. Government works

ISBN 13: 978-1-03-247706-0 (pbk)
ISBN 13: 978-1-4987-6080-5 (hbk)

DOI: 10.1201/9781315367644

Library of Congress Cataloging-in-Publication Data

Names: Hvidsten, Michael. | Hvidsten, Michael. Geometry with Geometry Explorer.
Title: Exploring geometry / Michael Hvidsten.
Other titles: Geometry with Geometry Explorer
Description: Second edition. | Boca Raton : CRC Press, 2017. | Series: Textbooks in mathematics series | Previous edtion: Geometry with Geometry Explorer / Michael Hvidsten (Dubuque, Iowa : McGraw-Hill, 2005).
Identifiers: LCCN 2016027443 | ISBN 9781498760805 (hardback : alk. paper)
Subjects: LCSH: Geometry--Textbooks.
Classification: LCC QA445 .H84 2017 | DDC 516--dc23
LC record available at https://lccn.loc.gov/2016027443

Visit the Taylor & Francis Web site at
http://www.taylorandfrancis.com

and the CRC Press Web site at
http://www.crcpress.com

Contents

Preface

It may well be doubted whether, in all the range of science, there is any field so fascinating to the explorer, so rich in hidden treasures, so fruitful in delightful surprises, as Pure Mathematics. – Lewis Carroll (Charles Dodgson), 1832–1898

Geometry is one of the most ancient subjects within mathematics. It is also one of the most vibrant and modern. One can simply walk down any street and see geometry come to life. The shapes that one sees in buildings, traffic signs, billboards, and paving stones are all based on some fundamental geometric design.

Geometry is also key to the development of computational "virtualities" that are now ubiquitous. These include the three-dimensional computer graphics animations used in television and film, virtual-reality displays, robotics, and the layout and design of architectural structures.

Geometry is a rich and rewarding area for mathematical exploration. The visual aspects of the subject make exploration and experimentation natural and intuitive. At the same time, the abstractions developed to explain geometric patterns and connections make the subject extremely powerful and applicable to a wide variety of physical situations.

The interplay between the practical and the abstract is one of the main features of this text. For students new to geometry, there are numerous concrete exploratory activities to aid in the understanding of geometric concepts. For students who have prior experience in geometry, the text provides detailed proofs and explanations and offers advanced topics such as elliptic and projective geometry. In general, the pedagogical approach used in this text is to give equal weight to the intuitive and imaginative exploration of geometry as well as to abstract reasoning and proofs.

DISCOVERY LEARNING

> Tell me and I forget.
> Teach me and I remember.
> Involve me and I learn. – Benjamin Franklin, 1706–1790

Ben Franklin's approach to teaching finds its current expression in the pedagogy of discovery learning. To truly learn, a student must be actively involved. In our modern era of instant communication it is a challenge to deeply engage students, especially in mathematical subjects such as geometry.

This textbook has been designed to promote student engagement with the beautiful ideas of geometry. Every major concept is introduced in its historical context and is explained in a way that connects the idea with some real-life basis. For example, in the first chapter on axiomatics, an analogy is made with playing a game. In a game it is important to have a solid understanding of the rules of play, so that the game can function correctly. This is the role of axiomatics in a mathematical system.

A system of experimentation followed by rigorous explanation and proof is central to the pedagogical approach of the text. Students conduct guided computer explorations of important foundational topics, followed by proofs of results they have observed. While many texts offer computer projects as part of homework assignments, or as supplementary activities, exploratory projects play an integral role in this text. They are the vehicle whereby students become actively engaged in learning. They help students develop a better sense of how to prove a result. The goal of this pedagogical approach is to enable students to master concepts at a deep level.

CHANGES IN SECOND EDITION

1. **Elliptic and Projective geometry** New chapters have been added on elliptic and projective geometry. The text now gives complete coverage of the three possible geometries (Euclidean, Hyperbolic, and Elliptic) based on variations of Euclid's fifth postulate. The new chapter on projective geometry provides a "universal" geometry that builds on the commonalities of these three geometries. The discovery learning approach is used extensively in both of these new chapters.

2. **Proofs** Supplemental material on proofs has been added as an

appendix. This helps students identify types of proofs and helps them learn strategies to use in writing proofs. There is also an expanded discussion of proofs in the first chapter.

3. **Computer Software** Almost all the computer projects are written in a neutral way so that any dynamic geometry software can be used for the projects. Supplementary material is available on the author's website for the support of various geometry software platforms, such as Geometer's Sketchpad and GeoGebra.

4. **Instructor's Manual** An instructor's manual, complete with worked out solutions for the homework and the projects, is available from the publisher.

5. **Student's manual** An on-line student manual is available with additional support for students. The manual also includes the solutions to all odd-numbered problems (except for project exercises).

6. **Consistency of Level** All chapters have been carefully reviewed for consistency of level. Material that is of a more advanced nature is now available separately as web chapters on the author's website at http://www.gac.edu/~hvidsten/geom-text.

COMPUTER PROJECTS

To encourage a playful appreciation of geometric ideas, we have incorporated many computer explorations in the text. These explorations can be accomplished using geometry software packages such as *Geometry Explorer, GeoGebra, Geometer's Skecthpad*, etc. These programs provide a virtual geometry laboratory where students can create geometric objects (like points, circles, polygons, areas, etc.), carry out transformations on these objects (dilations, reflections, rotations, and translations), and measure aspects of these objects (like length, area, radius, etc.). As such, it is much like doing geometry on paper (or sand) with a ruler and compass. However, on paper such constructions are static—points placed on the paper cannot be moved. When using a geometry program, all constructions are *dynamic*. One can draw a segment and then grab one of the endpoints and move it around the canvas, with the segment moving accordingly. Thus, one can construct a geometric figure and test out hypotheses by experimentation with the construction.

The goal of the computer projects in each chapter is to have students *actively* explore geometry through a three-fold approach. Students first see a topic introduced in the text. Then, they explore that topic using available geometry software or by means of in-class group projects. Finally, they review and report on their exploration, discussing what was discovered, conjectured, and proved during the course of the project.

The beginning of each project is designated by a special heading— the project title set between two horizontal lines. The conclusion of each project is designated by an ending horizontal line. Projects are illustrated with screen shots from the *Geometry Explorer* program. This is the recommended software for this text. *Geometry Explorer* can be downloaded from the author's website at http://www.gac.edu/~hvidsten/gex. Projects are designed to be general in scope, so that other, commonly used geometry software can also be used for exploration. Specific instructions for carrying out each project using different geometry programs can be found at the author's website http://www.gac.edu/~hvidsten/geom-text.

PROOFS AND FOUNDATIONS

The development of intuition is a critical first step in understanding geometric concepts. However, intuition alone cannot provide the basis for precise calculation and analysis of geometric quantities. For example, one may know experimentally that the sides of a right triangle follow the Pythagorean Theorem, but data alone do not show *why* this result is true. Only a logical proof of a result will give us confidence in using it in any given situation.

Throughout this text there is a focus on intuition/experimentation as a means to gain insight on explanations and proofs. This integration of exploration and explanation can be seen most clearly in the way computer projects are designed to help students develop abstract reasoning. For example, the first computer project explores the golden ratio and its amazing and ubiquitous properties. Students not only experimentally discover the properties of the golden ratio, but are then asked to dig deeper and analyze why these properties are true.

To aid students in the art of proof writing, there is a brief summary of proof techniques in Appendix A. Additionally, each chapter has a wide variety of completely worked-out proofs of major theorems. One unique feature of this text, in comparison to many other geometry texts, is the careful consideration of the foundations of mathematical reasoning found in Chapter 1. The chapter starts with a broad discussion of the historical development of the axiomatic method followed by a study of topics like completeness and independence. In keeping with the discovery learning focus of the text, the axiomatic foundation is accompanied by several concrete computer explorations.

A review of the main topics of Euclidean geometry in Chapter 2 follows this discussion of axiomatics. This review is prefaced with a brief discussion of axiomatic systems such as Hilbert's axioms and Birkhoff's axioms. However, a rigorous analysis of these axioms is put off so that students can fully explore more advanced notions of Euclidean geometry. A similar approach is taken in the chapters on constructions, transformations, hyperbolic geometry, and elliptic geometry.

A carefully developed consideration of geometric principles, based on Hilbert's axioms, can be found in a supplemental on-line chapter, which can be accessed at the website http://www.gac.edu/~hvidsten/geom-text. This chapter is a thorough axiomatic development of the geometric principles that serve as the foundation for all of the differing geometric systems that the book covers, including elliptic geometry.

INTENDED AUDIENCE

The design of the text makes it appropriate for college courses at several levels. The focus on exploration followed by proof makes the text accessible to prospective high school teachers. The author has successfully used this text to teach a geometry course for math education majors at Gustavus Adolphus College.

The rigorous analysis of all three classical geometries (Euclidean, Hyperbolic, and Elliptic) makes the text appropriate for courses designed for students with a strong mathematical background.

PREREQUISITES

This text is designed for use by mathematics students at the junior or senior collegiate level. Prior experience with proving mathematical results is recommended. A basic understanding of analytic geometry, including vectors, is needed throughout the text. This is reviewed in detail in Chapter 3. Some basic matrix algebra is used in Chapter 5. This material is at an elementary level and should not prove difficult for students who have not had linear algebra.

ARRANGEMENT OF TOPICS

The arrangement of topics in the text was designed to give as much flexibility as possible. While Chapters 1 and 2 are fundamental, many of the other chapters can be covered independently from one another.

Chapter 3 covers basic analytic geometry of vectors and angles, as well as complex numbers. Unless review of such matters is necessary, Chapter 3 can be viewed as optional foundational material.

Chapter 4 covers Euclidean constructions and depends only on the material in Chapters 1 and 2.

Chapters 5 and 6 are devoted to transformational geometry and require only a basic understanding of vectors and angles from Chapter 3, beyond the material covered in Chapters 1 and 2. Chapter 5 is a prerequisite for many subsequent chapters.

Chapters 7, 8, and 9 all cover various non-Euclidean geometries. Chapter 7 covers hyperbolic geometry, Chapter 8 covers elliptic geometry, and Chapter 9 covers projective geometry. While Chapters 7, 8, and 9 are quite independent of each other, it is advisable to cover hyperbolic geometry (Chapter 7) before covering the other non-Euclidean geometries.

The on-line chapter on Hilbert's axioms is a review of the foundations of neutral, elliptic, hyperbolic, and Euclidean geometry based on David Hilbert's axiomatic system. This chapter can be found at the author's website http://www.gac.edu/~hvidsten/geom-text.

Chapter 10 covers fractal geometry. This chapter is quite different from any other material in the text, as it is highly computational.

SUGGESTED COURSE SYLLABI

A suggested syllabus for a one-semester course for prospective high-school geometry teachers would include Chapters 1, 2, 4, sections 5.1–5.6, sections 6.1–6.2, and sections 7.1–7.6.

A suggested syllabus for a one-semester course for math majors would include Chapters 1, 2, 5, 7, 8, and as much of Chapter 9 as time permits.

If the focus of the course for math majors is on a thorough grounding of Euclidean geometry on an axiomatic foundation, then it would be advisable to start with Chapter 1 and then jump to the on-line chapter on Hilbert's axioms. A quick review of Chapter 2 can be followed by Chapters 5, 7, and as much of Chapter 9 as time permits.

A suggested syllabus for a one-semester course focusing on non-Euclidean geometry would include Chapters 1, 5, 7, 8, and 9.

Acknowledgments

This text, and the *Geometry Explorer* software, have evolved from the many geometry courses I have taught at Gustavus Adolphus College. I am deeply grateful to my students, who have graciously allowed me to experiment with different strategies in discovery-based learning and the integration of technology into the classroom. Their encouragement and excitement over these new approaches have been the primary motivating factors for writing an integrated learning environment for the *active* exploration of geometry.

I am especially grateful to those who helped out in the early phases of this project. Alicia Sutphen, a former student, provided valuable assistance in the early stages of designing the *Geometry Explorer* software. Special thanks go to those faculty members who field-tested early drafts of the text: Steve Benzel, Berry College; Jason Douma, University of Sioux Falls; George Francis, University of Illinois at Urbana-Champaign; Dan Kemp, South Dakota State University; Bill Stegemoller, University of Southern Indiana; Mary Wiest, Minnesota State University at Mankato; Matt Haines at Augsburg College, and Stephen Walk, St. Cloud State University.

I greatly appreciate the assistance of Bob Ross at CRC Press. He has been supportive of this project since it began with the first edition published in 2005. His commitment to this project has been continuous and I am thankful for his faith in the work. Ken Rosen, Series Editor and author of a fantastic book on discrete mathematics, provided many helpful suggestions on how to improve the text.

I would also like to thank the following reviewers: Nick Anghel, University of North Texas; Brian Beaudrie, Northern Arizona University; David Boyd, Valdosta State University; Anita Burris, Youngstown State University; Victor Cifarelli, University of North Carolina at Charlotte; Michael Dorff, Brigham Young University; Gina Foletta, Northern Kentucky University; Matthew Jones, California State University–Dominguez Hills; Tabitha Mingus, Western Michigan University; Chris Monico, Texas Tech University; F. Alexander Norman, University of

Texas at San Antonio; Ferdinand Rivera, San Jose State University; Craig Roberts, Southeast Missouri State University; Philippe Rukimbira, Florida International University; Don Ryoti, Eastern Kentucky University; Sherrie Serros, Western Kentucky University; and Wendy Hageman Smith at Radford University.

The design of the cover was accomplished with the assistance of Sarah Hickman, an artist and art therapist from Sacramento, California.

The material in this text is based upon work supported by the National Science Foundation under Grant No. 0230788.

Finally, I am grateful to my wife, Rebekah Richards, for her encouragement, her help in editing, and her understanding of the many hours needed to complete this project.

CHAPTER 1

Geometry and the Axiomatic Method

We owe geometry to the tax collector.

– J. L. Heilbron, *Geometry Civilized* [21]

Let no one ignorant of geometry enter here.

—Inscription over the doors to Plato's Academy

1.1 EARLY ORIGINS OF GEOMETRY

In a fundamental sense, geometry is a natural outgrowth of our exposure to the physical universe and in particular to the natural world. In our interactions with our environment, we encounter physical shapes, such as rocks and mountains, that we then organize by patterns into groups and classes. Rocks get put into the "round" category and mountains into a separate category. As our powers of perception become more refined, we notice other patterns of objects, such as the symmetries found in nature. An example of a natural symmetry is that of the rotational symmetry found in the California Poppy (Figure 1.1).

Figure 1.1 California Poppy, Mimi Kamp, Southwest School of Botanical Medicine

It is not surprising that human beings, being embedded in the natural world, should be inspired by and curious about geometrical shapes. For example, when constructing shelters our ancestors invariably chose to use precise geometric figures—most often circles or rectangles. There were very practical reasons to use these shapes; rectangular structures are easily laid out and circular huts provide a maximum of living space for the area they enclose.

While ancient peoples used geometric shapes for quite utilitarian purposes, they also surrounded themselves with patterns and designs that did not have any functional purpose.

In this classic Navajo rug, there is no *practical* need to decorate the fabric with such an intricate design. The decoration met a different need for the individual who created it, the need for beauty and abstraction. (Image courtesy of the National Park Service)

From the earliest times geometric figures and patterns have been used to represent abstract concepts, concepts that are expressed through the construction of objects having specific geometric shapes. A good example of this connection between the abstract and the concrete is that of the pyramids of ancient Egypt (Figure 1.2).

Figure 1.2 The Giza Pyramids and Giza Necropolis, Egypt, seen from above. Photo taken on 12 December 2008 by O.Mustafin (Creative Commons License)

The pyramids were built primarily as tombs for the pharaohs. However, a tomb for a pharaoh could not be just an ordinary box. The pharaoh was considered a god and as such his tomb was designed as a passageway connecting this life to the afterlife. The base of each pyramid represented the earth. It was laid out precisely with four sides oriented to face true north, south, east, and west. From the base, the sides reached a peak that symbolized the connection with the Egyptian sun god.

While the design and construction of the pyramids required very specific geometric knowledge—basic triangle geometry and formulas for the volume of four-sided pyramids—the Egyptians also developed simple geometric rules for handling a quite different task, that of surveying. The arable land of ancient Egypt lay close to the Nile and was divided into plots leased to local Egyptians to farm. Each year, after the Nile had flooded and wiped out portions of the land, tax collectors were forced to re-calculate how much land was left in order to levy the appropriate amount of rent. The Egyptians' study of land measurement was passed on to the Greeks and is evidenced by the word *geometry* itself, which in Greek means "earth measure."

A special class of Egyptian priests arose to handle these two types of geometrical calculations—practical surveying and the more abstract and spiritual design of monuments and tombs. The ancient Greek philosopher Aristotle believed that the existence of this priestly class motivated a more abstract understanding of geometry by the Egyptians than by any of their predecessors. However, this higher study of geometry was still quite primitive by our standards. A truly abstract and logical understanding of geometry would come with the Greeks' absorption of Egyptian ideas through the schools of Thales and Pythagoras.

1.2 THALES AND PYTHAGORAS

The Egyptians had remarkably good formulas for the volume of a truncated pyramid (not surprisingly) and had a good approximation (about $3\frac{1}{6}$) for the constant π. However, there was a serious problem with Egyptian geometry. They did not delineate between values that were approximations and those that were exact. Indeed, nowhere in Egyptian geometry is there a concern for what the "actual" value of a computation is. Their method of solving a problem was to take the numbers involved and follow a recipe of adding, subtracting, and so on, until they ended up with a final number. For example, they knew that certain triples of integers, say 3, 4, and 5, would form the lengths of a right triangle, but they had no notion of what the relationship between the sides of a right triangle were in general.

In the period between 900 and 600 BC, while the Egyptian empire was waning, a new seafaring and trading culture arose in Greece. As the Greeks traveled throughout the Mediterranean, they interacted with a diverse set of cultures, including the Egyptians. The entrepreneurial spirit of the Greeks was reflected in the creation of independent schools of learning led by master teachers such as Thales and Pythagoras. This was in stark contrast to the centralized monopoly of the priestly class in Egypt. The Greeks created a marketplace of ideas where theories were created and debated vigorously.

In this society of ideas, two notable schools arose—the Ionian school founded by Thales of Miletus and the Pythagorean school founded by Pythagoras of Samos. Much about both men is lost to history and comes down to us through myths and legends. However, the impact that their schools had on mathematics, science, music, and philosophy was revolutionary.

To understand the profound change that occurred in how we think of mathematics now versus before the Greeks, let us consider a simple problem—that of computing the area of a triangle. The Egyptians were aware that if the triangle were a right triangle, then one could "flip" the triangle across the longest side and get a rectangle. The triangle's area is then half the product of the two shortest sides.

They also knew that in some special triangles, with certain fixed lengths of sides, the area was half the length of one side times the height of the triangle. However, this is all that the Egyptians knew, or wanted to know, about the area of a triangle in general. Geometry to them was an *empirical* subject, needed only to analyze those triangles found in real objects.

The Greeks, on the other hand, viewed the abstraction of things as the ideal and perfect form of reality. Sides and edges in this world were imperfect. By abstracting edges into segments having no width and having *exact* length, the Greeks could talk about the exact answer to questions such as the area of a triangle. And not only that, they could discuss in general what the area should be for *any* triangle that is built of three abstract segments.

This abstraction, freed from the constraints of empirical foundation, allowed for the study of *classes* of objects, rather than the objects themselves. For example, if one supposed the earth was a sphere, then one could deduce many properties of the earth just by using what was known about abstract spheres.

But this way of thinking required a new way of determining the veracity of statements. For the Egyptians, the statement "a triangle with sides of length 3, 4, and 5 has a right angle" would be accepted as true if one drew a 3-4-5 triangle in the sand and measured the angle to be approximately a right angle. However, a statement about abstract triangles cannot be proved this way—one cannot draw a *perfect* 3-4-5 triangle, no matter how hard one tries. The greatest achievement of the Greeks was the development of a precise and logical way of reasoning called the *deductive method*, which provided sound rules of argument to be used for abstract systems of objects. The deductive method has formed the basis for scientific reasoning from the time of the Greeks until the modern age.

The Greeks' focus on abstraction and their creation of the deductive method can be traced to the two great schools of Thales and Pythagoras.

1.2.1 Thales

Thales (ca. 624–548 BC) lived in the coastal city of Miletus in the ancient region of Ionia (present day Turkey). He is reported to have visited Egypt in the first half of the sixth century BC. While there, he studied Egyptian mathematics and measured the height of the Egyptian pyramids by the use of shadows and similar triangles.

Thales was, in many respects, a bridge between the empirical and mythical world of the Egyptians and the abstract and rational world of Greek civilization. To the Egyptians, reality was infused with the actions of spiritual and mystical beings, and thus, while one could discover some basic facts about how things worked, the true nature of reality was not important—the gods would do as they wanted to with the world. Thales is known as the first scientist because he believed in a rational world where one could discover universal truths about natural phenomena by abstracting properties of the world into ideal terms. This was a fundamental shift in how nature was perceived. No longer were natural processes simply the whims of mythical beings; they were products of processes that could be described in terms that could be debated and proved true or false. Thales was perhaps the first thinker to seriously consider the question, What is matter composed of? His answer was *water*, which we know today to be incorrect, but it was an answer that Thales could back up with a logical argument and could, in principle, be shown to be true or false.

Thales is known for being the first mathematician, the first to use deductive reasoning to prove mathematical results. As best we can tell, his methods of proof actually involved a combination of deductive reasoning and inductive reasoning from examples. The full development of deductive reasoning from first principles (axioms) would come later, with its most complete expression found in the work of Euclid.

The five geometric theorems attributed to Thales are

1. A circle is bisected by a diameter.

2. The base angles of an isosceles triangle are equal.

3. The pairs of vertical angles formed by two intersecting lines are equal.

4. Two triangles are congruent if they have two angles and the included side equal.

5. An angle inscribed in a semicircle is a right angle.

The last result has become known as "The Theorem of Thales." It can be proved in several ways. In Figure 1.3 we suggest one method. Before moving on, try to develop your own proof (or explanation) of this theorem based on Figure 1.3. That is, show that $\beta + \gamma = 90$. In your proof you can use Thales' theorem on isosceles triangles (Theorem 2 in the list above) and the fact that the sum of the angles in a triangle is 180 degrees.

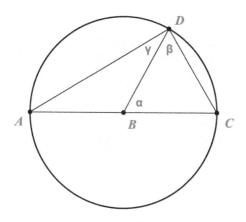

Figure 1.3

1.2.2 Pythagoras

According to legend, Pythagoras (ca. 580–500 BC) was a pupil of Thales. Like Thales, Pythagoras traveled widely, gathering not only mathematical ideas, but also religious and mystical teachings. To the Pythagoreans, philosophy and mathematics formed the moral basis of life. The words *philosophy* (love of wisdom) and *mathematics* (that which is learned) were coined by the Pythagoreans.

In the Pythagorean worldview, mathematics was the way to study the ideal, that which was truly harmonious and perfect. "Numbers rule the universe" was their motto. They believed that all of nature could be explained by properties of the *natural* numbers $1, 2, 3, \ldots$. The search for harmony and perfection were of utmost importance and translated directly to a focus on ratios and proportion—the harmonious balance of numbers.

The theorem historically attributed to Pythagoras, that *in a right triangle the square on the hypotenuse is equal to the sum of the squares on the two sides*, was known as an empirical fact to the Egyptians and others in the ancient Orient. However, it was the Pythagoreans who first provided a logical proof of this result.

The Pythagoreans' development of geometry and their focus on the theory of numbers in relation to music, astronomy, and the natural world had a profound effect on Greek culture. As Boyer and Merzbach state in their history of mathematics [7, page 57], "Never before or since has mathematics played so large a role in life and religion as it did among the Pythagoreans."

One of the most important features of the schools of both Pythagoras and Thales was their insistence that mathematical results be justified or proved true. This focus on proofs required a method of reasoning and argument that was precise and logical. This method had its origins with Pythagoras and Thales and culminated with the publication of *The Elements* by Euclid in about 300 BC. The method has become known as the "Axiomatic Method." Howard Eves in [12] records the invention of the axiomatic method as one of the very greatest moments in the history of mathematics. Before we investigate the axiomatic method, let's take a little side trip into one of the most elegant constructions used by the Pythagoreans—the construction of the golden section.

1.3 PROJECT 1 - THE RATIO MADE OF GOLD

Leonardo Da Vinci called it the "Sectio aurea." Luca Paccioli, an Italian mathematician of the 1500s, wrote a book, which Da Vinci illustrated, called *De Divine Proportione*. Johannes Kepler, in the late 1500s, said:

> Geometry has two great treasures: one is the theorem of Pythagoras; the other, the division of a line into extreme and mean ratio. The first we may compare to a measure of gold; the second we may name a precious jewel.

What these great thinkers were referring to is one of the simplest, yet most aesthetically pleasing, geometric constructions—that of the golden ratio.

The first study of the golden ratio is attributed to the Pythagoreans. In their search for harmony of number, they sought the ideal figure, one in which the dimensions were in perfect harmony of proportion. Proportion has to do with ratios of measurements. We say that two ratios are in *proportion* if they are equal to one another. The simplest geometric ratio is that of two segment lengths. The simplest way to create a proportion of ratios is to start with a single segment and cut it into two parts. We create one ratio by looking at the ratio of the total segment to the longer of the two sub-pieces, and we can create another ratio by that of the longer sub-piece to the smaller. The splitting of the segment will be most harmonious if these two ratios are equal, that is, if they are in proportion.

Our first project will be to construct geometric figures that have this perfect harmony of proportion. We will do so by using a dynamic geometry software environment. Before you start this project, review the brief notes on using geometry software found at http://www.gac.edu/~hvidsten/geom-text. In particular, review the notes on *creating* versus *constructing* geometric objects and the information about *attaching* points to objects.

You should work through all the constructions and exercises, keeping notes on how figures are constructed and writing down answers to the exercises. Your notes and answers will form the basis of the report you will create at the conclusion of the project.

1.3.1 Golden Section

Start up your geometry software. You should see a window in which you can create geometric objects. In Figure 1.4 we see the window opened using the *Geometry Explorer* software. Our first task will be to create a segment $\overline{AB}$ on the screen. Using the Segment tool in your software, click and drag on the screen to create a segment. Using the Point tool, *attach* a point C to segment $\overline{AB}$. Note that C is always "stuck" on $\overline{AB}$; if we drag C around, it will always stay on $\overline{AB}$. Now, measure the ratio of the length of segment $\overline{AB}$ to the length of segment $\overline{BC}$. Instructions for doing this measurement for various geometry programs can be found at http://www.gac.edu/~hvidsten/geom-text. In Figure 1.4 we see this ratio displayed in *Geometry Explorer*.

Next measure the ratio of the length of segment $\overline{BC}$ to $\overline{CA}$. Drag point C around and see if you can get these two ratios to match up; that is, see if you can create a *proportion* of ratios.

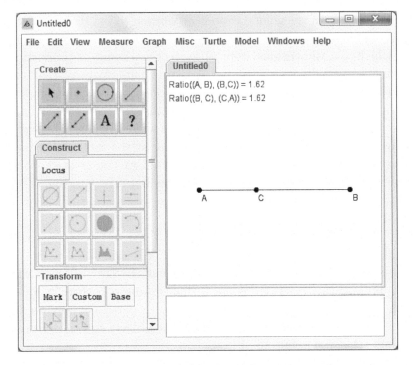

Figure 1.4 *Geometry Explorer* main window

Interesting!! The ratios seem to match at a magic ratio of about 1.62. Let's see why this is. For the sake of argument, let's set the length of $\overline{BC}$ equal to 1. Let x be the length of $\overline{AB}$. Then, what we are looking for is a value of x that satisfies the equation

$$\frac{x}{1} = \frac{1}{x-1}$$

If we solve this equation, we get that x must satisfy $x^2 - x - 1 = 0$. This has two roots $\frac{1 \pm \sqrt{5}}{2}$. The positive solution is $\frac{1+\sqrt{5}}{2}$, which is about 1.62. This ratio is the perfection of balance for which the Pythagoreans were searching. The segment subdivision having this ratio is what Da Vinci called the *Golden Section*.

Construction of the Golden Ratio

We can see from the previous discussion that it is not too hard to approximate the golden ratio by moving point C. However, in the true Pythagorean spirit, is it possible to construct two segments whose ratio of lengths is *exactly* the golden ratio?

Traditionally, the question of geometric construction of numerical values or geometric figures has played a key role in the development of geometry. (Euclid's notion of *construction* encompasses *both* of the ideas of creating and constructing used in many geometry programs.) Euclidean constructions are carried out by drawing lines (or segments) and circles and by finding the intersections of lines and circles. Such constructions are called *straightedge and compass* constructions as they represent pencil and paper constructions using a straightedge (line) and compass (circle). A review of constructions will come later in the text. For now, we point out that all of the tools available in the Create and Construct panels of buttons in *Geometry Explorer* are valid Euclidean constructions.

To construct the golden ratio of a segment, we will need to split a segment into two sub-segments such that the ratio of the larger to the smaller sub-segment is exactly $\frac{1+\sqrt{5}}{2}$. How can we do this? Since the fraction $\frac{1}{2}$ appears in the expression for the golden ratio, it might be useful to construct the midpoint of a segment. Also, an easy way to construct $\sqrt{5}$ would be to construct a right triangle with base lengths of 1 and 2. We'll keep these ideas in mind as we explore how to construct the golden ratio.

Golden Ratio Construction Step 1

To get started, we need a segment. Clear the screen and create segment $\overline{AB}$.

Since we already discussed the need for midpoints, let's construct the midpoint C of $\overline{AB}$ by using the Midpoint tool in your program. To make life easier for ourselves, let's assume the length of $\overline{AB}$ is 2. Then, we have one base of the triangle we discussed above.

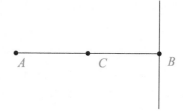

To get the other base, we need a right angle at B. Use the perpendicular line tool in your program to construct the perpendicular to $\overline{AB}$ at B (refer to the preceding figure).

Golden Ratio Construction Step 2

Now, we need to create a segment up from B along the vertical line that has length 1. This can be done using a circle centered at B of radius 1. To create this circle, use the Circle tool in your program. Click on point B. Drag the cursor until it is directly over point C. When the cursor is over C, that point will become highlighted. The cursor represents the radius point of the circle we want to construct, and we want this radius point to be *equal* to C so that the circle has radius exactly equal to 1. Release the mouse button and drag points A and B around the screen. Notice how the circle radius is always of constant length equal to the length of $\overline{BC}$.

The technique of dragging a point of a new circle or line until it matches an existing point will be a common technique used throughout the labs in this text. It is an easy way to create objects that are in synchronization with existing objects.

To construct the vertical leg of the desired right triangle, return A and B to the position shown. We will use the Point tool to construct the intersection point D. Hover the mouse where the intersection should be. The intersecting line and circle should be highlighted. Click the mouse to create D.

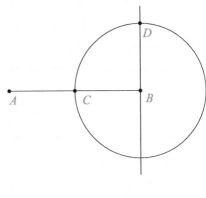

At this point, you should check that you have constructed all of the figures correctly. Drag points A and B around the screen and check that your circle moves accordingly, with center B and radius to C, and that your intersection points D and E move with the circle and the perpendicular.

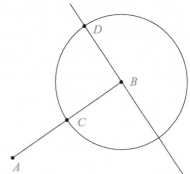

A great advantage in working with a geometry program, as compared with paper and pencil, is that you can dynamically move geometric figures, exploring how their properties change (or stay constant) as you do so.

Golden Ratio Construction Step 3

Point D is the point we are after, so let's hide the circle. Move $\overline{AB}$ back to a horizontal position and create a segment connecting A to D. To finish our triangle, create segment $\overline{BD}$.

By the Pythagorean Theorem we know that the length of $\overline{AD}$ will be $\sqrt{5}$. So we have constructed all the numbers that appear in the fraction for the golden ratio, but we have not actually found a point on $\overline{AB}$ that will subdivide this segment in this ratio.

Let's experiment a bit with what we are trying to find. Suppose G is a point between A and B that subdivides the segment into the golden ratio; that is $\frac{AB}{AG} = \frac{1+\sqrt{5}}{2}$. Let x be the length of $\overline{AG}$. (Draw a picture on a scrap piece of paper—all good mathematicians have paper on hand when reading math books!)

Exercise 1.3.1. *Given that the length of $\overline{AB}$ is 2, show that x must be equal to $\sqrt{5} - 1$.*

From this exercise, we see that to finish our construction we need to find a point G on $\overline{AB}$ such that the length of $\overline{AG}$ is $\sqrt{5} - 1$.

We make use of the fact that we already have a length of $\sqrt{5}$ in the hypotenuse of our right triangle. To cut off a length of 1 from the hypotenuse, create a circle centered at D with radius point B. Then, construct the intersection point F of this new circle with $\overline{AD}$.

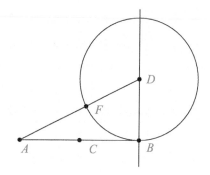

To construct G, we just need to transfer $\overline{AF}$ to $\overline{AB}$. Create a circle with center A and radius point F and construct the intersection point G of this new circle with $\overline{AB}$. Then, compute the ratio of the length of $\overline{AG}$ to the length of $\overline{GB}$. (Refer to the construction earlier in this project if you need a refresher on how to calculate this ratio)

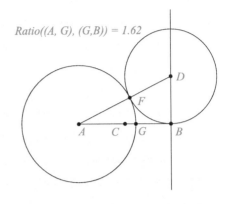

$Ratio((A, G), (G,B)) = 1.62$

It looks like our analysis was correct. We have constructed a golden section for $\overline{AB}$ at G!

1.3.2 Golden Rectangles

The golden ratio has been used extensively in art and architecture—not in the subdivision of a single segment but in the creation of rectangular shapes called golden rectangles. A *Golden Rectangle* is a rectangle where the ratio of the long side to the short side is exactly the golden ratio.

To construct a golden rectangle, we will again need to construct the numbers 1, 2, and $\sqrt{5}$. But this time we can interpret the numerator and denominator of the golden ratio fraction $\frac{1+\sqrt{5}}{2}$ as the separate side lengths of a rectangle. It makes sense to start out with a square of side length 2, as we can take one of its sides as the denominator and can split an adjacent side in two to get the 1 term in the numerator. To get the $\sqrt{5}$ term, we need to extend this smaller side appropriately.

Golden Rectangle Construction Step 1

To start the construction, clear the screen and create segment $\overline{AB}$. We will assume this segment is of length 2.

To construct a square on $\overline{AB}$, first construct the perpendicular to $\overline{AB}$ at A. Then construct a circle at A with radius $\overline{AB}$. Next, construct the intersection point C. Do a similar series of constructions at B to get point D as shown.

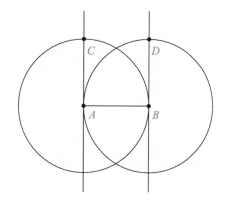

Points $A, B, C,$ and D will form a square. Hide all of the perpendiculars and circles and create the segments $\overline{AC}$, $\overline{CD}$, and $\overline{DB}$ to form a square.

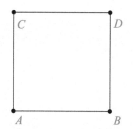

Golden Rectangle Construction Step 2

Construct the midpoint E of $\overline{CD}$. If we can extend $\overline{CD}$ to a point F so that $\overline{DF}$ has length $\sqrt{5} - 1$, then segment $\overline{CF}$ will have length $\sqrt{5} + 1$ and we will have the length ratio of $\overline{CF}$ to $\overline{AB}$ equal to the golden ratio.

To extend $\overline{CD}$, create a ray from C to D. Then, create a circle with center at E and radius point B and construct the intersection point F.

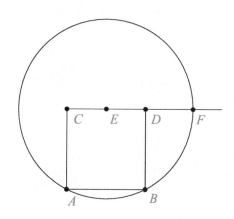

Exercise 1.3.2. *Use a right triangle to argue that $\overline{DF}$ has length $\sqrt{5} - 1$, given the construction of F done in the step above.*

Golden Rectangle Construction Step 3

Finally, to finish off the rectangle partially defined by B, D, and F, we construct a perpendicular to ray $\overrightarrow{CD}$ at F, then create a ray from A to B, and construct the intersection point G of the perpendicular with this ray. Measure the ratio of $\overline{AB}$ to $\overline{BG}$. We have created a segment ($\overline{AG}$) that is cut in the golden ratio at B, as well as a rectangle ($DFGB$) that is "golden."

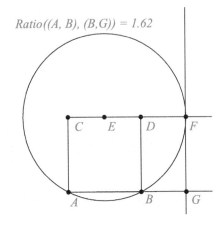

Ratio((A, B), (B,G)) = 1.62

Exercise 1.3.3. *Grab a ruler and a few of your friends and measure the proportions of the rectangles enclosing your friends' faces. Measure the ratio of the distance from the bottom of the chin to the top of the head to the distance between the ears. Make a table of these distances and find the average "face ratio." Does the average come close to the golden ratio? Find some magazine photos of actors and actresses considered beautiful. Are their faces "golden"?*

The Fibonacci sequence has a fascinating connection with the golden ratio. The sequence is defined as a sequence of numbers $u_1, u_2, u_3, \ldots$ where $u_1 = 1$, $u_2 = 1$, and $u_n = u_{n-1} + u_{n-2}$. Thus, $u_3 = 2$, $u_4 = 3$, $u_5 = 5$, and so on. The first ten terms of this sequence are $1, 1, 2, 3, 5, 8, 13, 21, 34, 55, 89$. These numbers come up in surprising places. They appear in the branching patterns of plants, in the numbers of rows of kernels in a sunflower head, and in many spiral patterns, such as the spiral found in the Nautilus shell. The ratio of successive Fibonacci numbers is also related to the golden ratio. Consider the first few ratios of terms in the sequence

$$\frac{2}{1} = 2, \quad \frac{3}{2} = 1.5, \quad \frac{5}{3} = 1.666..., \quad \frac{8}{5} = 1.6, \quad \frac{13}{8} = 1.625, \quad \frac{21}{13} = 1.61538...$$

It appears that these ratios are approaching the golden ratio. In fact, these ratios actually do converge *exactly* to the golden ratio (see [25] for a proof).

Exercise 1.3.4. *Find five objects in your environment that have dimensions given by two succeeding terms in the Fibonacci sequence. For example, a simple 3x5 index card is close to being a golden rectangle as $\frac{5}{3}$ is one of our Fibonacci ratios. Once you start looking, you will be amazed at how many simple, everyday objects are nearly golden.*

Project Report

This ends the exploratory phase of this project. Now it is time for the explanation phase—the final project report. The report should be carefully written and should include three main components: an introduction, a section discussing results, and a conclusion.

The introduction should describe what was to be accomplished in the project. It should focus on one or two major themes or ideas that were explored.

The main body of the report will necessarily be composed of the results of your investigation. This will include a (brief) summary of the constructions you carried out. This should not be a verbatim list or recipe of what you did, but rather a general discussion of what you were asked to construct and what you discovered in the process. Also, all results from exercises should be included in this section.

The conclusion of the report should document what you learned by doing the project and also include any interesting ideas or conjectures that you came up with while doing the project.

1.4   THE RISE OF THE AXIOMATIC METHOD

In the last section we looked at the ancient Greeks' search for the perfect harmony of proportion. In this section we consider another example of the Greeks' quest for perfection—that of perfection of *reasoning*. This quest culminated in the creation of a pattern of reasoning called the *axiomatic method*.

The axiomatic method is based on a system of *deductive* reasoning. In a deductive system, statements used in an argument must be derived, or based upon, prior statements used in the argument. These prior statements must themselves be derived from even earlier statements, and so on. There are two logical traps that one can fall into in such a system.

First, there is the trap of producing a never-ending stream of prior statements. Consider, for example, the definition of a word in the dictionary. If we want to define a word like *orange* we need to use other words such as *fruit* and *round*. To define *fruit* we need to use *seed*, and so forth. If every word required a different word in its definition, then we would need an infinite number of words in the English language!

Second, there is the trap of circular reasoning. In some dictionaries a *line* is defined as a type of curve, and a *curve* is defined as a line that deviates from being straight.

The Greeks recognized these traps and realized that the only way out of these logical dilemmas was to establish a base of concepts and statements that would be used *without* proof. This base consisted of *undefined terms* and *postulates/axioms.*

Undefined terms were those terms that would be accepted without any further definition. For example, in the original formulation of Euclid's geometry, the terms *breadth* and *length* are undefined, but a *line* is defined as "breadth-less length." One may argue whether this is a useful definition of a line, but it does allow Euclid to avoid an infinite, or circular, regression of definitions. In modern treatments of Euclidean geometry, the terms *point* and *line* are typically left undefined.

A postulate or axiom is a logical statement about terms that is accepted as true without proof. To the Greeks, postulates referred to statements about the specific objects under study and axioms (or *common notions*) referred to more universal statements about general logical systems. For example, the statement "A straight line can be drawn from any point to any point" is the first of Euclid's five postulates; whereas the statement "If equals be added to equals, the wholes are equal" is one of Euclid's axioms. The first is a statement about the specifics of a geometric system and the second is a general logical statement. In modern mathematical axiomatic systems, there is no distinction between these two types of mutually accepted statements, and the terms *axiom* and *postulate* are used interchangeably to refer to a statement accepted without proof.

Starting from a base of undefined terms and agreed upon axioms, we can define other terms and use our axioms to argue the truth of other statements. These other statements are called the *theorems* of the system. Thus, our deductive system consists of four components:

1. Undefined Terms

2. Axioms (or Postulates)

3. Defined Terms

4. Theorems

A system comprising these four components, along with some basic rules of logic, is an axiomatic system. (In Appendix B there is a complete listing of the axiomatic system Euclid used at the start of his first book on plane geometry.)

One way to think about an axiomatic system is by analogy with playing a game like chess. We could consider the playing pieces (as black and white objects) and the chessboard as undefined parts of the game. They just exist and we use them. A particular playing piece, for example the bishop, would be a defined term, as it would be a special kind of playing piece. The *rules* of chess would be axioms. The rules are the final say in what is allowed and what is not allowed in playing the game. Everyone (hopefully) agrees to play by the rules. Once the game starts, a player moves about the board, capturing his or her opponent's pieces. A particular configuration of the game, for example with one player holding another player in check, would be like a theorem in the game in that it is derived from the axioms (rules), using the defined and undefined terms (pieces) of the game, and it is a configuration of the game that can be verified as legal or not, using the rules. For example, if we came upon a chessboard set up in the starting position, but with all of a player's pawns behind all of the other pieces, we could logically conclude that this was not a legal configuration of the game.

It is actually very useful to have a game analogy in mind when working in an axiomatic system. Thinking about mathematics and proving theorems is really a grand game and can be not only challenging and thought-provoking like chess, but equally as enjoyable and satisfying.

As an example of playing the axiomatic game, suppose that we had a situation where students enrolled in classes. *Students* and *classes* will be left as undefined terms as it is not important for this game what they actually mean. Suppose we have the following rules (axioms) about students and classes.

A1 There are exactly three students.

A2 For every pair of students, there is exactly one class in which they are enrolled.

A3 Not all of the students belong to the same class.

A4 Two separate classes share at least one student in common.

What can be deduced from this set of axioms? Suppose that two classes shared more than one student. Let's call these classes C_1 and C_2. If they share more than one student—say students A and B are in both classes—then we would have a situation where A and B are in more than one class. This clearly contradicts the rule we agreed to in Axiom 2, that two students are in one and only one class. We will use a rule of logic that says that an assumption that contradicts a known result, or an axiom, cannot be true. The assumption that we made was that two classes could share more than one student. The conclusion we must make is that this assumption is false, and so two classes cannot share more than one student.

By Axiom 4 we also know that two separate classes share at least one student. Thus, two separate classes must have one and only one student in common. We can write these results down as a *theorem*. The set of explanations given above is called a *proof* of the theorem.

> **Theorem 1.1.** *Two separate classes share one and only one student in common.*

A proof is a logical explanation for why a theorem statement is true. A valid proof is based solely on statements that are either axioms or that have already been proven. For a brief review of proofs, consult Appendix A.

Here is another theorem and proof based on our axiomatic system. Note how the proof uses only the axioms and the theorem proved above.

> **Theorem 1.2.** *There are exactly three classes in our system.*

Proof: By Axiom 2 we know that for each pair of students, there is a class. By Axiom 3 all three students cannot be in a common class. Thus, there must be at least three classes, say C_1, C_2, C_3, as there are three different pairs of students. Suppose there is a fourth class, say C_4. By the theorem just proved, there must be a student shared by each pairing of C_4 with one of the other three classes. So C_4 has at least one student. It cannot contain all three students by Axiom 3. Also, it cannot have just one student since if it did, then classes C_1, C_2, and C_3 would be forced by Axiom 4 to share this student and, in addition, to have three other, different students among them, because the three classes must have different pairs of students. This would mean that there are at least four students and would contradict Axiom 1. Thus, C_4 must have exactly two students. But, since this pair of students must already be in one of the other three classes, we have a contradiction to Axiom 2. Thus, there cannot be a fourth class. □

Here is another example. Again, note the careful way the proof only uses axioms and previously proven theorems.

Theorem 1.3. *Each class has exactly two students.*

Proof: By the previous theorem we know that there are exactly three classes. By Axiom 4 we know that there is at least one student in a class. Suppose a class had just one student, call this student S. All classes would then have student S by Axiom 4. The other two students are in some class, call it class X, by Axiom 2. Class X must then have all three students as it also needs to have student S, the student common to all three classes. But, this contradicts Axiom 3. Thus, all classes must have at least two students and by Axiom 3 must have exactly two students. □

The careful reader will see that we are using basic properties of counting in this proof. While not axiomatically based, this will be a generally accepted rule of reasoning that we will use for proofs. We are also assuming that we can label objects with letters and that there is mutual agreement about what words mean. This will also be tacitly assumed for the readership of a proof. What cannot be assumed in a proof is any relationship between the *terms* (defined and undefined) of the axiomatic system, other than that given by the system.

Another rule of reasoning we are using in this proof is the idea of arguing by the contradiction principle. This is the idea that if we assume the opposite of a statement, and reach a result that contradicts a statement we know is true, then the original statement must be true.

In the proof above, we consider the opposite of the statement that each class has at least two students. One way to negate this statement is to say that some class has only one student. This leads to a contradiction to Axiom 3. Thus, we can conclude that the statement that each class has at least two students must be true. For further review of this principle, consult Appendix A.

It is important to point out that the precise meaning of the terms *students* and *classes* is not important in this axiomatic system. We could just as well have used the following axiom set:

A1 There are exactly three snarks.

A2 For every pair of snarks, there is exactly one bittle.

A3 Not all of the snarks belong to the same bittle.

A4 Two separate bittles have at least one snark in common.

By changing the labels in the theorems above, we would get equivalent theorems about snarks and bittles. The point of this silly little aside is that we are concerned about the *relationships* and *patterns* among the objects in an abstract axiomatic system and not about the objects themselves. This is the great insight of the Greeks—that it is the relationships that matter and not how we apply those relationships to objects.

The insistence on proofs in formal axiomatic mathematics can seem, at times, to be a tedious exercise in belaboring the obvious. This sentiment is actually as old as Greek geometry itself. J. L. Heilbron, in [21] describes how the Greek philosopher Epicurus (341–270 BC) criticized Euclid's proof that no side of a triangle can be longer than the sum of the other two sides. As Epicurus stated, "It is evident even to an ass." For if a donkey wanted to travel to a bale of hay, it would go directly there along a line and not go through any point not on that line (Figure 1.5).

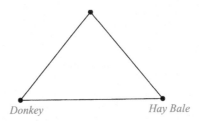

Donkey   *Hay Bale*

Figure 1.5 Donkey geometry

But, the Greek geometer Proclus (411–485 AD) refuted this criticism by arguing that something that seems evident to our senses cannot be relied on for scientific investigation. By training our minds in the most careful and rigorous forms of reasoning abstracted from the real world, we are preparing our minds for the harder task of reasoning about things that we cannot perceive. For Proclus, this type of reasoning "arouses our innate knowledge, awakens our intellect, purges our understanding, brings to light the concepts that belong essentially to us, takes away the forgetfulness and ignorance that we have from birth, and sets us free from the bonds of unreason" [21, page 8].

One clear illustration of Proclus' point is the development of modern views on the nature of the universe. From the time of Euclid, mathematicians believed that the universe was *flat*, that it was essentially a three-dimensional version of the flat geometry Euclid developed in the plane. In fact, Euclidean geometry was considered the only possible axiomatic geometric system. This was a reasonable extrapolation from these mathematicians' experience of the world. However, in the nineteenth century, a revolution occurred in the way mathematicians viewed geometry. *Non-Euclidean* geometries were developed that were just as logically valid as Euclidean geometry. The notion of a *curved* universe was now mathematically possible, and in 1854 George Frederich Riemann (1826–1866) set out the basic mathematical principles for analyzing curved spaces of not just three dimensions, but *arbitrary* dimensions. This mathematical theory must have seemed incredibly wild and impractical to non-mathematicians. However, Riemann's geometry of curved space was fundamental in Einstein's development of the theory of relativity and in his view of space-time as a four-dimensional object.

This revolution in the axiomatic basis of geometry laid the groundwork for a movement to formalize the foundations of *all* of mathematics. This movement reached its peak in the *formalist* school of the late 1800s and early 1900s. This school was led by David Hilbert (1862–1943) and had as its goal the axiomatic development of all of mathematics from first principles. Hilbert's *Grundlagen der Geometrie*, published in 1899, was a careful development of Euclidean geometry based on a set of 21 axioms.

Exercise 1.4.1. *Trace the following words through a dictionary until a circular chain of definitions has been formed for each: power, straight, real (e.g., power - strength - force - power). Why must dictionary definitions be inherently circular?*

Exercise 1.4.2. *In the game of Nim, two players are given several piles of coins, each pile having a finite number of elements. On each turn a player picks a pile and removes as many coins as he or she wants from that pile but must remove at least one coin. The player who picks up the last coin wins. (Equivalently, the player who no longer has coins to pick up loses.) Suppose that there are two piles with one pile having more coins than the other. Show that the first player to move can always win the game.*

Exercise 1.4.3. *Consider a system where we have children in a classroom choosing different flavors of ice cream. Suppose we have the following axioms:*

    A1 There are exactly five flavors of ice cream: vanilla, chocolate, strawberry, cookie dough, and bubble gum.

A2 Given any two different flavors, there is exactly one child who likes these two flavors.

A3 Every child likes exactly two different flavors among the five.

How many children are there in this classroom? Prove your result.

Exercise 1.4.4. *Using the ice cream axiom system, show that any pair of children likes at most one common flavor.*

Exercise 1.4.5. *In the ice cream system, show that for each flavor there are exactly four children who like that flavor.*

One of the most universal of abstract mathematical structures is that of a *group*. A group G consists of a set of undefined objects called *elements* and a *binary operation* "$\circ$" that relates two elements to a third. The axioms for a group are

A1 For all elements x and y, the binary operation on x, y is again an element of G. That is, for all $x, y \in G$, $x \circ y \in G$.

A2 For all $x, y, z \in G$, $(x \circ y) \circ z = x \circ (y \circ z)$. That is, the binary operation is *associative*.

A3 There is a special element $e \in G$ such that $x \circ e = e \circ x = x$ for all $x \in G$. The element e is called the *identity* of G.

A4 Given $x \in G$, there is an element $x^{-1} \in G$ such that $x \circ x^{-1} = x^{-1} \circ x = e$. The element x^{-1} is called the *inverse* to x.

The next three exercises are to be proven using just the group axioms. We will return to the idea of groups when we investigate transformations and symmetry in Chapters 5 and 6.

Exercise 1.4.6. *Show that if $x, y, z \in G$ and $x \circ z = y \circ z$, then $x = y$.*

Exercise 1.4.7. *Assume that the equation $x \circ y \circ z = e$ holds in a group G. Does it follow that $y \circ z \circ x = e$? Give a justification for your answer.*

Exercise 1.4.8. *Show that a group G can have only one identity.*

In 1889 Giuseppe Peano (1858–1932) published a set of axioms, now known as the *Peano axioms*, as a formal axiomatic basis for the natural numbers. Peano was one of the first contributors to the modern axiomatic and formalist view, and his system included five axioms based on the undefined terms *natural number* and *successor*. We will let N stand for the set of all natural numbers in the following listing of Peano's axioms.

A1 1 is a natural number.

A2 Every natural number x has a successor (which we will call x') that is a natural number, and this successor is unique.

A3 No two natural numbers have the same successor. That is, if x, y are natural numbers, with $x \neq y$, then $x' \neq y'$.

A4 1 is not the successor of any natural number.

A5 Let M be a subset of natural numbers with the property that

   (i) 1 is in M.

   (ii) Whenever a natural number x is in M, then x' is in M.

   Then, M must contain all natural numbers, that is, $M = N$.

The next four exercises are to be proven using just the Peano axioms.

Exercise 1.4.9. *The fifth Peano axiom is called the* Axiom of Induction. *Let M be the set of all natural numbers x for which the statement $x' \neq x$ holds. Use the axiom of induction to show that $M = N$, that is, $x' \neq x$ for all natural numbers x.*

Exercise 1.4.10. *Must every natural number be the successor of some number? Clearly, this is not the case for 1 (why not?), but what about other numbers? Consider the statement "If $x \neq 1$, then there is a number u with $u' = x$." Let M be the set consisting of the number 1 plus all other x for which this statement is true. Use the axiom of induction to show that $M = N$ and, thus, that every number other than 1 is the successor of some number.*

Exercise 1.4.11. *Define* addition *of natural numbers as follows:*

* *For every x, define $x + 1 = x'$.*
* *For every x and y, define $x + y' = (x + y)'$.*

Show that this addition is well-defined. That is, show that for all x and w, the value of $x + w$ is defined. [Hint: You may want to use induction.]

Exercise 1.4.12. *Think about how you might define multiplication in the Peano system. Come up with a two-part definition of multiplication of natural numbers.*

Exercise 1.4.13. *Hilbert once said that "mathematics is a game played according to certain simple rules with meaningless marks on paper." Lobachevsky, one of the founders of non-Euclidean geometry, once said, "There is no branch of mathematics, however abstract, which may not someday be applied to phenomena of the real world." Discuss these two views of the role of mathematics. With which do you most agree? Why?*

Exercise 1.4.14. *Historically, geometry had its origins in the empirical exploration of figures that were subsequently abstracted and proven deductively. Imagine that you are designing the text for a course on geometry. What emphasis would you place on discovery and empirical testing? What emphasis would you place on the development of proofs and axiomatic reasoning? How would you balance these two ways of exploring geometry so that students would become confident in their ability to reason logically, yet also gain an intuitive understanding of the material?*

1.5   PROPERTIES OF AXIOMATIC SYSTEMS

In this section we will look a bit deeper into the properties that characterize axiomatic systems. We will look at the axiomatic structure itself and at its properties.

In the last section, we looked at an axiomatic system consisting of classes and students. We saw that we could re-label these terms as snarks and bittles without any real change in the structure of the system or in the relationships between the terms of the system. In an axiomatic system, it does not matter what the terms represent. The only thing that matters is how the terms are related to each other. By giving the terms a specific meaning, we are creating an *interpretation* of the axiomatic system.

Suppose in our example of the last section that we replaced *class* and *student* by *line* and *point*.

We interpret *line* to be one of the three segments shown at right and *point* to be any one of the three points A, B, or C.

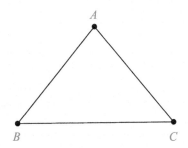

Then, our axioms would read as follows:

A1 There are exactly three points.

A2 Two distinct points belong to one and only one line.

A3 Not all of the points belong to the same line.

A4 Two separate lines have at least one point in common.

We should check that our axioms still make correct sense with this new interpretation and they do. In our new interpretation, the theorems above now say that there are exactly three lines, each pair of lines intersects in exactly one point, and each line has exactly two points, as can be seen in the figure above. This system is called *Three-Point geometry*.

This new interpretation of our original axiomatic system is called a *model* of the system.

> **Definition 1.1.** *A* model *is an interpretation of the undefined terms of an axiomatic system such that all of the axioms are true statements in this new interpretation.*

In any model of an axiomatic system, all theorems in that system are true when interpreted as statements about the model. This is the great power of the axiomatic method. A theorem of the abstract system needs to be proved once, but can be interpreted in a wide variety of models.

Three important properties of an axiomatic system are *consistency*, *independence*, and *completeness*.

1.5.1 Consistency

Consider the following axiomatic system:

A1 There are exactly three points.

A2 There are exactly two points.

One may consider this a rather stupid system, and rightly so. It is basically useless as Axiom 1 contradicts Axiom 2. It would be impossible to logically deduce theorems in this system as we start out with a fundamental contradiction. The problem here is that this system is *inconsistent*. One may argue that to avoid creating an inconsistent system, we just need to choose axioms that are not self-contradictory. This is not always easy to do. It may be the case that the axioms look fine, but there is some theorem that contradicts one of the axioms. Or, perhaps two theorems contradict one another. Any of these situations would be disastrous for the system.

> **Definition 1.2.** *An axiomatic system is* consistent *if no two statements (these could be two axioms, an axiom and theorem, or two theorems) contradict each other.*

To determine whether an axiomatic system is consistent, we would have to examine every possible pair of axioms and/or theorems in the system. In systems such as Euclidean geometry, this is not possible. The best we can do is to show *relative* consistency. If we can find a model for a system—let's call it system *A*—that is embedded in another axiomatic system *B*, and if we know that system *B* is consistent, then system *A* must itself be consistent. For, if there were two statements in *A* that were contradictory, then this would be a contradiction in system *B*, when interpreted in the language of *B*. But, *B* is assumed consistent, so there can be no contradictory statements in *B*.

If we believe that Euclidean geometry is consistent, then the axiomatic system for classes and students we have described must be consistent, as it has a model (Three-Point geometry) that is embedded in Euclidean geometry.

1.5.2 Independence

> **Definition 1.3.** *An individual axiom in an axiomatic system is called* independent *if it cannot be proved from the other axioms.*

For example, consider the following system for points and lines:

A1 There are exactly three points.

A2 Two distinct points belong to one and only one line.

A3 Not all of the points belong to the same line.

A4 Two separate lines have at least one point in common.

A5 A line has exactly two points.

In the preceding section, we saw that Axiom 5 could be *proved* from the first four. Thus, Axiom 5 is not independent. While this axiomatic system would be perfectly fine to use (it is just as consistent as the original), it is not as economical as it could be. A system with numerous axioms is difficult to remember and confusing to use as a basis for proving theorems. Just as physicists search for the most simple and elegant theories to describe the structure of matter and the evolution of the universe, so too mathematicians search for the most concise and elegant basis for the foundations of their subject. However, it is often the case that the more compact an axiomatic system is, the more work has to be done to get beyond elementary results.

How do we show that an axiom is independent of the other axioms in an axiomatic system? Suppose that we could find a model for a system including Axiom X. Suppose additionally that we could find a model for the system where we replaced Axiom X by its logical negation. If both models are consistent then Axiom X must be independent of the other axioms. For if it were dependent, then it would have to be provably true from the other axioms and its negation would have to be provably false. But, if consistent models exist with both X and the negation of X valid, then X must be independent of the other axioms.

As an example, consider the system for Three-Point geometry. Suppose we replaced Axiom 3 with "All of the points belong to the same line." A model for this would be just a line with three points. This would clearly be consistent with the first two axioms. Since there is only one line in the system, Axiom 4 would be vacuously true. (Here we use the logical rule that a statement is true if the hypothesis, the existence of two lines in this case, is false.) Thus, we can say that Axiom 3 is independent of Axioms 1, 2, and 4.

1.5.3 Completeness

The last property we will consider is that of completeness.

> **Definition 1.4.** *An axiomatic system is called* complete *if it is impossible to add a new consistent and independent axiom to the system. The new axiom can use only defined and undefined terms of the original system.*

An equivalent definition would be that every statement involving defined and undefined terms is provably true or false in the system. For if there were a statement that could not be proved true or false within the system, then it would be independent of the system, and we could add it to get an additional axiom.

As an example consider the following system for points and lines:

A1 There are exactly three points.

A2 Two distinct points belong to one and only one line.

Now consider the additional statement "Not all of the points belong to the same line." (This is our original Axiom 3 from the last section.) We have already looked at a consistent model for this larger system. Now, consider the logical opposite of this statement: "All of the points belong to the same line." A consistent model for a system with axioms A1, A2, and this additional statement is that of three points on a line. Thus, the statement "Not all of the points belong to the same line" is not provably true or false within the smaller system consisting of just axioms A1 and A2, and this smaller system cannot be complete.

What of the four axioms for Three-Point geometry? Do they form a consistent system? Suppose we label the points A, B, C. We know from Theorem 1.2 and Theorem 1.3 in section 1.4 that there must be exactly three lines in Three-Point geometry and that each line has exactly two points. The three lines can then be symbolized by pairs of points: $(A, B), (A, C), (B, C)$.

Suppose we had another model of Three-Point geometry with points L, M, N. Then, we could create a one-to-one correspondence between these points and the points A, B, C so that the lines in the two models could be put into one-to-one correspondence as well. Thus, any two models of this system are essentially the same, or *isomorphic*; one is just a re-labeling of the other. This must mean that the axiomatic system is complete. For if it were not complete, then a statement about points and lines that is independent of this system could be added as a new axiom. If this were the case, we could find a consistent model not only for this new augmented system, but also for the system we get by adding the *negation* of this new axiom, and obviously these two models could not be isomorphic.

An axiomatic system is called *categorical* if all models of that system are isomorphic to one other. Thus, Three-Point geometry is a categorical system. In Chapter 7 we will see a much more powerful display of this idea of isomorphism of systems when we look in detail at the system of non-Euclidean geometry called *Hyperbolic Geometry*.

1.5.4 Gödel's Incompleteness Theorem

We cannot leave the subject of completeness without taking a side-trip into one of the most amazing results in modern mathematics—the Incompleteness Theorem of Kurt Gödel.

At the second International Congress of Mathematics, held in Paris in 1900, David Hilbert presented a lecture entitled "Mathematical Problems" in which he listed 23 open problems that he considered critical for the development of mathematics. Hilbert was a champion of the *formalist* school of thinking in mathematics, which held that all of mathematics could be built as a logical axiomatic system.

Hilbert's second problem was to show that the simplest axiomatic system, the natural numbers as defined by Peano, was consistent; that is, the system could not contain theorems yielding contradictory results.

Ten years after Hilbert posed this challenge to the mathematical community, Bertrand Russell (1872–1970) and Alfred North Whitehead (1861–1947) published the first volume of the *Principia Mathematica*, an ambitious project to recast all of mathematics as an expression of formal logic. This line of reasoning has become known as *logicism*. Russell and Whitehead succeeded in deriving set theory and natural number arithmetic from a formal logical base, yet their development still begged the question of consistency. However, the *Principia* was very influential in that it showed the power of formal logical reasoning about mathematics.

In 1931, an Austrian mathematician named Kurt Gödel (1906–1978) published a paper titled "On Formally Undecidable Propositions of Principia Mathematica and Related Systems" in the journal *Monatshefte für Mathematik und Physik*. In this paper Gödel used the machinery of formal logic to show the impossibility of the universalist approach taken by the formalists and the logicists. This approach had as its goal the development of *consistent* and *complete* systems for mathematics.

To address the question of consistency, Gödel showed the following result:

Theorem 1.4. *Given a consistent axiomatic system that contains the natural number system, the consistency of such a system cannot be proved from within the system.*

Thus, Hilbert's second problem is impossible! If one built a formal system, even one based on pure logic like Whitehead and Russell, then

Gödel's theorem implies that it is impossible to show such a system has no internal contradictions.

Even more profound was the following result of Gödel's on the completeness of systems:

Theorem 1.5 (Incompleteness Theorem). *Given a consistent axiomatic system that contains the natural number system, the system must contain* undecidable *statements, that is, statements about the terms of the system that are neither provable nor disprovable.*

In other words, a system sufficiently powerful to handle ordinary arithmetic will necessarily be incomplete! This breathtaking result was revolutionary in its scope and implication. It implied that there might be a simple statement in number theory that is undecidable, that is impossible to prove true or false. Even with this disturbing possibility, most mathematicians have accepted the fact of incompleteness and continue to prove those theorems that are decidable.

Now that we have reviewed most of the important features of axiomatic systems, we will take a closer look in the next section at the first, and for most of history the most important, axiomatic system—that of Euclid's geometry.

Exercise 1.5.1. *Consider sets as collections of objects. If we allow a set to contain objects that are themselves sets, then a set can be an element of another set. For example, if sets A, B are defined as $A = \{\{a, b\}, c\}$ and $B = \{a, b\}$, then B is an element of A. Now, it may happen that a set is an element of itself! For example, the set of all mathematical concepts is itself a mathematical concept.*

Consider the set S which consists only of those sets which are not elements of themselves. Can an axiomatic system for set theory, that allows the existence of S, be a consistent system? [Hint: Consider the statement "S is an element of itself."] The set S has become known as the Russell Set *in honor of Bertrand Russell.*

Exercise 1.5.2. *Two important philosophies of mathematics are those of the* platonists *and the* intuitionists. *Platonists believe that mathematical ideas actually have an independent existence and that a mathematician only discovers what is already there. Intuitionists believe that we have an intuitive understanding of the natural numbers and that mathematical results can and should be derived by constructive methods based on the natural numbers, and such constructions must be* finite.

Of the four main philosophies of mathematics—platonism, formalism, logi-

cism, and intuitionism—which most closely matches the way you view mathematics? Which is the prevalent viewpoint taken in the teaching of mathematics?

Exercise 1.5.3. *Gödel's Incompleteness Theorem created a revolution in the foundations of mathematics. It has often been compared to the discovery of incommensurable magnitudes by the Pythagoreans in ancient Greece. Research the discovery of incommensurables by the Pythagoreans and compare its effect on Greek mathematics with the effect of Gödel's theorem on modern mathematics.*

The exercises that follow deal with various simple axiomatic systems. To gain facility with the abstract notions of this section, carefully work out these examples.

In Four-Point geometry we have the same types of undefined terms as in Three-Point geometry, but the following axioms are used:

A1 There are exactly four points.

A2 Two distinct points belong to one and only one line.

A3 Each line has exactly two points belonging to it.

Exercise 1.5.4. *Show that there are exactly six lines in Four-Point geometry.*

Exercise 1.5.5. *Show that each point belongs to exactly three lines.*

Exercise 1.5.6. *Show that Four-Point geometry is relatively consistent to Euclidean geometry. [Hint: Find a model.]*

Exercise 1.5.7. *A regular tetrahedron is a polyhedron with four sides being equilateral triangles (pyramid-shaped). If we define a point to be a vertex of the tetrahedron and a line to be an edge, will the tetrahedron be a model for Four-Point geometry? Why or why not?*

Exercise 1.5.8. *Consider an axiomatic system that consists of elements in a set S and a set P of pairings of elements (a, b) that satisfy the following axioms:*

A1 If (a, b) is in P, then (b, a) is not in P.

A2 If (a, b) is in P and (b, c) is in P, then (a, c) is in P.

Let $S = \{1, 2, 3, 4\}$ and $P = \{(1, 2), (2, 3), (1, 3)\}$. Is this a model for the axiomatic system? Why or why not?

Exercise 1.5.9. *In the previous problem, let S be the set of real numbers and let P consist of all pairs (x, y) where $x < y$. Is this a model for the system? Why or why not?*

Exercise 1.5.10. *Use the results of the previous two exercises to argue that the axiomatic system with sets S and P is not complete. Think of another independent axiom that could be added to the axioms in Exercise 1.5.8, for which S and P from Exercise 1.5.8 is still a model, but S and P from Exercise 1.5.9 is not a model.*

Several of the examples and exercises of this section deal with axiomatic systems having a finite number of points and lines. These are called *finite geometries.* An important class of finite geometries is the class of *projective planes of order n.*

Definition 1.5. *A* projective plane of order n *has undefined terms:* point, line, and incident. *Defined terms are:*

1. Collinear*: points incident with the same line are collinear.*

2. Concurrent*: lines incident with the same point are concurrent.*

We have the following axioms for a projective plane of order n:

A1 Given two distinct points, there is exactly one line incident with them both.

A2 Given two distinct lines, there is at least one point that is incident with both lines.

A3 There are at least four points, no three of which are collinear.

A4 There exists at least one line with exactly $n + 1(n > 1)$ distinct points incident with it.

Exercise 1.5.11. *Let the points in a projective plane be the seven points A, B, C, D, E, F, G as shown. Let a line be any Euclidean segment, or the arc from E to G to F. Show that this figure is a model for the projective plane of order 2.*

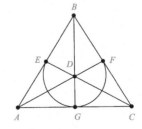

Exercise 1.5.12. *Show that for a projective plane of order n there are at least four lines, no three of which are concurrent. [Hint: By Axiom 3, there are at least four points, call them A, B, C, and D. By Axiom 1, there are four lines l_1, l_2, l_3, l_4 incident on pairs of points (A, B), (B, C), (C, D) and (A, D). Show that no three of these lines can be concurrent.]*

The last exercise shows that the *dual* to Axiom 3 holds in this finite geometry.

Definition 1.6. *The* dual *of a statement in the axiomatic system for a projective plane is obtained by replacing each occurence of the words* point *and* collinear *by* line *and* concurrent, *respectively.*

Exercise 1.5.13. *Write down and prove the dual of Axiom 1 for a projective plane of order n.*

Exercise 1.5.14. *Write down and prove the dual of Axiom 2 for a projective plane of order n.*

Exercise 1.5.15. *The dual of Axiom 4 would be: There exists at least one point with exactly $n + 1(n > 1)$ distinct lines incident with it. Prove this result. [Hint Axiom 4 guarantees the existence of a line l with $n + 1$ points, say P_1, $\ldots$, P_{n+1}. By Axiom 3, there must be a point Q that is not on l. Show there are $n + 1$ lines through Q that satisfy the dual statement of Axiom 4]*

1.6 EUCLID'S AXIOMATIC GEOMETRY

The system of deductive reasoning begun in the schools of Thales and Pythagoras was codified and put into definitive form by Euclid (ca. 325–265 BC) around 300 BC in his 13-volume work *Elements*. Euclid was a scholar at one of the great schools of the ancient world, the Museum of Alexandria, and was noted for his lucid exposition and great teaching ability. The *Elements* were so comprehensive in scope that this work superseded all previous textbooks in geometry.

What makes Euclid's exposition so important is its clarity. Euclid begins Book I of the *Elements* with a few definitions, a few rules of logic, and ten statements that are axiomatic in nature. Euclid divides these into five geometric statements which he calls *postulates*, and five *common notions*. These are listed in their entirety in Appendix B. We will keep Euclid's terminology and refer to the first five axiomatic statements—those that specifically deal with the geometric basis of his exposition—as postulates.

1.6.1 Euclid's Postulates

Euclid I To draw a straight line from any point to any point.

Euclid II To produce a finite straight line continuously in a straight line.

Euclid III To describe a circle with any center and distance (i.e., radius).

Euclid IV That all right angles are equal to each other.

Euclid V If a straight line falling on two straight lines makes the interior angles on the same side less than two right angles, the two straight lines, if produced indefinitely, meet on that side on which are the angles less than the two right angles.

The first three postulates provide the theoretical foundation for constructing figures based on a hypothetically *perfect* straightedge and compass. We will consider each of these three in more detail, taking care to point out exactly what they say and what they do *not* say.

The first postulate states that given two points, one can construct a line connecting these points. Note, however, that it does not say that there is *only* one line joining two points.

The second postulate says that finite portions of lines (i.e., segments) can be extended. It does not say that lines are *infinite* in extent, however.

The third postulate says that given a point and a distance from that point, we can construct a circle with the point as center and the distance as radius. Here, again, the postulate does not say anything about other properties of circles, such as continuity of circles.

As an example of Euclid's beautiful exposition, consider the very first theorem of Book I of the *Elements*. This is the construction of an equilateral triangle on a segment $\overline{AB}$ (Figure 1.6). The construction goes as follows:

1. Given segment $\overline{AB}$.

2. Given center A and distance equal to the length of $\overline{AB}$, construct circle c_1. (This is justified by Postulate 3.)

3. Given center B and distance equal to the length of $\overline{BA}$, construct circle c_2. (This is justified by Postulate 3.)

4. Let C be a point of intersection of circles c_1 and c_2.

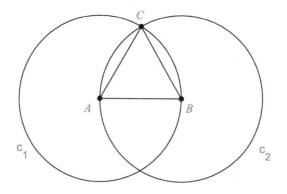

Figure 1.6 Euclid Book I, Proposition I

5. Construct segments from A to C, from C to B, and from B to A. (This is justified by Postulate 1.)

6. Since $\overline{AC}$ and $\overline{AB}$ are radii of circle c_1, and $\overline{CB}$ and $\overline{AB}$ are radii of c_2, then $\overline{AC}$ and $\overline{CB}$ are both equal to $\overline{AB}$ and, thus, must be equal to one another. (This is justified by use of the Common Notions.)

7. By definition, then, triangle ABC is equilateral.

The proof is an *almost* perfect example of a well-written mathematical argument. Euclid takes care to justify the steps in the construction in terms of the initial set of five postulates and five common notions, *except* for the step where the intersection point of the two circles is found. Euclid here assumed a principle of continuity of circles. That is, since circle $c1$ has points inside *and* outside of another circle ($c2$), then the two circles must intersect somewhere. In other words, there are no *holes* in the circles. To make this proof rigorous, we would have to add a postulate on circle continuity, or prove circle continuity as a consequence of the other five postulates.

We can see that from the very first proposition, Euclid was not logically perfect. Let's look at the last two postulates of Euclid.

Whereas the first three postulates are ruler and compass statements, the fourth postulate says that the rules of the game we are playing do not change as we move from place to place. In many of Euclid's theorems, he moves parts of figures on top of other figures. Euclid wants an axiomatic basis by which he can assume that segment lengths and angles remain unchanged when moving a geometric figure. In the fourth postulate, Euclid is saying that, at least, right angles are always equal no matter what configuration they are in.

In many of his proofs, Euclid assumes much more than Postulate 4 guarantees. For example, consider his proof of the Side-Angle-Side Congruence theorem for triangles.

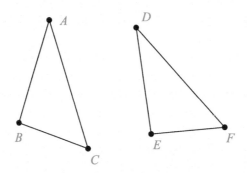

Figure 1.7 Euclid SAS theorem

Euclid starts with two triangles ABC and DEF with $\overline{AB}$ congruent to $\overline{DE}$, $\overline{AC}$ congruent to $\overline{DF}$, and angle A congruent to angle D, as shown in Figure 1.7. (For the time being, assume *congruent* to mean equal in magnitude.)

Euclid proceeds by moving angle A on top of angle D. Then, he says that since two sides are congruent and the angles at A and D are congruent, then point B must move to be on top of point E and point C must move to be on top of point F. Thus, the two triangles have to be congruent.

The movement or *transformation* of objects was not on a solid logical basis in Euclid's geometry. Felix Klein, in the late 1800s, developed an axiomatic basis for Euclidean geometry that *started* with the notion of an existing set of transformations and then constructed geometry as the set of figures that are unchanged by those transformations. We will look at transformations and transformational geometry in greater detail in Chapter 5.

The final postulate seems very different from the first four. Euclid himself waited until Proposition 29 of Book I before using the fifth postulate as a justification step in the proof of a theorem. Euclid's fifth postulate is often referred to as the *parallel postulate*, even though it is actually more of an "anti-parallel" postulate as it specifies a sufficient condition for two lines to intersect.

From the time of Euclid, the fifth postulate's axiomatic status has been questioned. An axiom should be a statement so obvious that it can be accepted without proof. The first four of Euclid's postulates are simple statements about the construction of figures, statements that resonate with our practical experience. We can draw lines and circles and can measure angles. However, we cannot follow a line indefinitely to see if it intersects another line.

Many mathematicians tried to find simpler postulates, ones that were more intuitively believable, to replace Euclid's fifth postulate, and then prove the postulate as a *theorem*. Others tried to prove Euclid's fifth postulate *directly* from the first four. That is, they tried to show that the fifth postulate was not *independent* of the other four.

One of the attempts to find a simpler postulate for Euclid's fifth was that of John Playfair (1748–1819). His substitute can be expressed as follows:

> Given a line and a point not on the line, it is possible to construct one and only one line through the given point parallel to the line.

This statement, which has become known as *Playfair's Postulate*, is certainly easier to read and understand when compared to Euclid's fifth postulate. However, it is not hard to show that this statement is logically equivalent to Euclid's fifth postulate and, thus, does not really *simplify* Euclid's system at all. A full discussion of this and other statements about parallelism can be found in the next chapter.

Other mathematicians tried to prove that Euclid's fifth postulate was actually a *theorem* that could be derived from the first four postulates. A popular method of attack was to assume the logical opposite of Euclid's fifth postulate and try to prove this new statement false, or find a contradiction to an already accepted result. Amazingly, no one could prove that the negation of the fifth postulate was false or produced a contradiction.

In hindsight, this persistent lack of success would seem to imply that there could be consistent *non-Euclidean* geometries, obtained by replacing the fifth postulate by its opposite. However, this possibility ran counter to the generally held belief that Euclidean geometry was the only consistent geometry possible. This belief had been a bedrock of philosophy from the time of Euclid, and it was not until the 1800s that non-Euclidean geometry was fully explored. Even then, Carl Frederich Gauss (1777–1855), who was undoubtedly the first to recognize the consistency of non-Euclidean geometry, refrained from publishing results in this area. It wasn't until the work of Janos Bolyai (1802–1860) and Nicolai Lobachevsky (1793–1856) that theorems resulting from non-Euclidean replacements for Euclid's fifth postulate appeared in print. We will look at the fascinating development of non-Euclidean geometry in more detail in Chapter 7.

To end this discussion of Euclid's original axiomatic system, let's consider the system in terms of the properties discussed in the last section. Perhaps the most important question is whether this system is complete. A careful reading of the postulates shows that Euclid *assumed* the existence of points with which to construct his geometry. However, none of the postulates implies that points exist! For example, the statement "There exists at least two points" cannot be proven true or false from the other postulates. Thus, Euclid's axiomatic system is not complete.

Euclid's work contains quite a few *hidden* assumptions that make his axiomatic system far from complete. However, we should not be too hard on Euclid. It was only in the late 1800s that our modern understanding of abstract axiomatic systems and their properties was developed. No theorem that Euclid included in the *Elements* has ever been shown to be false, and various attempts to provide better axioms for Euclidean geometry have served to shore up the *foundations* of the subject, but the results that Euclid proved thousands of years ago are just as true today.

The first geometric axiomatic system to be shown logically consistent (at least relative to the consistency of natural number arithmetic) was the system developed by David Hilbert in 1899 [22]. Other modern axiomatic systems for geometry have been developed since then and it would be interesting to compare and contrast these. However, having a good axiomatic foundation is only the first step in exploring geometry. One must also *do* geometry, that is, one must experiment with geometric concepts and their patterns and properties. In the next chapter we will start with a basic set of geometric results and assumptions that will get us up and running in the grand game of Euclidean geometry.

Exercise 1.6.1. *Euclid's* Elements *came to be known in Europe through a most circuitous history. Research this history and describe the importance of*

Arabic scholars to the development of Western mathematics of the late Middle Ages.

Exercise 1.6.2. *The 13 books of Euclid's* Elements *contain a wealth of results concerning not only plane geometry, but also algebra, number theory, and solid geometry. Book II is concerned with geometric algebra, algebra based on geometric figures. For example, the first proposition of Book II is essentially the distributive law of algebra, but couched in terms of areas. Show, using rectangular areas, how the distributive law $(a(b+c) = ab + ac)$ can be interpreted geometrically.*

Exercise 1.6.3. *Show how the algebraic identity $(a + b)^2 = a^2 + 2ab + b^2$ can be established geometrically.*

Exercise 1.6.4. *Euclid's* Elements *culminates in Book XIII with an exploration of the five* Platonic Solids. *Research the following questions:*

- *What are the Platonic Solids? Briefly discuss why there are only five such solids.*

- *The solids have been known for thousands of years as* Platonic *solids. What is the relationship to Plato and how did Plato use these solids in his explanation of how the universe was constructed?*

Exercise 1.6.5. *The* Elements *is most known for its development of geometry, but the 13 books also contain significant non-geometric material. In Book VII of the* Elements, *Euclid explores basic number theory. This book starts out with one of the most useful algorithms in all of mathematics, the* Euclidean Algorithm. *This algorithm computes the* greatest common divisor *(gcd) of two integers. A* common divisor *of two integers is an integer that divides both. For example, 4 is a common divisor of 16 and 24. The* greatest common divisor *is the largest common divisor. The Euclidean Algorithm works as follows:*

- *Given two positive integers a and b, assume $a \geq b$.*

- *Compute the quotient q_1 and remainder r_1 of dividing a by b. That is, find integers q_1, r_1 with $a = q_1 b + r_1$ and $0 \leq r_1 < b$. If $r_1 = 0$, we stop and b is the $\gcd(a, b)$.*

- *Otherwise, we find the quotient of dividing b by r_1, that is, find q_2, r_2 with $b = q_2 r_1 + r_2$ and $0 \leq r_2 < r_1$. If $r_2 = 0$, we stop and r_1 is the $\gcd(a, b)$.*

- *Otherwise, we iterate this process of dividing each new divisor $(r_1, r_2,$ etc.) by the last remainder $(r_2, r_3,$ etc.), until we finally reach a new remainder of 0. The last non-zero remainder is then the gcd of a, b.*

Use this algorithm to show that the $\gcd(36, 123) = 3$.

Exercise 1.6.6. *Prove that the Euclidean Algorithm works, that it does find the gcd of two positive integers.*

Suppose we re-interpreted the term *point* in Euclid's postulates to mean a point on a sphere and a *line* to be a part of a great circle on the sphere. A *great circle* is a circle on the sphere cut by a plane passing through the center of the sphere.

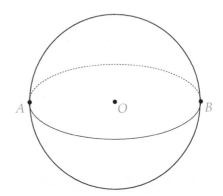

To gain an intuitive feel for this geometry, one could take a ball (beach ball, bowling ball, soccer ball, etc.) and a string and experiment with constructing lines (segments) on the sphere. Distance in this new geometry will be measured along lines (segments) on the sphere. With this notion of distance, circles will be defined analogously with the Euclidean definition of circles. Angles will be defined by tangent directions defined by lines on the sphere.

With this basis of definitions and interpretations for lines, points, circles, and angles, we will consider whether Euclid's postulates hold in this new *spherical geometry*.

Exercise 1.6.7. *Show that Euclid's first postulate holds for* most *pairs of points, but not all. Show that the second, third, and fourth postulates do hold in spherical geometry. Show that the fifth postulate (or equivalently Playfair's Postulate) does* not *hold. (Give general explanations for your answers — we do not yet have the resources to do rigorous proofs in spherical geometry)*

In the following exercises, experiment to determine if each statement is most likely true or false in spherical geometry. Give short explanations (not proofs) for your answers.

Exercise 1.6.8. *The sum of the angles of a triangle is 180 degrees.*

Exercise 1.6.9. *Given a line and a point not on the line, there is a perpendicular to the line through that point.*

Exercise 1.6.10. *A four-sided figure with three right angles must be a rectangle.*

Exercise 1.6.11. *One can construct a triangle with three right angles.*

1.7 PROJECT 2 - A CONCRETE AXIOMATIC SYSTEM

We have seen that the axioms in an axiomatic system are the statements about the terms of the system that must be accepted without proof. Axioms are generally statements that are written down at the start of the development of some mathematical area.

When exploring geometry using a dynamic geometry program, there are similarly constructions and tools that one uses *without proof*. That is, we draw lines and circles, find intersections, and construct perpendiculars using the tools built into the software, assuming that these constructions are valid. We might say that the abstract Euclidean axiomatic system has been made *concrete* within the software world of the program.

When one first starts up a dynamic geometry program, constructions are based on the Euclidean axiomatic system. However, there is another axiomatic system that one can explore in most geometry programs. This is the *Poincaré Disk* model of Hyperbolic Geometry. The most natural implementation of this model is found in *Geometry Explorer*, but there are versions of this model available in all major dynamic geometry software packages. Before you start this project, go to http://www.gac.edu/~hvidsten/geom-text for instructions on how to download the hyperbolic geometry system for your particular software environment.

To start this project, open the *Poincaré Disk* model for your particular geometry software. The world for hyperbolic geometry will be the *interior* of the disk that you see on the screen. The Euclidean points on the boundary circle are not points in this new geometry. If one lived in this two-dimensional geometry, one could approach the boundary circle, but never actually reach it.

The dynamic software environment you are using will provide a set of tools for this new geometry, just as we had for Euclidean geometry. In Figure 1.8, we see the environment for the *Poincaré Disk* model as you would see it in *Geometry Explorer*.

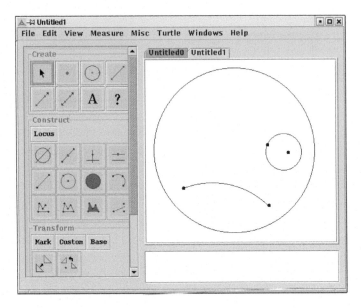

Figure 1.8 A concrete axiomatic system

Try creating some segments, lines, rays, and circles using the new tools for the *Poincaré Disk* model.

How is this system like Euclidean geometry and how is it different? Let's review Euclid's first postulate—that between any two points we can draw a segment. Experiment with creating segments on the screen. We know that points in this new geometry are restricted to points within the boundary disk. Thus, it does appear that given any two points *inside* this disk, we can always create a *segment* connecting those points. It is true that segments do not look "normal" in the sense that they are not straight. However, we must use the axiomatic view in this new geometry. That is, a segment is whatever gets created when we use the Segment tool. Thus, it appears that Euclid's first postulate holds in this new geometry.

Euclid's second postulate states that segments can always be extended. Before we explore the validity of this postulate in our new geometry, let's consider what is meant by *extending* segments. In Euclidean geometry we can extend a segment $\overline{AB}$ by *moving* one of the endpoints, say B, a certain distance in the direction determined by the segment $\overline{AB}$. As has been mentioned before, the movement of objects in Euclidean geometry is not specified in the postulates, but is nevertheless assumed by Euclid in many of his proofs.

Create a segment on the screen and drag one of the endpoints towards the boundary. It appears that we get "stuck" at the boundary, but this is merely a limitation of the finite nature of the software (mouse movement and pixel resolution) we are using. Since the boundary can never be reached, then, no matter how close we are, we can always move a little bit closer. So, it appears that Euclid's second postulate holds.

As one approaches the boundary, strange things happen in this geometry. If one creates a segment and then selects and drags that segment towards the boundary, it appears to change length. Using your hyperbolic geometry tools, measure the length of the segment and then drag the segment around inside the disk. The length does not change!

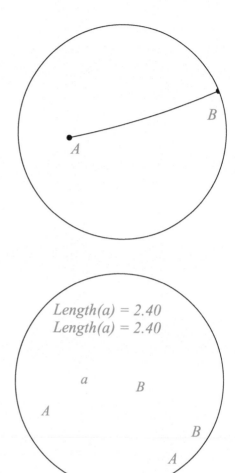

Even though the shape and length appear to change (to us outsiders looking at the geometry), the shape and length actually do *not* change, as viewed internally to the geometry. To try and make sense of this, imagine you were a two-dimensional creature living entirely within this geometry and you had no idea there was an *outside* to your world. Imagine dragging a piece of rope (a segment) around your world. You would notice no difference in the rope's shape and length, even though from the outside things look much different.

Let's consider Euclid's third postulate on the constructibility of circles.

Clear the screen and create a variety of circles on the screen using the circle construction tool for your software. For later use, let's measure the radius of one of these circles.

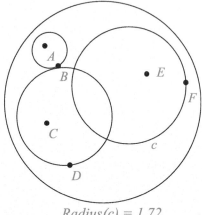

Radius(c) = 1.72

If we drag one of these circles towards the boundary, we see that the circle appears to change shape and size. However, the measured radius does not change. We also note that since our circles seem to shrinking to nothing as we approach the boundary, then we can cram circles of ever larger radii into the space near the boundary.

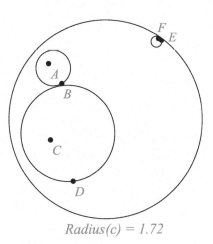

Radius(c) = 1.72

Thus, it appears that circles with any center and any specified length can be constructed in this geometry.

Let's now consider the fourth Euclidean axiom—that all right angles are congruent. Let's consider not just a right angle, but *any* angle.

Clear the screen and create a
segment $\overline{AB}$ near the center of
the disk. Then, create another
segment $\overline{AC}$ originating from A.
Next, measure the angle created
by A, B, C.

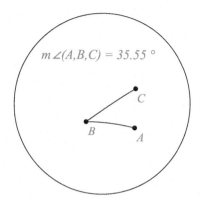

Now, select the three points A, B,
and C in a group and drag them
with the mouse, moving the angle around the screen. Verify that
movement of the angle has no effect on the angle measurement.

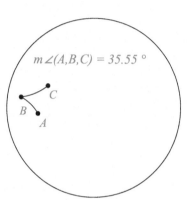

We see that the first four of Euclid's axioms still hold in this strange
new geometry. What about the parallel axiom?

Clear the screen and create a line
l near the bottom. Then, create a
point P above l. As shown in the
figure at the right, one can create many lines through P that do
not intersect l, that is, are *parallel*
to l. In fact, there should be two
unique parallels through P that
are *just* parallel to l. These are
lines n and m in the figure.

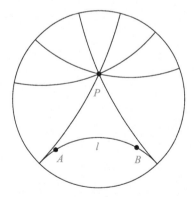

The geometry we have been discussing in this project is called *hyperbolic* geometry. From the last construction above on parallel lines we
see that Euclid's fifth postulate does *not* hold in this geometry, and so

we have had our first exploration of a *non-Euclidean* geometry. We will cover this geometry in more depth in Chapter 7.

Exercise 1.7.1. *Determine which of the following Euclidean properties holds in the Hyperbolic geometry of* Geometry Explorer.

- *Rectangles can be constructed. (Try any construction you remember from high school geometry.)*

- *The sum of the angles of a triangle is 180 degrees.*

- *Euclid's construction of an equilateral triangle.*

- *Given a line and a point not on the line there is a perpendicular to the line through that point.*

Project Report

Your report should include a general discussion of how this new system of geometry is similar to Euclidean geometry and how it is different. This discussion should reference the five postulates as we have explored them in this chapter, and should include your analysis from Exercise 1.7.1.

CHAPTER 2

Euclidean Geometry

Euclid alone has looked on Beauty bare.
Let all who prate of Beauty hold their peace,
And lay them prone upon the earth and cease
To ponder on themselves, the while they stare
At nothing, intricately drawn nowhere
In shapes of shifting lineage; let geese
Gabble and hiss, but heroes seek release
From dusty bondage into luminous air.
O blinding hour, O holy, terrible day,
When first the shaft into his vision shone
Of light anatomized! Euclid alone
Has looked on Beauty bare. Fortunate they
Who, though once only and then but far away,
Have heard her massive sandal set on stone.

– Edna St. Vincent Millay (1892–1950)

In the first chapter we saw that the origins of our modern views on geometry can be traced back to the work of Euclid and earlier Greek mathematicians. However, Euclid's axiomatic system is not a *complete* axiomatic system. Euclid makes many assumptions that are never formalized as axioms or theorems. These include assumptions about the existence of points, the unboundedness and continuity of space, and the transformation of figures.

Since the time of Euclid, mathematicians have attempted to "complete" Euclid's system by either supplementing Euclid's axioms with additional axioms or scrapping the whole set and replacing them with an entirely new set of axioms for Euclidean geometry.

Hilbert's *Grundlagen der Geometrie* [22] is the best example of a consistent formal development of geometry in the spirit of Euclid. Hilbert uses six undefined terms (*point, line, plane, incident* (or *on*), *between, congruent*) and a set of 21 axioms as a foundation for Euclidean geometry. Hilbert's axioms for plane geometry are included in Appendix D. Some of Euclid's axioms are repeated in Hilbert's axioms and others are provable as theorems. Is Hilbert's system complete? Is it consistent? Axioms V-1a and V-2b in Appendix D can be used to show that any model of Hilbert's axiomatic system is *isomorphic* to the real Cartesian Plane [19, page 70]. Assuming real arithmetic is consistent, then Hilbert's system is consistent as well. Also, completeness of Hilbert's system translates to the question of completeness of an axiomatic system for the real numbers.

A very different axiomatic development of Euclidean geometry was created in 1932 by G. D. Birkhoff [6]. This system departs significantly from Euclid's style of using purely *geometric* constructions in axioms and theorems and assumes at the outset an arithmetic structure of the real numbers. The system has only four axioms and two of these postulate a correspondence between numbers and geometric angles and lengths. From this minimal set of axioms (fewer than Euclid!), Birkhoff develops planar Euclidean geometry, and his development is as logically consistent as Hilbert's. Birkhoff achieves his economy of axioms by assuming the existence of the real numbers and the vast array of properties that entails, whereas Hilbert has to construct the reals from purely geometric principles.

We will not take the time at this point to go through the sequence of proofs and definitions needed to base Euclidean geometry firmly and carefully on Hilbert's axioms or Birkhoff's axioms. Hilbert's approach to geometry is covered in detail in the on-line chapter on Hilbert's axioms found at http://www.gac.edu/~hvidsten/geom-text. The reader who wishes to have this foundational material covered first, should jump to that chapter before continuing with the current chapter.

Birkhoff's axiomatic system is described in Appendix C and also discussed at the end of Chapter 3. If one wants to cover the analytic, coordinate system basis for geometry, one should jump to that appendix and chapter before continuing.

In this chapter we will first review some of the basic results from classical Euclidean geometry, referencing those theorems that are most important for doing geometry. Again, for complete proofs and a careful development of this material, the reader is referred to the on-line chapter on Hilbert's axioms. We will then look at some interesting results not often covered in high school geometry. We will take care to provide precise definitions of terms and to point out areas where the work of Hilbert and Birkhoff has filled in the "gaps" of Euclid described above.

2.1 ANGLES, LINES, AND PARALLELS

The simplest figures in geometry are points, lines, and angles. The terms *point* and *line* will remain undefined, and an *angle* will be defined in terms of *rays*, which are themselves defined in terms of *segments*. We used many of these terms in the last chapter without really giving them a precise definition. To begin our review of Euclidean geometry, it is critical to have a solid set of definitions for these basic terms.

Definition 2.1. *A* segment $\overline{AB}$ *consists of points A and B and all those points C on the line through A and B such that C is between A and B.*

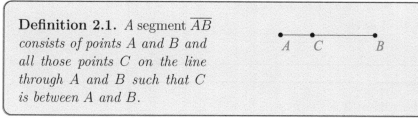

Note that this definition relies on the first of Euclid's postulates, the postulate on the existence of lines. Also, this definition uses an undefined term—*between*. The notion of the ordering of points on a line is a critical one, but one which is very hard to define. Hilbert, in his axiomatic system, leaves the term *between* undefined and establishes a set of axioms on how the quality of *betweenness* works on points. We will assume that *betweenness* works in the way our intuition tells us it should.

Also, note that this definition is not very useful unless we actually have points to work with and can tell when they lie on a line. Here, again, Hilbert introduces a set of axioms about the existence of points and how the property of being *on* a line works, the term *on* being left undefined.

Definition 2.2. *A* ray $\overrightarrow{AB}$ *consists of the segment $\overline{AB}$ together with those points C on the line through A and B such that B is between A and C.*

Definition 2.3. *The* angle with vertex *A consists of the point A together with two rays* $\overrightarrow{AB}$ *and* $\overrightarrow{AC}$ *(the* sides *of the angle). We denote an angle with vertex A and sides* $\overrightarrow{AB}$ *and* $\overrightarrow{AC}$ *by* $\angle BAC$.

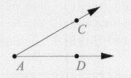

Definition 2.4. *Two angles with a common vertex and whose sides form two lines are called* vertical *angles.*

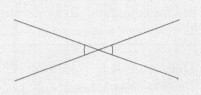

Definition 2.5. *Two angles are* supplementary *if they share a common side and the other sides lie in opposite directions on the same line. (Here* $\angle BAC$ *and* $\angle CAD$ *are supplementary.)*

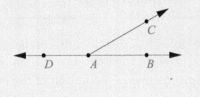

We will often talk of *congruent* angles and segments. Congruence is another of those terms that is difficult to define explicitly. For Euclid, congruence of angles or segments meant the ability to move an angle or segment on top of another angle or segment so that the figures coincided. This transformational geometry pre-supposes an existing set of transformations that would itself need an axiomatic basis. Hilbert leaves the term *congruence* undefined and provides a set of axioms as to how the quality of *congruence* works for angles, as well as for segments.

When speaking of two angles being congruent, we will assume that we have a way of associating an angle with a real number, called the *degree measure* of the angle, in such a way that

- The measure of an angle is a real number between 0 and 180 degrees.

- Congruent angles have the same angle measure.

- The measure of a right angle is 90.

- If two angles are supplementary, then the sum of the measures of the angles is 180.

Definition 2.6. *The* degree measure *of* $\angle ABC$ *will be denoted by* $m\angle ABC$.

Similarly, for any segment $\overline{AB}$ we will assume that we have a way of associating a real number, called the *length* of $\overline{AB}$, satisfying the usual notions of Euclidean length. For example, congruent segments will have the same *length*, and if B is between A and C, then the length of $\overline{AC}$ is equal to the sum of the lengths of $\overline{AB}$ and $\overline{BC}$.

Definition 2.7. *The* length *of* $\overline{AB}$ *will be denoted by* AB.

In classical Euclidean geometry, the *orientation* of an angle is not significant. Thus, $\angle ABC$ is congruent to $\angle CBA$, and both angles have the same measure, which is less than or equal to 180 degrees. This is distinctly different from the notion of angle commonly found in trigonometry, where orientation is quite significant. Be careful of this distinction between oriented and unoriented angles, particularly when using *Geometry Explorer*, as the program measures angles as *oriented* angles.

Definition 2.8. *A* right angle *is an angle that has a supplementary angle to which it is congruent.*

Definition 2.9. *Two lines that intersect are* perpendicular *if one of the angles made at the intersection is a right angle.*

Definition 2.10. *Two lines are* parallel *if they do not intersect.*

> **Definition 2.11.** *A* bisector *of a segment* $\overline{AB}$ *is a point C on the segment such that* $\overline{AC}$ *is congruent to* $\overline{CB}$. *A bisector of an angle* $\angle BAC$ *is a ray* $\overrightarrow{AD}$ *such that* $\angle BAD$ *is congruent to* $\angle DAC$.

Now that we have a good base of precise definitions, let's review some of the basic results on points, lines, and angles from the first 28 propositions of Book I of Euclid's *Elements*. We will review these propositions without proof. Proofs can be found in the on-line chapter on Hilbert's axioms.

> **Theorem 2.1.** *(Prop. 9, 10 of Book I) Given an angle or segment, the bisector of that angle or segment can be constructed.*

We point out that to *construct* a geometric figure we mean a very specific geometric process, a sequence of steps carried out using only a straightedge and compass. Why just these two types of steps? Clearly, if one does geometry on paper (or sand), the most basic tools would be a straightedge, for drawing segments, and a compass, for drawing circles. Euclid, in his first three axioms, makes the assumption that there are *ideal* versions of these tools that will construct *perfect* segments and circles. Euclid is making an abstraction of the concrete process we carry out to draw geometric figures. By doing so, he can prove, without a shadow of a doubt, that properties such as the apparent equality of vertical angles are *universal* properties and not just an artifact of how we draw figures.

For example, how would we use a straightedge and compass to bisect an angle? The following is a list of steps that will bisect $\angle BAC$ (see Figure 2.1). The proof is left to the reader as an exercise at the end of the next section.

- At A construct a circle of radius $\overline{AB}$, intersecting $\overrightarrow{AC}$ at D.

- At B construct a circle of radius $\overline{BD}$ and at D construct a circle of radius $\overline{BD}$. These will intersect at E.

- Construct ray $\overrightarrow{AE}$. This will be the bisector of the angle.

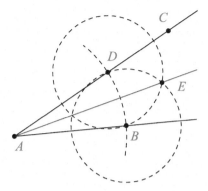

Figure 2.1 Bisector of an angle

A more complete discussion of constructions can be found in Chapter 4. You may wish to review some of the simple constructions in the first section of that chapter before continuing.

Theorem 2.2. *(Prop. 11, 12 of Book I) Given a line and a point either on or off the line, the perpendicular to the line through the point can be constructed.*

Theorem 2.3. *(Prop. 13, 15 of Book I) If two straight lines cross one another they make the vertical angles congruent. Also, the angles formed by one of the lines crossing the other either make two right angles or will have angle measures adding to two right angles. In Figure 2.2, ∠BEC and ∠AED are congruent and* $m\angle BEC + m\angle CEA = 180.$

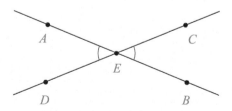

Figure 2.2 Angles made by lines

Theorem 2.4. *(Prop. 14 of Book I) If the measures of two adjacent angles add to two right angles (180 degrees), then the angles form a line.*

Theorem 2.5. *(Prop. 23 of Book I) Given an angle, a line (or part of a line), and a point on the line, an angle whose vertex is at the point, and whose measure is equal to the measure of the given angle, can be constructed.*

For the next set of theorems, we need a way of specifying types of angles created by a line crossing two other lines.

Definition 2.12. *Let t be a line crossing lines l and m and meeting l at A and m at A', with A ≠ A' (Figure 2.3). Choose points B and C on either side of A on l and B' and C' on either side of A' on m, with C and C' on the same side of t.*

Then ∠CAA', ∠C'A'A, ∠BAA', and ∠B'A'A are called interior *angles (the angles having $\overrightarrow{AA'}$ as a side). Also, ∠CAA' and ∠B'A'A are called* alternate interior angles, *as are ∠C'A'A and ∠BAA'. All other angles formed are called* exterior *angles. Pairs of nonadjacent angles, one interior and one exterior, on the same side of the crossing line t are called* corresponding *angles.*

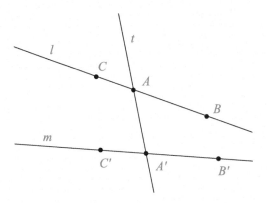

Figure 2.3 Interior-exterior angles

Definition 2.13. *We will say an angle is* greater *than another angle if its angle measure is greater than the other angle's measure.*

Theorem 2.6. *(The Exterior Angle Theorem, Prop. 16 of Book I) Given* $\triangle ABC$, *if one of the sides* $(\overline{AC})$ *is extended, then the exterior angle produced* $(\angle DAB)$ *is greater than either of the two interior and opposite angles* $(\angle BCA$ *or* $\angle ABC)$ *as shown in Figure 2.4.*

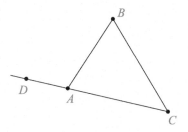

Figure 2.4

In the proof of this result found in the on-line chapter on Hilbert's axioms, we note that the proof uses the idea of extending a segment along a line, which is the focus of Euclid's second axiom. Geometers have often intuitively understood this axiom to imply that a line has infinite extent, or length. A careful reading of the second Euclidean axiom does not support this understanding.

In fact, consider the "spherical" geometry introduced in Exercise 1.6.7 of Chapter 1. Here we interpreted the term *point* to mean a point on a sphere and a *line* to be a part of a great circle on the sphere. In this geometry segments can always be extended (you can always move around the great circle, even though you may re-trace your path), but lines are not infinite in extent.

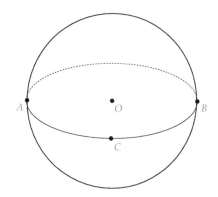

We will return to this question of the boundedness of lines in Chapter 8, where we develop a non-Euclidean geometry akin to that of the sphere.

Theorem 2.7. *(Prop. 27 of Book I) If a line n falling on two lines l and m makes the alternate interior angles congruent to one another, then the two lines l and m must be parallel (Figure 2.5).*

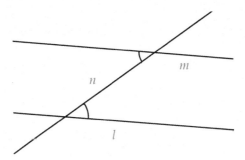

Figure 2.5 Alternate interior angles

Theorem 2.8. *(Prop. 28 of Book I) If a line n falling on two lines l and m makes corresponding angles congruent, or if the sum of the measures of the interior angles on the same side equal two right angles, then l and m are parallel (Figure 2.6).*

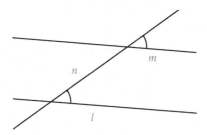

Figure 2.6 Angles and parallels

We note here that all of the preceding theorems are *independent* of the fifth Euclidean postulate, the parallel postulate. That is, they can be proved from an axiom set that does not include the fifth postulate. Such results form the basis of what is called *absolute* or *neutral* geometry. All of the theorems and results found in the on-line chapter on Hilbert's axioms, except the section on Euclidean Geometry, are independent of the parallel postulate and thus form the basis for neutral geometry.

The first time that Euclid actually used the fifth postulate in a proof was for Proposition 29 of Book I. Recall that the fifth postulate states:

> If a straight line falling on two straight lines make the interior angles on the same side less than two right angles, the two straight lines, if produced indefinitely, meet on that side on which are the angles less than the two right angles.

As was noted in the last chapter, many mathematicians have attempted to prove this postulate or replace it with a more palatable alternative. One of these alternatives is Playfair's Postulate, which we restate here:

> Given a line and a point not on the line, it is possible to construct one and only one line through the given point parallel to the line.

Let's see how Playfair's Postulate can be used to prove Proposition 29, which is essentially the converse of Propositions 27 and 28.

Theorem 2.9. *(Prop. 29 of Book I) If a line n falls on two parallel lines l and m, then alternate interior angles are congruent, corresponding angles are congruent, and the sum of the measures of the interior angles on the same side of n is equal to two right angles (see Figure 2.7).*

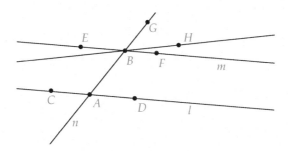

Figure 2.7 Proposition 29

Proof: Let n be the line through A and B and let the two parallel lines be ones through A, D and B, F. We will prove the statement about the corresponding angles and leave the other two parts of the proof as an exercise.

Assume $\angle FBG$ is not congruent to $\angle DAB$. We may assume that $\angle FBG$ is greater than $\angle DAB$. By Theorem 2.5 we can create a new angle $\angle HBG$ on $\overrightarrow{BG}$ so that $\angle HBG$ is congruent to $\angle DAB$. By Theorem 2.8 the line through H and B will be parallel to $\overleftrightarrow{AD}$. But this means that there are two different lines through B that are parallel to $\overleftrightarrow{AD}$, which contradicts Playfair's Postulate. □

In Euclid's original proof of Proposition 29, he bases the proof on the fifth Euclidean postulate, whereas we have used Playfair's Postulate. Actually, these two postulates are *logically equivalent*. Either postulate, when added to Euclid's first four postulates, produces an equivalent axiomatic system. How can we prove this?

The simplest way to show logical equivalence is to prove that each postulate is a *theorem* within the other postulate's axiomatic system. To show this, we need to prove two results: first, that Playfair's Postulate can be derived using the first five Euclidean postulates and second, that Euclid's fifth postulate can be derived using the first four Euclidean postulates plus Playfair's Postulate. These derivations are shown in the exercises.

Exercise 2.1.1. *Finish the proof of Theorem 2.9. That is, show that $\angle DAB$ is congruent to $\angle EBA$ and that $m\angle DAB + m\angle ABF = 180$.*

Exercise 2.1.2. *Use the Exterior Angle Theorem to show that the sum of the measures of two interior angles of a triangle is always less than 180 degrees.*

It is critical to have a clear understanding of the terms of an axiomatic system when working on a mathematical proof. One must use *only* the facts given in the definitions and not impose preconceived notions on the terms.

Exercise 2.1.3. *Each of the following statements are about a specific term of this section. Determine, solely on the basis of the definition of that term, if the statement is true or false.*

(a) A right angle *is an angle whose measure is 90 degrees.*

(b) An angle *is the set of points lying between two rays that have a common vertex.*

(c) An exterior angle *results from a line crossing two other lines.*

(d) A line *is the union of two opposite rays.*

Exercise 2.1.4. *In this exercise we will practice defining terms. Be careful to use only previously defined terms, and take care not to use imprecise and colloquial language in your definitions.*

(a) Define the term midpoint *of a segment.*

(b) Define the term perpendicular bisector *of a segment.*

(c) Define the term triangle *defined by three non-collinear points* A, B, C.

(d) Define the term equilateral triangle.

In the next set of exercises, we consider the logical equivalence of Euclid's fifth postulate with Playfair's Postulate. You may use any of the first 28 Propositions of Euclid (found in Appendix B) and/or any of the results from this section for these exercises.

Exercise 2.1.5. *In this exercise you are to show that Euclid's fifth postulate implies Playfair's Postulate. Given a line* l *and a point* A *not on* l, *we can copy* ∠CBA *to* A *to construct a parallel line* n *to* l. *(Which of Euclid's first 28 Propositions is this based on?) Suppose that there was another line* t *through* A *that was not identical to* n. *Use Euclid's fifth postulate to show that* t *cannot be parallel to* l *(Figure 2.8).*

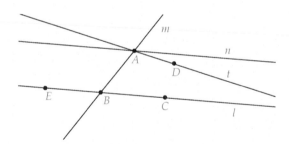

Figure 2.8

Exercise 2.1.6. *Now, we will prove the converse of the preceding exercise, that Playfair's Postulate implies Euclid's fifth postulate. Consider Figure 2.8. Suppose line* m *intersects lines* t *and* l *such that the measures of angles* ∠CBA *and* ∠BAD *add up to less than* 180 *degrees. Copy* ∠CBA *to* A, *creating line* n, *and then use Playfair's Postulate to argue that lines* t *and* l *must intersect. We now need to show that the lines intersect on the same side of* m *as* D *and* C. *We will prove this by contradiction. Assume that* t *and* l *intersect on the other* side of m *from point* C, *say at some point* E. *Use the exterior angle* ∠CBA *to triangle* △ABE *to produce a contradiction.*

Exercise 2.1.7. *Show that Playfair's Postulate is equivalent to the statement, Whenever a line is perpendicular to one of two parallel lines, it must be perpendicular to the other.*

Exercise 2.1.8. *Given triangle $\triangle ABC$, construct a parallel to $\overleftrightarrow{BC}$ at A. (How would we do this?) Use this construction to show that Playfair's Postulate (or Euclid's fifth) implies that the angle sum in a triangle is 180 degrees, namely, equal to two right angles. (The converse to this statement is also true: If the angle sum of a triangle is always 180 degrees, then Playfair's Postulate is true. For a proof see [43], [pages 21–23].)*

Exercise 2.1.9. *Show that Playfair's Postulate is equivalent to the statement, Two lines that are parallel to the same line are coincident (the same) or themselves parallel.*

Exercise 2.1.10. *Show that Playfair's Postulate is equivalent to the statement, If a line intersects but is not coincident with one of two parallel lines, it must intersect the other.*

2.2 CONGRUENT TRIANGLES AND PASCH'S AXIOM

One of the most useful tools in a geometer's toolbox is that of congruence, especially triangle congruence. In this section we will review the basic triangle congruence results found in Propositions 1–28 of Book I of *Elements*.

> **Definition 2.14.** *Two triangles are congruent if and only if there is some way to match vertices of one to the other such that corresponding sides are congruent and corresponding angles are congruent.*

If $\triangle ABC$ is congruent to $\triangle XYZ$, we shall use the notation $\triangle ABC \cong \triangle XYZ$. We will use the symbol "$\cong$" to denote congruence in general for segments, angles, and triangles. Thus, $\triangle ABC \cong \triangle XYZ$ if and only if

$$\overline{AB} \cong \overline{XY}, \overline{AC} \cong \overline{XZ}, \overline{BC} \cong \overline{YZ}$$

and

$$\angle BAC \cong \angle YXZ, \angle CBA \cong \angle ZYX, \angle ACB \cong \angle XZY$$

Let's review a few triangle congruence theorems.

Theorem 2.10. (SAS: Side-Angle-Side, *Prop. 4 of Book I) If in two triangles there is a correspondence such that two sides and the included angle of one triangle are congruent to two sides and the included angle of another triangle, then the triangles are congruent.*

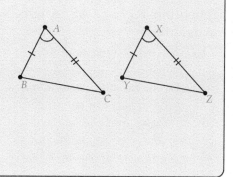

This proposition is one of the *axioms* in Hilbert's axiomatic basis for Euclidean geometry. Hilbert chose to make this result an axiom rather than a theorem to avoid the trap that Euclid fell into in his proof of the SAS result. In Euclid's proof, he *moves* points and segments so as to overlay one triangle on top of the other and thus prove the result. However, there is no axiomatic basis for such transformations in Euclid's original set of five postulates. Most modern treatments of Euclidean geometry assume SAS congruence as an axiom. Birkhoff chooses a slightly different triangle comparison result, the SAS condition for triangles to be *similar*, as an axiom in his development of Euclidean geometry.

Theorem 2.11. (ASA: Angle-Side-Angle, *Prop. 26 of Book I) If in two triangles there is a correspondence in which two angles and the included side of one triangle are congruent to two angles and the included side of another triangle, then the triangles are congruent.*

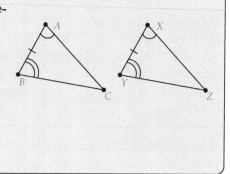

Theorem 2.12. *(AAS: Angle-Angle-Side, Prop. 26 of Book I)* *If in two triangles there is a correspondence in which two angles and the side subtending one of the angles are congruent to two angles and the side subtending the corresponding angle of another triangle, then the triangles are congruent.*

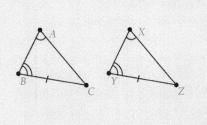

Theorem 2.13. *(SSS: Side-Side-Side, Prop. 8 of Book I)* *If in two triangles there is a correspondence in which the three sides of one triangle are congruent to the three sides of the other triangle, then the triangles are congruent.*

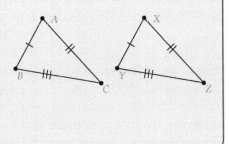

We note here for future reference that the four fundamental triangle congruence results are independent of the parallel postulate; that is, their proofs do not make reference to any result based on the parallel postulate.

Let's see how triangle congruence can be used to analyze isosceles triangles.

Definition 2.15. *An* isosceles triangle *is a triangle that has two sides congruent. The two congruent sides are called the* legs *of the triangle and the third side is called the* base. *The* base angles *of the triangle are those angles sharing the base as a side.*

Isosceles triangles were a critical tool for many of Euclid's proofs, and he introduced the next result very early in the *Elements*. It followed immediately after SAS congruence (Proposition 4).

Theorem 2.14. *(Prop. 5 of Book I) In an isosceles triangle, the two base angles are congruent.*

Proof: Let triangle $\triangle ABC$ have sides $\overline{AB}$ and $\overline{BC}$ congruent, as shown in Figure 2.9. Let $\overrightarrow{BD}$ be the bisector of $\angle ABC$, with D inside the triangle. Let $\overrightarrow{BD}$ intersect $\overline{AC}$ at E. Then, by SAS we have triangles $\triangle ABE$ and $\triangle CBE$ congruent and thus $\angle EAB \cong \angle ECB$ and we're done. □

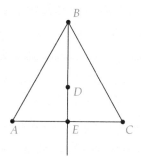

Figure 2.9 Base angles congruence

This proof seems okay—it relies solely on definitions, on the existence of an angle bisector, and on the SAS congruence result. It is the proof one finds in many modern geometry texts, but is not the proof Euclid used. Why not?

If we were true to Euclid's development, we would have a major problem with this proof. First, the construction of angle bisectors comes in Proposition 9 of the *Elements*, which itself depends on Proposition 5, the result we are trying to prove. To use angle bisectors for this proof would be circular reasoning, if we assumed Euclid's development of axioms and theorems. Circular reasoning is an often subtle error that can creep into our attempts to prove a result.

Beyond the problem of circular reasoning, there is an even more fundamental axiomatic problem with this proof, as seen from within Euclid's system. The use of SAS congruence is fine, except for a hidden assumption that has slipped into the proof—that the bisector actually intersects the third side of the triangle. This seems intuitively obvious to us, as we *see* triangles having an *inside* and an *outside*; that is, we see the triangle separating the plane into two regions. In fact, Euclid assumes this *separation* property without proof and does not include it as one of his axioms.

Moritz Pasch (1843–1930) was the first to notice this hidden assumption of Euclid, and in 1882 he published *Vorlesungen uber neure Geometrie*, which for the first time introduced axioms on separation and ordering of points in the plane. One of these axioms specifically addresses the issue of triangle separation and has become known as *Pasch's Axiom*:

> Let A, B, C be three non-collinear points and let l be a line that does not pass through A, B, or C. If l passes through side $\overline{AB}$, it must pass through either a point on $\overline{AC}$ or a point on $\overline{BC}$, but not both.

Hilbert built on Pasch's work in his own axiomatic system. Hilbert's four axioms of "order" or "betweenness" are essentially the axioms Pasch used in 1882. Using Pasch's Axiom, the hidden assumption in Euclid's proof of Theorem 2.14 can be justified. (See the Crossbar Theorem in the on-line chapter covering Hilbert's axioms).

In fact, the proof given above for the isosceles triangle theorem is perfectly correct within Hilbert's axiomatic system, as SAS congruence is an *axiom* in this system, and thus needs no proof, and the triangle intersection and angle bisection properties are proved before this theorem appears.

We see, then, that the same proof can be valid in one system and not valid in another. What makes the difference is how the two systems progressively build up results that then are used for proving later results. Can we give a valid proof in Euclid's system? We know that we can safely use SAS as it is Proposition 4 of Book I. If we consider $\triangle ABC$ in comparison to *itself* by the ordering $\triangle ABC$ with $\triangle CBA$, then we get the base angles congruent by a simple application of SAS. (Convince yourself of this.) This may seem like cheating, but if we look carefully at the statement of SAS, it never stipulated that the two triangles for comparison needed to be different.

This concludes our whirlwind review of the highlights of basic planar geometry (the first 28 Propositions of Book I of *Elements*). This basic material concerns points, lines, rays, segments and segment measure, angles and angle measure, parallels, and triangles as covered in many elementary geometry courses. We summarized these areas without taking the time to carefully derive the results from first principles, although we did take care to point out where more foundational work was needed.

Our brief treatment of formal issues is not an indication that careful axiomatic exposition of basic geometry is unimportant. Hilbert's axiomatic system is an outstanding achievement in the foundations of geometry and we thoroughly cover that system in the on-line chapter covering Hilbert's development of basic geometry. However, it is also important to cover the power and complexity of more advanced geometric concepts, which we do in the rest of this chapter and in the following chapters.

Exercise 2.2.1. *Is it possible for a line to intersect all three sides of a triangle? If so, does this contradict Pasch's Axiom?*

Exercise 2.2.2. *Use Pasch's Axiom to show that a line intersecting one side of a rectangle, at a point other than one of the vertices of the rectangle, must intersect another side of the rectangle. Will the same be true for a pentagon? A hexagon? An n-gon? If so, develop a proof for the case of a line intersecting a side of a regular n-gon (all sides and angles equal).*

Exercise 2.2.3. *If a line intersects three (or more) sides of a four-sided polygon (any shape made of four connected segments), must it contain a vertex of the polygon? If so, give a proof; if not, give a counter-example.*

Let l be a line and A, B two points not on l. If $A = B$, or if segment $\overline{AB}$ contains no points on l, we will say that A and B are on the *same side* of l. Otherwise, they are on *opposite sides* (i.e., $\overline{AB}$ must intersect l). We say that the set of points all on one side of l are in the same *half-plane*.

Use Pasch's Axiom to prove the following pair of results, which Hilbert termed the *Plane Separation Property*. This property is sometimes used to replace Pasch's Axiom in modern axiomatic developments of geometry.

Exercise 2.2.4. *For every line l and every triple of points A, B, C not on l, if A, B are on the same side of l and B, C are on the same side of l, then A, C are on the same side of l.*

Exercise 2.2.5. *For every line l and every triple of points A, B, C not on l, if A, B are on opposite sides of l and B, C are on opposite sides of l, then A, C are on the same side of l.*

Definition 2.16. *Define a point D to be in the* interior *of an angle* $\angle CAB$ *if D is on the same side of* $\overleftrightarrow{AC}$ *as B and also D is on the same side of* $\overleftrightarrow{AB}$ *as C. (The interior is the intersection of two half-planes.)*

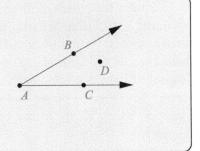

Exercise 2.2.6. *Define precisely what is meant by the* interior *of a triangle.*

The next set of exercises deals with triangle congruence.

Exercise 2.2.7. *Prove the converse to Theorem 2.14. That is, show that if a triangle has two angles congruent, then it must be isosceles.*

Exercise 2.2.8. *Use congruent triangles to prove that the angle bisector construction discussed in section 2.1 is valid.*

Exercise 2.2.9. *Show that if two sides of a triangle are not congruent, then the angles opposite those sides are not congruent, and the larger angle is opposite the larger side of the triangle. [Hint: Use isosceles triangles.]*

Exercise 2.2.10. *Using SAS congruence, prove Angle-Side-Angle congruence.*

Exercise 2.2.11. *Show that for two right triangles, if the hypotenuse and leg of one are congruent to the hypotenuse and leg of the other, then the two triangles are congruent. [Hint: Try a proof by contradiction.]*

A *quadrilateral ABCD* is composed of segments $\overline{AB}$, $\overline{BC}$, $\overline{CD}$, and $\overline{DA}$ such that no three of the points of the quadrilateral are collinear and no pair of segments intersects, except at the endpoints. *Saccheri quadrilaterals* are quadrilaterals having two congruent sides perpendicular to a third side, called the *base*. The remaining side is called the *summit*. Saccheri quadrilaterals are named for Giovanni Saccheri (1667–1773 who studied them in detail, and almost discovered non-Euclidean geometry (see Chapter 7).

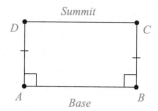

Figure 2.10

Exercise 2.2.12. *Show that the summit angles in a Saccheri Quadrilateral (Figure 2.10) are always congruent.*

Exercise 2.2.13. *Given a triangle △ABC, Let E, and F be the midpoints of sides $\overline{AB}$ and $\overline{AC}$, as shown in Figure 2.11. (We show two possible configurations of the triangle for assisting in the proof for this exercise).*

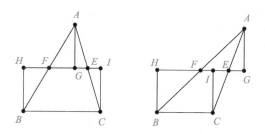

Figure 2.11

Construct the line through F and E and drop perpendiculars to this line from A, B, and C meeting the line at G, H, and I, respectively. Show that BHIC is a Saccheri Quadrilateral and that the angle sum of the triangle is equal to the sum of the summit angles.

Exercise 2.2.14. *Use the preceding exercise, Exercise 2.1.8, and Theorem 2.9 to show that Playfair's Postulate is equivalent to the statement, The summit angles of a Saccheri quadrilateral are right angles. [Remember: This requires two proofs]*

Exercise 2.2.15. *State and prove a SASAS (Side-Angle-Side-Angle-Side) congruence result for quadrilaterals.*

2.3 PROJECT 3 - SPECIAL POINTS OF A TRIANGLE

Triangles are the simplest two-dimensional shapes that we can construct with segments. The simplicity of their construction masks the richness of relationships exhibited by triangles. In this project we will explore several interesting points of intersection that can be found in triangles.

We will be using dynamic geometry software for this project. Notes on how to use particular software packages for this project can be found at http://www.gac.edu/~hvidsten/geom-text.

2.3.1 Circumcenter

Start up your geometry software and use connected segments to create $\triangle ABC$ on the screen.

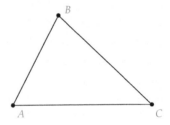

Construct the midpoints D, E, F of the sides using the midpoint tool of your software.

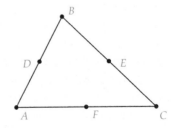

Now, construct the perpendicular bisectors of $\overline{AB}$ and $\overline{BC}$ by using the perpendicular construction tool of your software. Then, construct the intersection point G where these perpendiculars meet. Create a segment from F to G and measure $\angle CFG$.

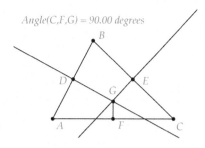

Angle(C,F,G) = 90.00 degrees

It appears that $\overleftrightarrow{GF}$ is perpendicular to $\overleftrightarrow{AC}$ and thus all three perpendicular bisectors meet at G. Drag the vertices of the triangle around. Does the common intersection property persist? This suggests a theorem.

Theorem 2.15. *The perpendicular bisectors of the sides of a triangle intersect at a common point called the* circumcenter *of the triangle.*

Let's see if we can prove this.

Consider the two perpendicular bisectors we constructed, the ones at D and E. Since $\overleftrightarrow{AB}$ and $\overleftrightarrow{BC}$ are not parallel, we can find their intersection at point G.

Exercise 2.3.1. *Show that triangles $\triangle DGB$ and $\triangle DGA$ are congruent, as are triangles $\triangle EGB$ and $\triangle EGC$. Use this to show that the line through F and G will be a perpendicular bisector for the third side, and thus all three perpendicular bisectors intersect at point G. (Make sure your labels match the ones shown above.)*

Create a circle with center at the circumcenter (point G above) and radius point equal to one of the vertices of the triangle. (Be careful that you drag the radius point onto one of the triangle vertices, thus *attaching* the circle's radius point to a triangle vertex.) What do you notice about the circle in relation to the other vertices of the triangle?

Corollary 2.16. *The circle with center at the circumcenter of a triangle and radius out to one of the vertices will pass through the other vertices of the triangle. This is called the* circumscribed *circle of the triangle (Figure 2.12).*

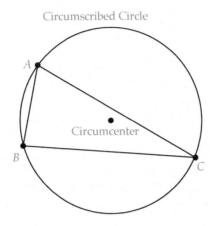

Figure 2.12 Circumscribed circle of a triangle

Exercise 2.3.2. *Prove Corollary 2.16.*

2.3.2 Orthocenter

An *altitude* of a triangle will be a perpendicular to a side of the triangle that passes through the opposite vertex. In Figure 2.13, line a is an altitude to $\triangle ABC$ through vertex B.

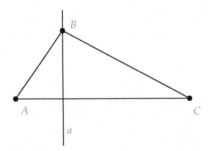

Figure 2.13 Altitude

What can we say about the three altitudes of a given triangle? Do they intersect at a common point?

We will show that the altitudes of a triangle do intersect at a common point by showing that the altitudes are also the perpendicular bisectors of an associated triangle to the given triangle.

Clear the screen and create a triangle $\triangle ABC$. Use the Parallel tool of your software to construct three parallels, each one a parallel to a side of the triangle passing through an opposite triangle vertex. Since the original sides of $\triangle ABC$ were not parallel, each pair of the new parallels will intersect. These three intersection points will form a new triangle, $\triangle DEF$. We will call this the *associated* triangle to $\triangle ABC$.

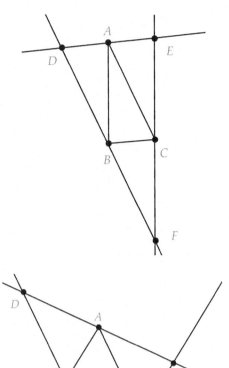

Drag the vertices around and notice the relationship among the four triangles $\triangle BAD$, $\triangle ABC$, $\triangle FCB$, and $\triangle CEA$. What would your conjecture be about the relationship among these four triangles?

Exercise 2.3.3. *Use Theorem 2.9 to verify the angle congruences illustrated in Figure 2.14 for $\triangle ABC$ and its associated triangle $\triangle DEF$. Then, show that all four sub-triangles are congruent. That is,*

$$\triangle ABC \cong \triangle BAD \cong \triangle FCB \cong \triangle CEA$$

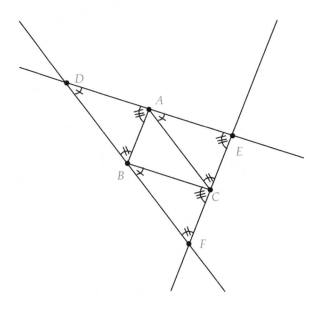

Figure 2.14

This leads directly to the following result:

Theorem 2.17. *The altitudes of a triangle are the perpendicular bisectors of the associated triangle as we have described.*

Exercise 2.3.4. *Prove Theorem 2.17. [Hint: Use the preceding exercise.]*

Since the altitudes of a triangle are the perpendicular bisectors of the associated triangle, and since we have already shown that the three perpendicular bisectors of any triangle have a common intersection point, then we have the following.

Corollary 2.18. *The altitudes of a triangle intersect at a common point. This point is called the* orthocenter *of the triangle.*

Corollary 2.19. *The orthocenter of a triangle is the circumcenter of the associated triangle.*

Does the orthocenter have a nice circle like the circumscribed circle?

Create a circle with center at the orthocenter and radius point equal to one of the vertices of triangle $\triangle ABC$. What do you notice that is different about this situation in comparison to the preceding circumscribed circle?

2.3.3 Incenter

Clear the screen and create $\triangle ABC$ one more time. In this part of our project, we will find the angle bisectors of each of the three interior angles of the triangle. Use the Angle Bisector tool of your software to bisect each of the triangles angles.

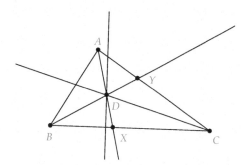

It appears that the three angle bisectors have a common intersection point.

Theorem 2.20. *The angle bisectors of triangle $\triangle ABC$ intersect at a common point. This point is called the* incenter *of the triangle.*

Proof: First note that the bisector at A will intersect side $\overline{BC}$ at some point X. The bisector at B will intersect $\overline{AC}$ at some point Y and $\overline{AX}$ at some point D that is interior to the triangle.

In the next exercise, we will show that the points on an angle bisector are equidistant from the sides of an angle and, conversely, if a point is interior to an angle and equidistant from the sides of an angle, then it is on the bisector. Thus, the distance from D to $\overrightarrow{AB}$ equals the distance from D to $\overrightarrow{AC}$. Also, the distance from D to $\overrightarrow{AB}$ equals the distance from D to $\overrightarrow{BC}$, as D is on the bisector at B. Thus, the distance from D to $\overrightarrow{AC}$ equals the distance from D to $\overrightarrow{BC}$, and D must be on the bisector at C. $\square$

Exercise 2.3.5. *Use congruent triangles to show that the points on an angle bisector are equidistant from the sides of the angle and, conversely, if a point is interior to an angle and equidistant from the sides of an angle, then it is on the bisector. [Hint: Distance from a point to a line is measured by dropping a perpendicular from the point to the line. You may also use the Pythagorean Theorem if you wish.]*

Drop a perpendicular line from the incenter to one of the sides of the triangle and find the intersection of this perpendicular with the side. (For example, we get point Z in Figure 2.15 by dropping a perpendicular from D to $\overleftrightarrow{BC}$.) Create a circle with center at the incenter and radius point equal to the intersection point.

> **Definition 2.17.** *The circle just constructed is called the* inscribed *circle to the triangle. A circle is inscribed in a triangle (or other polygon) if it intersects each side of the triangle (polygon) in a single point.*

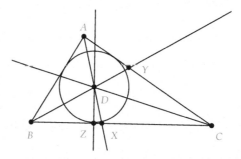

Figure 2.15

Project Report

It is amazing what structure exists in a simple shape like a triangle! In your report discuss the similarities and differences you discovered in exploring these three special points of a triangle. Also, take care that all of your work on proofs and exercises is written in a logical and careful fashion.

2.4 MEASUREMENT AND AREA IN EUCLIDEAN GEOMETRY

Euclid, in his development of triangle congruence, had a very restrictive notion of what congruence meant for segments and angles. When we hear the word *congruence* our modern mathematical background prompts us to immediately think of *numbers* being equal. For example, we immediately associate a number with a segment and consider two segments congruent if their numerical lengths are equal.

This is very different from the way that Euclid viewed congruence. In fact, Euclid's notion of congruence was one of *coincidence* of figures. That is, two segments were congruent if one could be made to be coincident with the other, by movement of one onto the other. Euclid never used segments as substitutes for numerical values, or vice versa. If one looks at Book II of Euclid, where he develops what we would now call algebraic relationships, we see that these relationships are always based on geometric figures. When Euclid mentions the *square* on segment $\overline{AB}$, he literally means the geometric square constructed on $\overline{AB}$ and not the product of the length of $\overline{AB}$ with itself.

Modern axiomatic treatments of Euclidean geometry resolve this problem of dealing with numerical measure by *axiomatically* stipulating that there is a way to associate a real number with a segment or angle. Hilbert, staying true to Euclid, does not explicitly tie numbers to geometric figures, but instead provides two axioms of continuity that allow for the *development* of measurement through an ingenious but rather tedious set of theorems. Birkhoff, on the other hand, directly specifies a connection between real numbers and lengths in his first axiom and between numbers and angles in his third axiom (see Appendix C).

In the following project, we will continue to assume the standard connection between geometric figures and real numbers via lengths and angle degree measures. This connection will lead naturally to a definition of area based on segment length.

2.4.1 Mini-Project - Area in Euclidean Geometry

In this project we will create a definition for area, based on properties of four-sided figures. Our investigation of area will follow Euclid's use of the idea of *equality of figures*, beginning in his proof of Prop. 35 of Book I. Euclid never uses the terminology of area to describe how two figures can be considered "equal", but it is clear from his proofs that his reasoning about shapes can be directly compared to our modern notion of area.

As you work through this project you will notice that it is different than our previous computer-based activities. Whereas earlier projects

used computer software, for this project we will primarily be using human "grayware," that is, the canvas of our minds.

We start our investigation where Euclid did, with the study of four-
sided shapes.

> **Definition 2.18.** *A quadrilateral ABCD is a figure comprised of
> segments $\overline{AB}$, $\overline{BC}$, $\overline{CD}$, and $\overline{DA}$ such that no three of the points
> of the quadrilateral are collinear and no pair of segments intersects,
> except at the endpoints. We call $\overline{AB}$, $\overline{BC}$, $\overline{CD}$, and $\overline{DA}$ the* sides
> *of the quadrilateral and $\overline{AC}$, $\overline{BD}$ the* diagonals *of the quadrilateral.*

> **Definition 2.19.** *A parallelogram is a quadrilateral ABCD, where
> $\overleftrightarrow{AB}$ is parallel to $\overleftrightarrow{CD}$ and $\overleftrightarrow{BC}$ is parallel to $\overleftrightarrow{AD}$. That is, opposite
> sides are parallel.*

Here's the first set of exercises to work through. The tools that will
prove most helpful to you include the theorems covered so far on parallels
and triangle congruence.

Exercise 2.4.1. *(Prop. 34 of Book
I) Show that in parallelogram ABCD,
opposite sides are congruent, opposite angles are congruent and the
diagonal bisects the parallelogram.
(That is, the diagonal divides the
parallelogram into two congruent triangles.).*

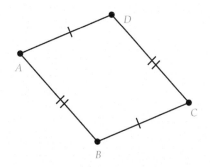

Now that we have covered the basic properties of a parallelogram we
will consider Euclid's first analysis of how two parallelogram shapes can
be compared.

Prop. 35 of Book I: Two parallelograms having the same base and
defined between the same parallels are *equal.*

Euclid never defines what "equals" means, but from his proofs it is
clear that it has to do with being able to dissect two figures into pairs of
congruent triangles. We will define two ways of dissecting figures —one
additive and one subtractive.

Definition 2.20. *We will call two geometric figures P and P' equidecomposable or* congruent by addition *if it is possible to decompose each into a finite number of non-overlapping triangles*

$$P = T_1 \cup \cdots \cup T_n$$
$$P' = T_1' \cup \cdots \cup T_n'$$

where for each i, triangle T_i is congruent to T_i' [19, page 197].

For the proof of **Prop. 35 of Book I**, let parallelograms $ABCD$ and $EBCF$ be two parallelograms sharing base $\overline{BC}$ with $\overleftrightarrow{AF}$ parallel to $\overleftrightarrow{BC}$.

There are two possible configurations for the parallelograms: either D will be between E and F as shown in the top configuration, or D will not be on segment $\overline{EF}$ as shown in the second configuration.

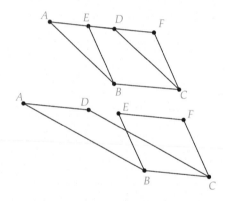

Exercise 2.4.2. *In the first case, where D is between E and F, use a SAS argument to show that triangles $\triangle EAB$ and $\triangle FDC$ are congruent. Then, show that these two parallelograms are congruent by addition.*

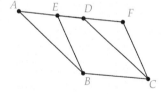

Definition 2.21. *We say that two geometric figures P and P' have* equal content *or are* congruent by subtraction *if there are figures Q, Q', that are congruent by addition with P and Q non-overlapping and P', Q' non-overlapping such that*

$$R = P \cup Q$$
$$R' = P' \cup Q'$$

are congruent by addition.

Exercise 2.4.3. *Let's consider the second case for* **Prop. 35 of Book I.** *We have parallelograms $ABCD$ and $EBCF$ sharing base $\overline{BC}$ with $\overleftrightarrow{AF}$ parallel to $\overleftrightarrow{BC}$ and D not on segment $\overline{EF}$.*

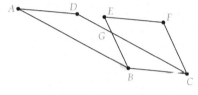

Let $Q = Q'$ be the figure defined by triangle DEG. Show that the two parallelograms are congruent by subtraction, using figure Q.

It is clear that if two figures are congruent by addition, then they are congruent by subtraction (Q can be empty). What is not so clear is the converse. If we take the axiomatic system of Hilbert and assume a reasonable method of assigning numerical areas to figures, then the converse is true. However, a geometry built on all of Hilbert's axioms, except the Archimedean axiom, is possible, and in this geometry the converse is *not* true [19, pages 198–199].

It is reasonable to take as a replacement for Euclid's vague notion of equality of figures the concept of congruence by subtraction, as it is intuitively clear that two figures that are congruent by subtraction should have the same area. With this replacement, and the results of the previous exercises, we have established a proof of **Prop. 35 of Book I.**

We now are in a position to define area in terms of rectangles!

Definition 2.22. *A* rectangle *is a quadrilateral in which all the angles defined by the vertices are right angles. Given a rectangle $ABCD$, we can identify two adjacent sides and call the length of one of the sides a* base *of the rectangle and the length of the other a* height. *The* area *of the rectangle will be defined as the product of a base and height.*

Exercise 2.4.4. *Show that this definition of rectangular area is not ambiguous. That is, it does not depend on how we choose the sides for the base and height. [Be careful—this is not as trivial as it may appear. You need to base your argument only on known geometric theorems and not on analytic or numerical reasoning.]*

Definition 2.23. *If two figures are congruent by subtraction, we will say that they have the same area.*

Definition 2.24. *If we identify a side of a parallelogram to be a* base side, *then the* height *of the parallelogram, relative to that base side, is the perpendicular distance between the base side and the opposite side.*

Exercise 2.4.5. *Show that the area of a parallelogram ABCD is the product of its base length and its height by showing that the parallelogram is congruent by subtraction (or addition) to a rectangle with those side lengths (Figure 2.16). [Hint: Given parallelogram ABCD, with base AB, drop perpendiculars from A to $\overleftrightarrow{CD}$ at E and from B to $\overleftrightarrow{CD}$ at F. Show that ABFE is a rectangle with the desired properties.]*

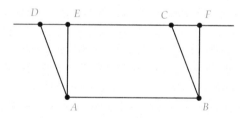

Figure 2.16

The next set of theorems follows immediately from your work.

Theorem 2.21. *The area of a right triangle is one-half the product of the lengths of its legs, the legs being the sides of the right angle.*

> **Theorem 2.22.** *(Essentially Prop. 41 of Book I) The area of any triangle is one-half the product of a base of the triangle with the height of the triangle.*

Project Report

In this project we have developed the notion of area from simple figures such as parallelograms, rectangles, and triangles. From these simple beginnings, we could expand our idea of area to arbitrary polygonal figures, and from there find areas of curved shapes by using approximating rectilinear areas and limits, as is done in calculus when defining the integral.

Provide a summary of the proofs for each of the exercises in this project. If you finish early, discuss what aspects of area are still not totally defined. That is, were there hidden assumptions made in our proofs that would necessitate area *axioms*?

2.4.2 Cevians and Areas

We will now apply our results on areas to a very elegant development of the intersection properties of cevians. A full account of the theory of cevians can be found in the monograph *Geometry Revisited*, by Coxeter and Greitzer [10].

> **Definition 2.25.** *Given a triangle, a cevian is a segment from a vertex to a point on the opposite side.*

> **Theorem 2.23.** *Given $\triangle ABC$, and cevians $\overline{AX}$, $\overline{BY}$, and $\overline{CZ}$ (refer to Figure 2.17), if the three cevians have a common intersection point P, then*
>
> $$\frac{BX}{XC}\frac{CY}{YA}\frac{AZ}{ZB} = 1$$

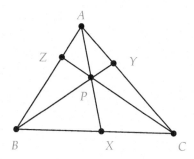

Figure 2.17

Proof: Since two triangles with the same altitudes have areas that are proportional to the bases, then

$$\frac{area(\Delta ABX)}{area(\Delta AXC)} = \frac{BX}{XC} \quad \text{and} \quad \frac{area(\Delta PBX)}{area(\Delta PXC)} = \frac{BX}{XC}$$

A little algebra shows

$$\frac{area(\Delta ABX) - area(\Delta PBX)}{area(\Delta AXC) - area(\Delta PXC)} = \frac{area(\Delta ABP)}{area(\Delta APC)} = \frac{BX}{XC}$$

Similarly,

$$\frac{area(\Delta BCP)}{area(\Delta APB)} = \frac{CY}{YA} \quad \text{and} \quad \frac{area(\Delta CAP)}{area(\Delta BPC)} = \frac{AZ}{ZB}$$

Thus,

$$\frac{BX}{XC}\frac{CY}{YA}\frac{AZ}{ZB} = \frac{area(\Delta ABP)}{area(\Delta APC)}\frac{area(\Delta BCP)}{area(\Delta ABP)}\frac{area(\Delta APC)}{area(\Delta BCP)}$$

Cancellation of terms yields the desired result. □

This theorem has become known as *Ceva's Theorem*, in honor of Giovanni Ceva (1647–1734), an Italian geometer and engineer. The term *cevian* is derived from Ceva's name, in recognition of the significance of his theorem in the history of Euclidean geometry.

The converse to Ceva's theorem is also true.

Theorem 2.24. *If the three cevians satisfy*

$$\frac{BX}{XC}\frac{CY}{YA}\frac{AZ}{ZB} = 1$$

then they intersect at a common point (refer to Figure 2.18).

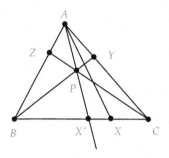

Figure 2.18

Proof: $\overline{BY}$ will intersect $\overline{CZ}$ at some point P which is interior to the triangle. The ray $\overrightarrow{AP}$ will intersect $\overline{BC}$ at some point X' that is between B and C. By the previous theorem we have

$$\frac{BX'}{X'C}\frac{CY}{YA}\frac{AZ}{ZB} = 1$$

Thus, $\frac{BX'}{X'C} = \frac{BX}{XC}$.

Suppose that X' is between B and X. Then

$$\frac{BX}{XC} = \frac{BX' + X'X}{X'C - X'X}$$

Since $\frac{BX'}{X'C} = \frac{BX}{XC}$, we have

$$\frac{BX'}{X'C} = \frac{BX' + X'X}{X'C - X'X}$$

Cross-multiplying in this last equation gives

$$(BX')(X'C - X'X) = (X'C)(BX' + X'X)$$

Simplifying we get

$$-(BX')(X'X) = (X'C)(X'X)$$

Since X' is between B and C, the only way that this equation can be satisfied is for $X'X$ to be zero. Then $X' = X$.

A similar result would be obtained if we assumed that X' is between X and C. □

> **Definition 2.26.** *A* median *of a triangle is a segment from the midpoint of a side to the opposite vertex.*

Exercise 2.4.6. *Show that the medians of a triangle intersect at a common point. This point is called the* centroid *of the triangle.*

Exercise 2.4.7. *Use an area argument to show that the centroid must be the balance point for the triangle. [Hint: Consider each median as a "knife edge" and argue using balancing of areas.]*

Exercise 2.4.8. *The medians of a triangle split the triangle into six sub-triangles. Show that all six have the same area.*

Exercise 2.4.9. *Use Exercise 2.4.8 and an area argument to show that the medians divide one another in a 2:1 ratio.*

2.5 SIMILAR TRIANGLES

One of the most useful tools for a more advanced study of geometric properties (e.g., the geometry used in surveying) is that of *similar* triangles.

> **Definition 2.27.** *Two triangles are* similar *if and only if there is some way to match vertices of one to the other such that corresponding sides are in the same ratio and corresponding angles are congruent.*

If $\triangle ABC$ is similar to $\triangle XYZ$, we shall use the notation $\triangle ABC \sim \triangle XYZ$. Thus, $\triangle ABC \sim \triangle XYZ$ if and only if

$$\frac{AB}{XY} = \frac{AC}{XZ} = \frac{BC}{YZ}$$

and

$$\angle BAC \cong \angle YXZ, \angle CBA \cong \angle ZYX, \angle ACB \cong \angle XZY$$

Similar figures have the same *shape* but are of differing (or the same) *size*. One is just a scaled-up version of the other. The scaling factor is the constant of proportionality between corresponding lengths of the two figures. Thus, similarity has to do with the equality of ratios of corresponding measurements of two figures.

Let's review some basic similarity and proportionality results.

Theorem 2.25. *Suppose that a line is parallel to one side of a triangle and intersects the other two sides at two different points. Then, this line divides the intersected sides into proportional segments (refer to Figure 2.19).*

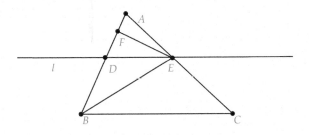

Figure 2.19

Proof: Let line l be parallel to $\overleftrightarrow{BC}$ in $\triangle ABC$. Let l intersect sides $\overline{AB}$ and $\overline{AC}$ at points D and E. Construct a perpendicular from point E to $\overleftrightarrow{AB}$ intersecting at point F. Consider the ratio of the areas of triangles $\triangle BED$ and $\triangle AED$:

$$\frac{Area(\triangle BED)}{Area(\triangle AED)} = \frac{\frac{1}{2}(BD)(EF)}{\frac{1}{2}(AD)(EF)} = \frac{BD}{AD}$$

Now, consider dropping a perpendicular from D to $\overleftrightarrow{AC}$, intersecting this line at point G (Figure 2.20).

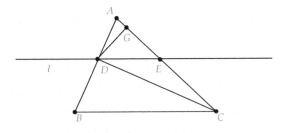

Figure 2.20

Then,

$$\frac{Area(\triangle CED)}{Area(\triangle AED)} = \frac{\frac{1}{2}(CE)(DG)}{\frac{1}{2}(AE)(DG)} = \frac{CE}{AE}$$

Now, triangles BED and CED share base $\overline{DE}$ and so the heights of these two triangles will be the same, as both heights are the lengths of perpendiculars dropped to $\overleftrightarrow{DE}$ from points on another line parallel to $\overleftrightarrow{DE}$. Thus, these two triangles have the same area and

$$\frac{Area(\triangle BED)}{Area(\triangle AED)} = \frac{Area(\triangle CED)}{Area(\triangle AED)}$$

Thus,

$$\frac{BD}{AD} = \frac{CE}{AE}$$

This completes the proof. □

Corollary 2.26. *Given the assumptions of Theorem 2.25, it follows that*
$$\frac{AB}{AD} = \frac{AC}{AE}$$

Proof: Since $AB = AD + BD$ and $AC = AE + CE$, we have

$$\frac{AB}{AD} = \frac{AD + BD}{AD} = 1 + \frac{BD}{AD} = 1 + \frac{CE}{AE} = \frac{AE + CE}{AE} = \frac{AC}{AE}$$

□

The converse to this result is also true.

Theorem 2.27. *If a line intersects two sides of a triangle so that the segments cut off by the line are proportional to the original sides of the triangle, then the line must be parallel to the third side.*

Proof: Let line l pass through $\triangle ABC$ at sides $\overline{AB}$ and $\overline{AC}$, intersecting at points D and E (Figure 2.21).

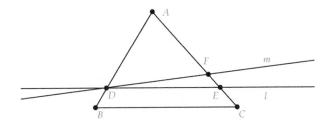

Figure 2.21

It is given that

$$\frac{AB}{AD} = \frac{AC}{AE}$$

By Playfair's Axiom there is a unique parallel m to $\overleftrightarrow{BC}$ through D. Since m is parallel to $\overleftrightarrow{BC}$ and it intersects side $\overline{AB}$, then it must intersect side $\overline{AC}$ at some point F. But, Corollary 2.26 then implies that

$$\frac{AB}{AD} = \frac{AC}{AF}$$

Thus,

$$\frac{AC}{AE} = \frac{AC}{AF}$$

and $AE = AF$. This implies that points E and F are the same and lines l and m are the same. So, l is parallel to $\overleftrightarrow{BC}$. □

We now have enough tools at our disposal to prove the following similarity condition.

Theorem 2.28. *(AAA Similarity) If in two triangles there is a correspondence in which the three angles of one triangle are congruent to the three angles of the other triangle then the triangles are similar (Figure 2.22).*

Proof: Let $\triangle ABC$ and $\triangle DEF$ be two triangles with the angles at $A, B,$ and C congruent to the angles at $D, E,$ and F, respectively. If segments $\overline{AB}$ and $\overline{DE}$ are congruent, then the two triangles are congruent by the AAS congruence theorem and thus are also similar.

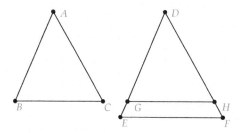

Figure 2.22 AAA similarity

Suppose that $\overline{AB}$ and $\overline{DE}$ are not congruent. We can assume that one is larger than the other, say $\overline{DE}$ is larger than $\overline{AB}$. Then, there is a point G between D and E such that $\overline{DG} \cong \overline{AB}$. Also, there must be a point H along the ray $\overrightarrow{DF}$ such that $\overline{DH} \cong \overline{AC}$. Then, since the angles at A and D are congruent, we have that $\triangle ABC \cong \triangle DGH$ (SAS) and $\angle DGH \cong \angle ABC$. Since $\angle ABC$ and $\angle DEF$ are assumed congruent, then $\angle DGH \cong \angle DEF$ and $\overline{GH} \parallel \overline{EF}$ (Theorem 2.7). This implies that H is between D and F. Applying Corollary 2.26, we get that

$$\frac{DG}{DE} = \frac{DH}{DF}$$

Since $\overline{DG} \cong \overline{AB}$ and $\overline{DH} \cong \overline{AC}$, we get

$$\frac{AB}{DE} = \frac{AC}{DF}$$

A similar argument can be used to show that

$$\frac{AB}{DE} = \frac{BC}{EF}$$

☐

Another important similarity condition is the SAS condition.

Theorem 2.29. *(SAS Similarity) If in two triangles there is a correspondence in which two sides of one triangle are proportional to two sides of the other triangle and the included angles are congruent, then the triangles are similar (Figure 2.23).*

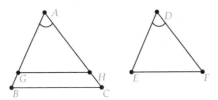

Figure 2.23 SAS similarity

Proof: Let triangles $\triangle ABC$ and $\triangle DEF$ be as specified in the theorem with

$$\frac{AB}{DE} = \frac{AC}{DF}$$

If $\overline{AB}$ and $\overline{AC}$ are congruent to their counterparts in $\triangle DEF$, then the triangles are congruent by SAS and thus are similar as well. We may then assume that $\overline{AB}$ and $\overline{AC}$ are greater than their counterparts in $\triangle DEF$. On $\overline{AB}$ and $\overline{AC}$ there must be points G and H such that $AG = DE$ and $AH = DF$. It is left as an exercise to show that $\overleftrightarrow{GH}$ and $\overleftrightarrow{BC}$ are parallel and thus the angles at B and C are congruent to the angles at E and F. □

Exercise 2.5.1. *In $\triangle ABC$ let points D and E be the midpoints of sides $\overline{AB}$ and $\overline{AC}$. Show that DE must be half the length of $\overline{BC}$.*

Exercise 2.5.2. *Fill in the gap in Theorem 2.29. That is, show in Figure 2.23 that $\overleftrightarrow{GH}$ and $\overleftrightarrow{BC}$ are parallel.*

Exercise 2.5.3. *(SSS Similarity Condition) Suppose that two triangles have three sides that are correspondingly proportional. Show that the two triangles must be similar.*

Exercise 2.5.4. *Let $\triangle ABC$ be a right triangle with the right angle at C. Let $a, b,$ and c be the side lengths of this triangle. Let $\overline{CD}$ be the altitude to side $\overline{AB}$ (the hypotenuse). Use similarity to prove the Pythagorean Theorem; that is, show that $a^2 + b^2 = c^2$.*

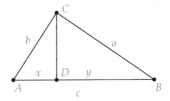

Exercise 2.5.5. *The basic results on similar triangles allow us to define the standard set of trigonometric functions on angles as follows:*

Given an acute angle $\angle BAC$, construct a right triangle with one angle congruent to $\angle BAC$ and right angle at C. Then, define

- $\sin(\angle A) = \frac{BC}{AB}$
- $\cos(\angle A) = \frac{AC}{AB}$

Show that these definitions are well defined; that is, they do not depend on the construction of the right triangle.

Exercise 2.5.6. *We can extend the functions defined in the last exercise to* obtuse *angles (ones with angle measure greater than 90 degrees) by defining*

- $\sin(\angle A) = \sin(\angle A')$
- $\cos(\angle A) = -\cos(\angle A')$

where $\angle A'$ is the supplementary angle to $\angle A$. We can extend the definitions to right angles by defining $\sin(\angle A) = 1$ and $\cos(\angle A) = 0$ if $\angle A$ is a right angle.

Using this extended definition of the trig functions, prove the Law of Sines, *that in any triangle $\triangle ABC$, if a and b are the lengths of the sides opposite A and B, respectively, then*

$$\frac{a}{b} = \frac{\sin(\angle A)}{\sin(\angle B)}$$

[Hint: Drop a perpendicular from C to AB at D and use the two right triangles $\triangle ADC$ and $\triangle BDC$.]

Exercise 2.5.7. *In this exercise we will prove Menelaus's Theorem: Given triangle ABC, let R be a point outside of the triangle on the line through A and B. From R draw any line intersecting the other two sides of the triangle at points P, Q. Then*

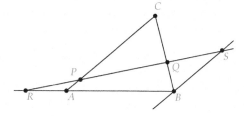

$$\frac{CP}{PA}\frac{AR}{RB}\frac{BQ}{QC} = 1$$

Prove this result using similar triangles. Specifically, construct a parallel to $\overleftrightarrow{AC}$ through B and let this intersect $\overrightarrow{RP}$ at point S. (Why must these lines intersect?) Then, show that $\triangle RBS$ and $\triangle RAP$ are similar, as are $\triangle PCQ$ and $\triangle SBQ$, and use this to prove the result. (Menelaus was a Roman mathematician who lived in Alexandria from about 70–130 AD.)

Exercise 2.5.8. *Blaise Pascal (1623–1662) did fundamental work in many areas of science and mathematics, including the physics of hydrostatics, the geometry of conic sections, and the foundations of probability and philosophy. He is also credited with inventing the first digital calculator. The following result is due to Pascal.*

Given $\triangle ABC$, construct a line $\overleftrightarrow{DE}$, intersecting sides $\overline{AB}$ and $\overline{AC}$ at D, E, and parallel to $\overleftrightarrow{BC}$. Pick any point F on $\overline{BD}$ and construct $\overline{FE}$. Then, construct $\overleftrightarrow{BG}$ parallel to $\overleftrightarrow{FE}$, intersecting side $\overline{AC}$ at G. Then, $\overleftrightarrow{FC}$ must be parallel to $\overleftrightarrow{DG}$.

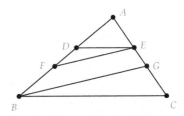

Prove Pascal's Theorem. [Hint: Use the theorems in this section on parallels and similar triangles to show that $\frac{AD}{AG} = \frac{AF}{AC}$.] Hilbert [22, page 46] used Pascal's Theorem extensively to develop his "arithmetic of segments," by which he connected segment length with ordinary real numbers.

2.5.1 Mini-Project - Finding Heights

A *surveyor* is one who measures things like distances and elevations. While modern surveyors use sophisticated equipment like laser range finders, the basic mathematics of surveying is that of triangle geometry and in particular similar triangles.

Consider the situation shown in Figure 2.24. At C we have a water tower. At a particular time of the day, the sun will cast a shadow of the tower that hits the ground at A. Suppose that a person stands at exactly point B, where his or her shadow will match the shadow of the tower at A.

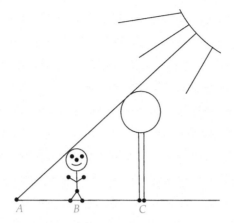

Figure 2.24 Finding heights

For this project, choose a partner and find a tall building (or tree) to measure in this fashion. You will need a long tape measure (or you can use a yard stick) and a pencil and paper for recording your measurements.

The report for this part of the project will include (1) a discussion of how to calculate the height using similar triangles, (2) a table of measurements needed to find the height of the building (or tree), and (3) the calculation of the height of your tall object.

Now, let's consider a second method of finding the height, as depicted in Figure 2.25.

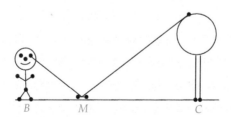

Figure 2.25 Finding heights by mirror

For this method you will need a measuring device and a small mirror. Place the mirror flat on the ground some distance from the tall object whose height you wish to measure. Then, walk back from the mirror and stop when the top of the object is just visible in the mirror. Describe how you can find the height of the object using this mirror method. Include in your project report a table of measurements for this second method and a short discussion of which method is more accurate and/or practically useful.

2.6 CIRCLE GEOMETRY

One of the most beautiful of geometric shapes is the circle. Euclid devoted Book III of the *Elements* to a thorough study of the properties of circles.

To start our discussion of circles, we begin with the following fundamental result on the existence of circles defined by three points.

> **Theorem 2.30.** *Given three distinct points, not all on the same line, there is a unique circle through these three points.*

Proof: Let the three points be A, B, C. Let l_1 and l_2 be the perpendicular bisectors of $\overline{AB}$ and $\overline{BC}$, with M_1, M_2 the midpoints of $\overline{AB}$ and $\overline{BC}$ (Figure 2.26). We first show that l_1 and l_2 must intersect.

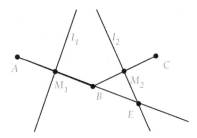

Figure 2.26

Suppose l_1 and l_2 were parallel. Then, since $\overleftrightarrow{AB}$ intersects l_1, it must intersect l_2 at some point E (by Exercise 2.1.10). Since A, B, C are not collinear, then E cannot be M_2. Since l_1 and l_2 are parallel, then by Theorem 2.9 $\angle BEM_2$ must equal the right angle at M_1. But, this is impossible as then triangle $\triangle BEM_2$ would have an angle sum greater than 180 degrees. This would contradict Exercise 2.1.8, which showed that the angle sum for a triangle is 180 degrees.

Thus l_1 and l_2 intersect at some point O (Figure 2.27). By SAS we know that $\triangle AM_1O$ and $\triangle BM_1O$ are congruent, as are $\triangle BM_2O$ and $\triangle CM_2O$, and thus $\overline{AO} \cong \overline{BO} \cong \overline{CO}$ and the circle with center O and radius $\overline{AO}$ passes through these three points.

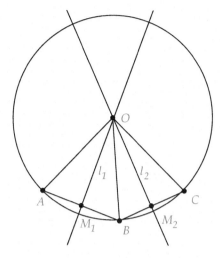

Figure 2.27

Why is this circle unique? Suppose there is another circle c' with center O' passing through A, B, C. Then, O' is equidistant from A and B and thus must be on the perpendicular bisector l_1. (This can be seen from a simple SSS argument.) Likewise O' is on l_2. But, l_1 and l_2 already meet at O and thus $O = O'$. Then, $\overline{O'A} \cong \overline{OA}$ and c' must be equal to the original circle. □

2.6.1 Chords and Arcs

It will be useful to have terminology to refer to *pieces* of circles.

Definition 2.28. *Let c be a circle with center O. If A and B are distinct points on c, we call the segment $\overline{AB}$ a* chord *of the circle. If a chord passes through the center O, we say it is a di-ameter of the circle.*

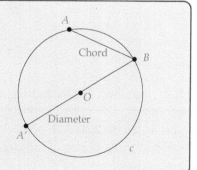

Definition 2.29. *A chord* $\overline{AB}$ *of a circle c with center O will divide the points of a circle c (other than A and B) into two parts, those points of c on one side of* $\overleftrightarrow{AB}$ *and those on the other side. Each of these two parts is called an* open arc *of the circle. An open arc determined by a diameter is called a* semi-circle. *If we include the endpoints A, B we would call the arc or semi-circle* closed.

Definition 2.30. *Note that one of the two open arcs defined by a (non-diameter) chord* $\overline{AB}$ *will be within the angle* $\angle AOB$. *This will be called a* minor *arc. This is arc a1 in Figure 2.28. The arc that is exterior to this angle is called a* major *arc. This is arc a2 in the figure.*

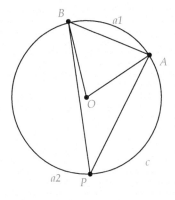

Figure 2.28

2.6.2 Inscribed Angles and Figures

Definition 2.31. *A* central angle *of a circle is one that has its vertex at the center of the circle. For example,* $\angle AOB$ *in Figure 2.28 is a central angle of the circle c. An* inscribed angle *of a circle within a major arc is one that has its vertex on the circle at a point on the major arc and has its sides intercepting the endpoints of the arc (*$\angle APB$ *in the figure).*

Let's look at some basic properties of inscribed angles.

> **Theorem 2.31.** *The measure of an angle inscribed in a circle is half that of its intercepted central angle.*

Proof: Let $\angle APB$ be inscribed in circle c having center O. Let $\overrightarrow{PO}$ intersect circle c at Q. Then, either A and B are on opposite sides of $\overleftrightarrow{PO}$ (as shown in Figure 2.29), or A is on the diameter through $\overline{OP}$, or A and B are on the same side of $\overleftrightarrow{PO}$. We will prove the result in the first case and leave the remaining cases as an exercise.

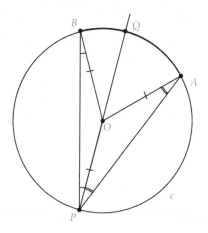

Figure 2.29

Now, $\overline{OP} \cong \overline{OB} \cong \overline{OA}$ and $\triangle POB$ and $\triangle POA$ are isosceles. Thus, the base angles are congruent (Theorem 2.14). Let $\alpha = m\angle PBO$ and $\beta = m\angle PAO$. Then, $m\angle POB = 180 - 2\alpha$ and $m\angle POA = 180 - 2\beta$. Also, $m\angle QOB = 2\alpha$ and $m\angle QOA = 2\beta$.

Clearly, the measure of the angle at P, which is $\alpha + \beta$, is half the measure of the angle at O, which is $2\alpha + 2\beta$. □

The following two results are immediate consequences of this theorem.

> **Corollary 2.32.** *If two angles are inscribed in a circle such that they share the same arc, then the angles are congruent.*

Corollary 2.33. *An inscribed angle in a semi-circle is always a right angle.*

Definition 2.32. *A polygon is* inscribed *in a circle if its vertices lie on the circle. We also say that the circle* circumscribes *the polygon in this case.*

Corollary 2.34. *If quadrilateral ABCD is inscribed in a circle, then the opposite angles of the quadrilateral are supplementary.*

The proof is left as an exercise.

Theorem 2.35. *(Converse to Corollary 2.33) If an inscribed angle in a circle is a right angle, then the endpoints of the arc of the angle are on a diameter.*

Proof (refer to Figure 2.30): Let $\angle BAC$ be a right angle inscribed in a circle c. We need to show that $\overline{BC}$ is a diameter of the circle.

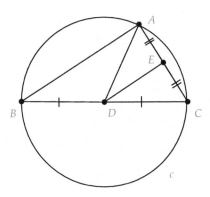

Figure 2.30

Let D and E be the midpoints of segments $\overline{BC}$ and $\overline{AC}$. Then, since $\overline{BD}$ and $\overline{AE}$ are both in the same proportion to $\overline{BC}$ and $\overline{AC}$, we have by Theorem 2.27 that $\overleftrightarrow{DE}$ is parallel to $\overleftrightarrow{AB}$ and thus the angle at E is a right angle. Then, $\triangle AED \cong \triangle CED$ by SAS and $\overline{DC} \cong \overline{DA}$.

Similarly, $\overline{BD} \cong \overline{DA}$. Thus, the circle with center D and radius equal to $\overline{DA}$ passes through A, B, C. But, c is the unique circle through these points and thus $\overline{DA}$ is the radius for the circle and $\overline{BC}$ is a diameter. □

2.6.3 Tangent Lines

Among all lines that pass through a circle, we will single out for special consideration those lines that intersect the circle only once.

> **Definition 2.33.** *A line l is said to be* tangent *to a circle c if l intersects the circle at a single point T, called the* point of tangency.

What properties do tangents have?

> **Theorem 2.36.** *Given a circle c with center O and radius $\overline{OT}$, a line l is tangent to c at T if and only if l is perpendicular to $\overleftrightarrow{OT}$ at T.*

Proof (refer to Figure 2.31): First, suppose that l is tangent to c at T. If l is not perpendicular to $\overleftrightarrow{OT}$ at T, then let P be the point on l where $\overleftrightarrow{OP}$ is perpendicular to l.

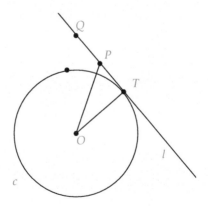

Figure 2.31

On l we can construct a point Q opposite of T from P such that

$\overline{PQ} \cong \overline{PT}$. Then, by SAS $\triangle OPT \cong \triangle OPQ$. Then, $\overline{OT} \cong \overline{OQ}$. But, this would imply that Q is on the circle, which contradicts T being the point of tangency.

On the other hand, suppose that l is perpendicular to $\overleftrightarrow{OT}$ at T. Let P be any other point on l. By the exterior angle theorem, the angle at P in $\triangle OPT$ is smaller than the right angle at T. Thus, since in any triangle the greater angle lies opposite the greater side and vice versa (Propositions 18 and 19 of Book I), we have that $OP > OT$ and P is not on the circle. Thus, l is tangent at T. □

Do tangents always exist? Theorem 2.36 shows us how to construct the tangent to a circle from a point on the circle. What if we want the tangent to a circle from a point outside the circle?

Theorem 2.37. *Given a circle c with center O and radius $\overline{OA}$ and given a point P outside of the circle, there are exactly two tangent lines to the circle passing through P.*

Proof (refer to Figure 2.32): Let M be the midpoint of $\overline{OP}$ and let c' be the circle centered at M with radius $\overline{OM}$. Then, since P is outside circle c and O is inside, we know that c' and c will intersect at two points, T_1 and T_2. Since $\angle PT_1O$ is inscribed in a semi-circle of c', then it must be a right angle. Thus, $\overleftrightarrow{PT_1}$ is perpendicular to $\overline{OT_1}$ at T_1 and by the previous theorem T_1 is a point of tangency. Likewise, T_2 is a point of tangency.

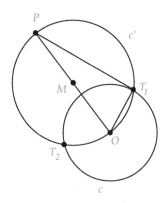

Figure 2.32

Can there be any more points of tangency passing through P? Suppose that W is another point of tangency on c such that the tangent line at W passes through P. Then, $\angle PWO$ is a right angle. Let c'' be the circle through P, W, O. Then, by Corollary 2.33 we know that $\overline{OP}$ is a diameter of c''. But, then M would be the center of c'' and the radius of c'' would be $\overline{OM}$. This implies that c' and c'' are the same circle and that W must be one of T_1 or T_2. □

2.6.4 General Intercepted Arcs

Given an arc on a circle and two (non-parallel) lines that intercept that arc, there are three different possible configurations for the lines in relation to the circle. Let P be the point of intersection for the lines.

Then, P is either outside, on, or inside the circle.

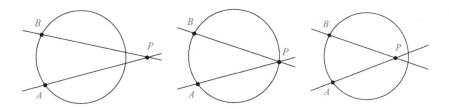

Figure 2.33

To analyze these three possibilities we need to distinguish between *secant* and tangent lines.

Definition 2.34. *A line l is said to be* secant *to a circle c if l intersects the circle at two points.*

Theorem 2.38. *Suppose two non-parallel secant lines $\overleftrightarrow{PA}$ and $\overleftrightarrow{PB}$ intersect the arc $\overarc{AB}$ of circle c, with P outside c. Let O be the center of the circle. Let C and D be the other points of intersection of the secant lines. Then, the angle created by the two secant lines is equal to the difference in the central angles to the two intercepted arcs. That is, $m\angle BPA = \frac{1}{2}(m\angle BOA - m\angle COD)$.*

Proof: Construct $\overline{AD}$.

By Euclid's Proposition 32, the measure of the exterior angle of a triangle is equal to the sum of the measures of the opposite interior angles. Thus, $m\angle BDA = m\angle CAD + m\angle BPA$, or $m\angle BPA = m\angle BDA - m\angle CAD$.

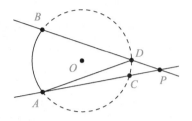

By Theorem 2.31, we know that $m\angle BDA = \frac{1}{2}m\angle BOA$. Likewise, $m\angle CAD = \frac{1}{2}m\angle COD$. Thus, $m\angle BPA = \frac{1}{2}m\angle BOA - \frac{1}{2}m\angle COD$. □

Theorem 2.39. *Suppose that a tangent line $\overleftrightarrow{PA}$ and a secant line $\overleftrightarrow{PB}$ intersect the arc $\overset{\frown}{AB}$ of circle c (center O), with P outside c. Let D be the other point of intersection of the secant line. Then, the angle created by the two lines is equal to the difference in the central angles to the two intercepted arcs. That is, $m\angle BPA = \frac{1}{2}(m\angle BOA - m\angle AOD)$.*

Proof: Construct $\overline{AD}$.

As in the proof of the previous theorem, we have $m\angle BDA = m\angle PAD + m\angle BPA$, or $m\angle BPA = m\angle BDA - m\angle PAD$. By Theorem 2.31, we know that $m\angle BDA = \frac{1}{2}m\angle BOA$. By Exercise 2.6.10 we know that $m\angle PAD = \frac{1}{2}m\angle AOD$. Thus, $m\angle BPA = \frac{1}{2}m\angle BOA - \frac{1}{2}m\angle AOD$. □

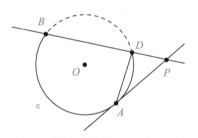

Theorem 2.40. *Suppose that lines $\overleftrightarrow{PT}$ and $\overleftrightarrow{PU}$ are tangent to circle c (center O), with P outside c. Then, the angle created by the two lines is equal to the difference in the central angles to the two intercepted arcs. That is, $m\angle UPT = \frac{1}{2}(m\angle UOT - m\angle TOU)$.*

(Here, $\angle UOT$ will be the angle defined by major arc $\overset{\frown}{UT}$.)

Proof: Construct $\overline{UT}$. Let Q be a point on the other of side of U from P on $\overleftrightarrow{PU}$.

As in the proof of the previous theorems, we have $m\angle QUT = m\angle PTU + m\angle UPT$, or $m\angle UPT = m\angle QUT - m\angle PTU$.

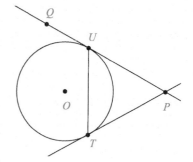

By Exercise 2.6.10 we know that $m\angle QUT = \frac{1}{2}m\angle UOT$ and $m\angle PTU = \frac{1}{2}m\angle TOU$. The result follows immediately. $\square$

These three theorems completely characterize the situation where two lines, intersecting at a point P, intercept an arc of a circle c. What about the case where P is inside the circle? One might think that there are again three cases to consider, but in the exercises it is shown that neither line can be a tangent line to the circle if P is inside c.

Theorem 2.41. *Suppose two intersecting secant lines $\overleftrightarrow{PA}$ and $\overleftrightarrow{PB}$ intersect the arc $\overparen{AB}$ of circle c, with P inside c. Let O be the center of the circle. Let C and D be the other points of intersection of the lines. Then, the angle created by the two lines is equal to the sum of the central angles to the two intercepted arcs. That is, $m\angle BPA = \frac{1}{2}(m\angle BOA + m\angle COD)$.*

Proof: Construct $\overline{AD}$.

By using similar reasoning as was used in the three preceding proofs, one can show the result. The proof is left as an exercise.

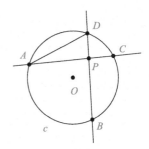

$\square$

We end this section with a nice result which is essentially the converse to Theorem 2.31.

Theorem 2.42. *Let c be a circle with center O. Let $\overleftrightarrow{PA}$ and $\overleftrightarrow{PB}$ intersect c at A and B. If the measure of the angle made by the lines at P is always half the measure of the intercepted arc's central angle, then P is on c.*

Proof: We are given that $m\angle APB = \frac{1}{2}m\angle AOB$. Suppose P is outside c. There are three cases to consider:

If the two lines are secant lines, then by Theorem 2.31 we have $m\angle ADB = \frac{1}{2}m\angle AOB$. Thus, $m\angle APB = m\angle ADB$. But this contradicts the Exterior Angle Theorem as applied to $\angle ADB$ and $\triangle ADP$.

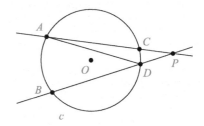

If one of the lines, say $\overleftrightarrow{PB}$ is a tangent line and the other line is a secant line, then as in the previous case we get $m\angle APB = m\angle ACB$. But this contradicts the Exterior Angle Theorem as applied to $\angle ACB$ and $\triangle BCP$.

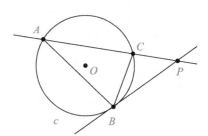

If both lines are tangent lines, let Q and R be the intersections of $\overleftrightarrow{OP}$ with c. Since $\overline{OA} \cong \overline{OB}$, then by Exercise 2.3.5 $\overrightarrow{PO}$ will bisect $\angle APB$ and also bisect $\angle AOB$. Thus, $m\angle APO = \frac{1}{2}m\angle AOR$. But this configuration was already shown to lead to a contradiction in the previous case.

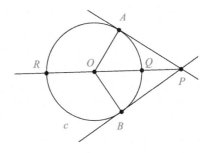

The proof for P inside c is similar and is left as an exercise. We conclude that P must be on c. $\square$

Exercise 2.6.1. *Finish the proof for the remaining two cases in Theorem 2.31. That is, prove the result for the case where A is on the diameter through $\overline{OP}$ and for where A and B are on the same side of $\overleftrightarrow{PO}$.*

Exercise 2.6.2. *Prove Corollary 2.34.*

Exercise 2.6.3. *Two circles meet at points P and Q. Let $\overline{AP}$ and $\overline{BP}$ be diameters of the circles. Show that $\overline{AB}$ passes through the other intersection Q.*

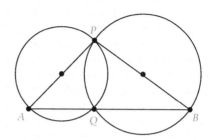

Exercise 2.6.4. *Let c be the circumscribed circle of $\triangle ABC$ and let P be the point on c where the bisector of $\angle ABC$ meets c. Let O be the center of c. Prove that the radius $\overline{OP}$ meets $\overline{AC}$ at right angles.*

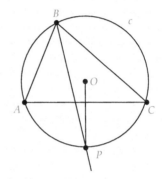

Exercise 2.6.5. *Show that the line passing through the center of a circle and the midpoint of a chord is perpendicular to that chord, provided the chord is not a diameter.*

Exercise 2.6.6. *Let $\overline{AD}$ and $\overline{BC}$ be two chords of a circle that intersect at P. Show that $(AP)(PD) = (BP)(PC)$. [Hint: Use similar triangles.]*

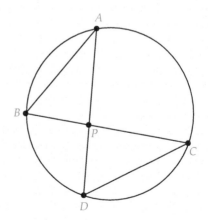

Exercise 2.6.7. *In the last project we saw that every triangle can be circumscribed. Show that every rectangle can be circumscribed.*

Exercise 2.6.8. *Prove Theorem 2.41.*

Exercise 2.6.9. *Finish the proof of Theorem 2.42. That is, show that the case where P is inside c is impossible.*

Exercise 2.6.10. *Let line $\overleftrightarrow{PT}$ be tangent to circle c at T. Let O be the center of c. Let $\overline{TA}$ be a chord of c. If $\angle PTA$ is an acute angle, show that $m\angle PTA = \frac{1}{2}m\angle TOA$.*

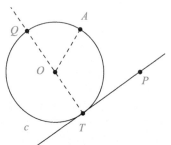

[Hint: Let $\overline{QT}$ be the diameter through T. Then, $\angle PTA = 90 - \angle ATO$. Use isosceles triangle $\triangle OTA$ and the fact that the angle sum of a triangle is 180 degrees.]

Definition 2.35. *Two distinct circles $c_1 \neq c_2$ are mutually tangent at a point T if the same line through T is tangent to both circles.*

Exercise 2.6.11. *Show that two circles that are mutually tangent must have the line connecting their centers passing through the point of tangency.*

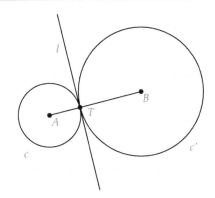

Exercise 2.6.12. *Show that mutually tangent circles intersect in only one point. [Hint: Suppose they intersected at another point P. Use the previous exercise and isosceles triangles to yield a contradiction.]*

Exercise 2.6.13. *Show that two circles that are mutually tangent at T must either be*

1. *on* opposite *sides of the common tangent line at T, in which case we will call the circles* externally *tangent, or*

2. *on the* same *side of the tangent line, with one inside the other, in which case we will call the circles* internally *tangent.*

Exercise 2.6.14. *Given two circles externally tangent at a point T, let $\overline{AB}$ and $\overline{CD}$ be segments passing through T with A, C on c_1 and B, D on c_2. Show that $\overleftrightarrow{AC}$ and $\overleftrightarrow{BD}$ are parallel. [Hint: Show that the alternate interior angles at C and D are congruent.]*

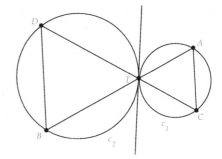

Exercise 2.6.15. *Show that the line from the center of a circle to an outside point bisects the angle made by the two tangents from that outside point to the circle. [Hint: Use Exercise 2.2.11.]*

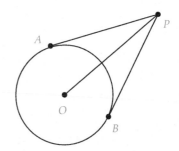

Exercise 2.6.16. *Show the converse to the preceding exercise, that is, that the bisector of the angle made by two tangents from a point outside a circle to the circle must pass through the center of the circle. [Hint: Try a proof by contradiction.]*

Exercise 2.6.17. *Let c and c' be externally tangent at T. Show that there are two lines that are tangent to both circles (at points other than T). [Hint: Let m be the line through the centers. Consider the two radii that are perpendicular to m. Let l be the line through the endpoints of these radii on their respective circles. If l and m are parallel, show that l is a common line of tangency for both circles. If l and m intersect at P, let n be a tangent from P to one of the circles. Show n is tangent to the other circle.]*

2.7 PROJECT 4 - CIRCLE INVERSION AND ORTHOGONALITY

In this project we will explore the idea of inversion through circles. Circle inversion will be a critical component of our construction of non-Euclidean geometry in Chapter 7.

We will be using dynamic geometry software for this project. Notes on how to use particular software packages for this project can be found at http://www.gac.edu/~hvidsten/geom-text.

We start out with the notion of the *power* of a point with respect to a given circle.

Start up your geometry software and create a circle c with center O and radius point A, and create a point P not on c.

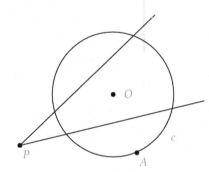

Now create two lines originating at P that pass through the circle. Find the two intersection points of the first line with the circle (call them P_1 and P_2) and the two intersection points of the second line with the circle (call them Q_1 and Q_2).

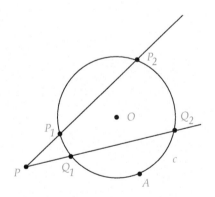

Now we will compare the product of PP_1 and PP_2 to the product of PQ_1 and PQ_2.

To do this, first measure the four distances PP_1, PP_2, PQ_1, and PQ_2. Next, use the Calculator tool which is a component of your geometry software to compute the product of PP_1 and PP_2 and the product of PQ_1 and PQ_2. Here we see the measurements as displayed when using the *Geometry Explorer* software.

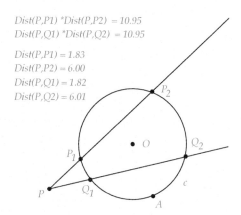

Dist(P,P1) *Dist(P,P2) = 10.95
Dist(P,Q1) *Dist(P,Q2) = 10.95

Dist(P,P1) = 1.83
Dist(P,P2) = 6.00
Dist(P,Q1) = 1.82
Dist(P,Q2) = 6.01

Interesting! It appears that these two products are the same. Drag point P around and see if this conjecture is supported.

Exercise 2.7.1. *Our first task in this project is to* prove *that these two products are always the same. [Hint: Consider some of the inscribed angles formed by P_1, P_2, Q_1, Q_2. Use Corollary 2.32 to show that $\triangle PP_1Q_2$ is similar to $\triangle PQ_1P_2$ and thus show the result.]*

Exercise 2.7.2. *Show that the product of PP_1 and PP_2 (or PQ_1 and PQ_2) can be expressed as $PO^2 - r^2$, where r is the radius of the circle.*

Definition 2.36. *Given a circle c with center O and radius r and given a point P, we define the* Power *of P with respect to c as:*

$$\textit{Power of } P = PO^2 - r^2.$$

Note that by Exercise 2.7.2 the Power of P is also equal to the product of PP_1 and PP_2 for any line l from P, with P_1 and P_2 the intersections of l with the circle c.

Also note that the Power of P can be used to classify whether P is inside (*Power* < 0), on (*Power* = 0), or outside (*Power* > 0) the circle.

Now we are ready to define circle inversion.

Definition 2.37. *The* inverse *of P with respect to c is the unique point P' on ray $\overrightarrow{OP}$ such that $OP' = \frac{r^2}{OP}$ (or $(OP')(OP) = r^2$).*

Note that if the circle had unit radius ($r = 1$), and if we considered O as the origin in Cartesian coordinates with $OP = x$, then the inverse P' of P can be interpreted as the usual multiplicative inverse; that is, we would have $OP' = \frac{1}{x}$.

How do we construct the inverse point?

Clear the screen and create a circle c with center O and radius point A and then create a point P inside c. Create the ray $\overrightarrow{OP}$. At P construct the perpendicular to $\overrightarrow{OP}$ and find the intersection points (T and U) of this perpendicular with the circle. Create segment $\overline{OT}$ and find the perpendicular to $\overline{OT}$ at T. Let P' be the point where this second perpendicular intersects $\overrightarrow{OP}$.

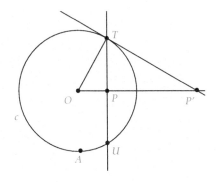

Measure the distances for segments $\overline{OP}$ and $\overline{OP'}$ and measure the radius of the circle. Use the Calculator to compute the product of OP and OP' and the square of the radius as shown in the figure.

Dist(O,P) = 1.24
Dist(O,P') = 5.10
Radius(c) = 2.52

Dist(O,P) *Dist(O,P') = 6.33
Radius(c) ^2 = 6.33

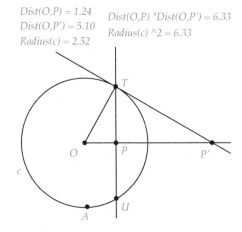

It appears that we have constructed the inverse!

Exercise 2.7.3. *Prove that this construction actually gives the inverse of P. That is, show that $(OP)(OP') = r^2$.*

In the last part of this lab, we will use the notion of circle inversion to construct a circle that meets a given circle at right angles.

Definition 2.38. *Two circles c and c′ that intersect at distinct points A and B are called* orthogonal *if the tangents to the circles at each of these points are perpendicular.*

Suppose we have a circle c and two points P and Q inside c, with P not equal to Q and neither point equal to the center O of the circle. The goal is to construct a circle through P and Q that meets c at right angles.

Using the ideas covered earlier in this project, construct the inverse P' of P with respect to c. Then, use the Circle tool in your geometry program to construct the unique circle c' through the three points P, P', and Q. The claim is that c' is orthogonal to c.

To see if this is the case, let's first find the center of c'. Let R be the intersection of $\overleftrightarrow{TP}$ with circle c' (Fig 2.34).

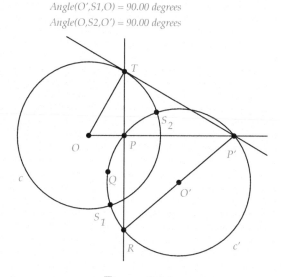

Figure 2.34

Then $\angle RPP'$ is a right angle in circle c' as $\angle OPT$ is a right angle. Thus, by Theorem 2.33 $\overline{RP'}$ is a diameter of c'. The midpoint O' of $\overline{RP'}$ will be the center of c'. Let S_1 and S_2 be the intersection points of c with c'. Measure $\angle O'S_1O$ and $\angle OS_2O'$ and check that they are right angles. Since the tangents to c and c' are orthogonal to $\overline{OS_1}$, $\overline{OS_2}$, $\overline{O'S_1}$, and $\overline{O'S_2}$, then the tangents to the circles at S_1 and S_2 must also be orthogonal and the circles are orthogonal. Note that this *evidence* of the orthogonality of c and c' is not a rigorous proof. The proof will be covered when we get to Theorem 2.43.

Exercise 2.7.4. *What do you think will happen to circle c' as one of the points P or Q approaches the center O of circle c? Try this out and then explain why this happens.*

Project Report

The ability to construct orthogonal pairs of circles is crucial to developing a model of Hyperbolic geometry, where parallels to a line through a point are "abundant." We will look at this model in detail in Chapter 7.

For the project report, provide a detailed analysis of the constructions used in this project and all answers to the exercises.

2.7.1 Orthogonal Circles Redux

Here is a proof of orthogonality of the circles constructed in the text preceding Exercise 2.7.4.

> **Theorem 2.43.** *Given a circle c with center O and radius $\overline{OA}$ and given two points P and Q inside c, with P not equal to Q and neither point equal to O, there exists a unique circle c' (or line) that passes through P and Q that is orthogonal to the given circle (Figure 2.35).*

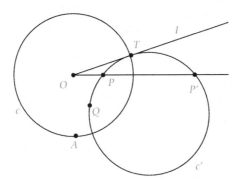

Figure 2.35

Proof: It is clear that if P and Q lie on a diameter of c, then there is a unique line (coincident with the diameter) that is orthogonal to c. So, in the rest of this proof we assume that P and Q are not on a diameter of c.

Suppose that one or the other of P or Q, say P, is strictly inside c. As above construct the inverse P' to P and let c' be the unique circle passing through Q, P, P'. Construct a tangent l to circle c' that passes through O. Let l be tangent to c' at T. (To construct l, use the construction discussed in Theorem 2.37.) We claim that T is also on circle c. To see this, consider the power of O with respect to circle c':

$$Power\ of\ O = (OP)(OP') = (OT)^2$$

But, $(OP)(OP') = r^2$ (r being the radius of c) since P' is the inverse point to P with respect to c. Thus, $(OT)^2 = r^2$ and T is on circle c, and the circles are orthogonal at T.

To see that this circle is unique, suppose there was another circle c'' through P and Q that was orthogonal to c. Let P'' be the intersection of $\overrightarrow{OP}$ with c''. Let T'' be a point where c and c'' intersect. Then $(OP)(OP'') = (OT'')^2$. But, $(OT'')^2 = r^2$ and thus, P'' must be the inverse P' to P, and c'' must then pass through Q, P, P' and must be the circle c.

The final case to consider is when both P and Q are on the boundary of c (Figure 2.36). Then, any circle through P and Q that is orthogonal to c must have its tangents at P and Q lying along $\overline{OP}$ and $\overline{OQ}$. Thus, the diameters of this circle must lie along tangent lines to c at P and Q. Thus, the center of the orthogonal circle must lie at the intersection of these tangents, which is a unique point. □

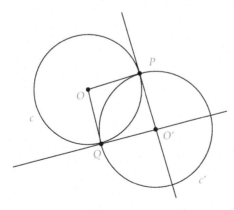

Figure 2.36

We conclude this section on orthogonal circles with two results that will prove useful when we study non-Euclidean geometry in Chapter 7.

Theorem 2.44. *Let c and c' be two circles and let P be a point that is not on c and is not the center O of c. Suppose that c' passes through P. Then, the two circles are orthogonal if and only if c' passes through the inverse point P' to P with respect to c.*

Proof: First, suppose that c' passes through the inverse point P' (refer to Figure 2.37). We know from the proof of Theorem 2.30 that the center O' of c' lies on the perpendicular bisector of $\overline{PP'}$. Since P and P' are inverses with respect to c, then they both lie on the same side of ray $\overrightarrow{OP}$. Thus, O is not between P and P' and we have that $O'O > O'P$. Thus, O is outside of c'. We then can construct two tangents from O to c' at points T_1 and T_2 on c'. Using the idea of the power of points with respect to c', we have $(OT_1)^2 = (OP)(OP')$. But, $(OP)(OP') = r^2$ by assumption, where r is the radius of c. Therefore, $(OT_1)^2 = r^2$, and T_1 is on c. A similar argument shows that T_2 is also on c. This implies that the two circles are orthogonal.

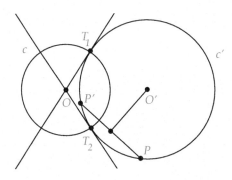

Figure 2.37

Conversely, suppose that c and c' are orthogonal at points T_1 and T_2. The tangent lines to c' at these points then pass through O, which implies that O is outside c'. Thus, $\overrightarrow{OP}$ must intersect c' at another point P'. Using the power of points, we have $r^2 = (OT_1)^2 = (OP)(OP')$, and thus P' is the inverse point to P with respect to circle c. □

Corollary 2.45. *Suppose circles c and c' intersect. Then c' is orthogonal to c if and only if the circle c' is mapped to itself by inversion in the circle c.*

Proof: Suppose the circles are orthogonal, and let P be a point on c'. If P is also on c, then it is fixed by inversion through c. If P is not on c, then by the proof of Theorem 2.44, we know that P is also not the center O of c. Thus, Theorem 2.44 implies that the inverse P' of P with respect to c is on c'. Thus, for all points P on c', we have that the inverse point to P is again on c'.

Conversely, suppose c' is mapped to itself by inversion in the circle c. Let P be a point on c' that is not on c and which is not the center O of c. Then, the inverse point P' with respect to c is again on c'. By Theorem 2.44, the circles are orthogonal. □

CHAPTER 3

Analytic Geometry

There once was a very brilliant horse who mastered arithmetic, plane geometry, and trigonometry. When presented with problems in analytic geometry, however, the horse would kick, neigh, and struggle desperately. One just couldn't put Descartes before the horse.

– Anonymous

In 1637 Réne Descartes (1596–1650) published the work *La Geometrie*, in which he laid out the foundations for one of the most important inventions of modern mathematics—the Cartesian coordinate system and analytic geometry.

In classical Euclidean geometry, points, lines, and circles exist as idealized objects independent of any concrete context. In this strict *synthetic* geometry, algebraic relations can be discussed, but *only* in relation to underlying geometric figures.

For example, the Pythagorean Theorem, in Euclid's geometry, reads as follows: "The square on the hypotenuse equals the squares on the two sides." This statement literally means that the square constructed on the hypotenuse equals the other two squares, and Euclid's proof of the Pythagorean Theorem amounts to showing how one can rearrange the square figures to make this equivalence possible.

From the time of Euclid until Descartes, there was always this insistence on tying algebraic expressions precisely to geometric figures. The Arabs were the first to introduce symbols, such as x^2, for algebraic quantities, but they too insisted on strict geometric interpretations of algebraic variables—x^2 literally meant the *square* constructed on a segment of length x.

Descartes was the first person to assume that algebraic relationships need not be tied to geometric figures. For Descartes, expressions like x^2, x^3, xy, and the like, were all *numbers*, or lengths of segments, and algebraic expressions could be thought of as either arithmetic or geometric expressions.

This was a great leap forward in the level of mathematical abstraction, in that it opened up mathematical avenues of study that were artificially closed. For example, equations involving arbitrary powers of x were now possible, since one was no longer restricted to segments (x), squares (x^2), and cubes (x^3).

Descartes' geometry was groundbreaking, although by no means what we think of today as *Cartesian* geometry. There were no coordinate axes, and Descartes did not have algebraic expressions for such simple figures as straight lines.

A much more modern-looking attempt to merge algebra with geometry was that of Pierre de Fermat (1601–1655), a contemporary of Descartes working in Toulouse, who had a system of perpendicular axes and coordinate equations describing lines, quadratics, cubics, and the conic sections. Fermat also developed a general method for finding tangents and areas enclosed by such algebraic expressions. In this regard, his work foreshadowed the development of calculus by Newton and Leibniz.

The great insight of Descartes and Fermat was to embed the study of geometric figures in a grid system, where a point is precisely located by its distances from two fixed lines that are perpendicular to one another. These two distances are called the *coordinates* of a point and are customarily labeled "x" and "y."

By studying the set of coordinates for a geometric figure, one can identify patterns in these coordinates. For example, a line is a set of points (x, y) where x and y have a relationship of the form $ax+by+c = 0$, with a, b, and c constants. Similarly, the points making up a circle have their own x-y relationship.

The coordinate geometry of Descartes and Fermat ultimately led to the notion of *functions* and to the creation of calculus by Newton and Leibniz toward the end of the seventeenth century. The great achievement of analytic geometry is that it allows one to enrich the traditional synthetic geometry of figures by the study of the equations of the x-y relationships for those figures.

In this chapter we will construct analytic geometry from first principles. Initially, we will develop this geometry from a synthetic geometric base, similar to the development of Descartes and Fermat. Later in the chapter, we will develop analytic geometry from a modern *axiomatic* basis. In the chapters following, we will make optimum use of both synthetic and analytic geometry, using whichever approach to Euclidean geometry is most transparent in devising proofs.

3.1 THE CARTESIAN COORDINATE SYSTEM

In analytic geometry we create a *coordinate system* in the plane so that we can have a way of uniquely referencing points by numerical values. The simplest such coordinate system is the *rectangular* coordinate system (Figure 3.1).

Let l be a line in the plane and let O be a point on that line. Let m be a perpendicular to l through O. We will call these two lines the *axes* of our coordinate system, with O being the *origin*.

By the continuity properties of the line, we know that for every real number x there is a unique point X on l such that the length of $\overline{OX}$ is x. We pick one side (ray) of l from O to be the positive side where the length of $\overline{OX}$ is $x \geq 0$ and let $-x$ be the coordinate of $\overline{OX'}$ on the opposite ray such that the length of $\overline{OX'}$ is also x. Likewise, there is a similar correspondence on m. We call x and y the *coordinates* of the points on the axes.

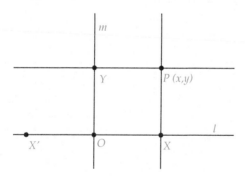

Figure 3.1

Now, let P be any other point in the plane, not on l or m. We can drop (unique) perpendiculars down to l and m at points X and Y, represented by coordinates x and y. The lines will form a rectangle $OXPY$. Conversely, given coordinates x and y, we can construct perpendiculars that meet at P. Thus, there is a one-to-one correspondence between pairs of coordinates (x, y) and points in the plane. This system of identification of points in the plane is called the *rectangular* (or *Cartesian*) coordinate system.

To see the power of this method, let's look at the equation of a circle, that is, the algebraic relationship between the x and y coordinates of points on a circle. The circle is defined by starting with two points O, R and consists of points P such that $\overline{OP} \cong \overline{OR}$.

Let's create a coordinate system with origin at O and x-axis along $\overline{OR}$. Let r be the length of $\overline{OR}$. Given any point P on the circle, let (x, y) be its coordinates (Figure 3.2). The angle $\angle OXP$ is a right angle. By the Pythagorean Theorem the length of $\overline{OP}$ is $\sqrt{x^2 + y^2}$, and this must equal the length of $\overline{OR}$, which is r. Thus, the circle is the set of points (x, y) such that $x^2 + y^2 = r^2$, a familiar equation from basic analysis.

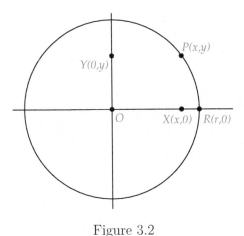

Figure 3.2

In the preceding discussion on the equation of a circle, we saw that the length between a point (x, y) on a circle and the center of the circle is given by $\sqrt{x^2 + y^2}$, assuming the origin of the coordinate system is at the center of the circle. If the coordinate system is instead constructed so that the center of the circle is at (x_0, y_0), then clearly the new length formula will be given by

$$\sqrt{(x - x_0)^2 + (y - y_0)^2}$$

In general this will give the distance between the points (x, y) and (x_0, y_0) and is called the *distance formula in the plane*.

Here are some other useful algebraic-geometric facts.

Theorem 3.1. *Let $A = (x, y)$ and $B = (-x, -y)$. Then, the line through A, B passes through the origin $(0, 0)$ with A and B on opposite sides of the origin on this line.*

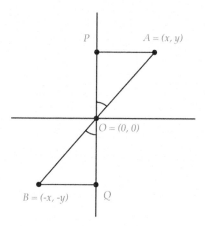

Figure 3.3

Proof (refer to Figure 3.3): If A is on either of the two axes, then so is B, and the result is true by the definition of positive and negative coordinates on the axes.

Otherwise, we can assume that A (and thus B) is not on either axis. Drop perpendiculars from A and B to the y-axis at P and Q. Then, the distance from P to the origin will be y as is the distance from Q to the origin. By the distance formula $\overline{PA} \cong \overline{QB}$ and $\overline{AO} \cong \overline{BO}$. Thus, by SSS $\triangle AOP \cong \triangle BOQ$ and the angles at O in these triangles are congruent. Since $\angle BOQ$ and $\angle BOP$ are supplementary, then so are $\angle AOP$ and $\angle BOP$. Thus, A, O, and B lie on a line.

If A and B were on the same side of O on the line through A,B, then since this line intersects the x-axis only at O, they must be on the same side of the x-axis. But, then P and Q would have to be on the same side of the x-axis, which is not the case. $\square$

> **Theorem 3.2.** *Let $A = (x, y)$ and k be a number. Then, $B = (kx, ky)$ is on the line through A and the origin, and the distance from B to the origin is equal to k times the distance from A to the origin.*

Proof: This can be proved using similar triangles and is left as an exercise. □

> **Theorem 3.3.** *If $A = (x, y) \neq (0, 0)$ and $B = (x_1, y_1)$ are points on the same line through the origin, then $(x_1, y_1) = k(x, y)$ for some number k.*

The proof of this theorem is left as an exercise.

> **Theorem 3.4.** *If $A = (x, y)$ and $B = (x_1, y_1)$ are not collinear with the origin, then the point $C = (x + x_1, y + y_1)$ forms, with O, A, B, a parallelogram where $\overline{OC}$ is the diagonal.*

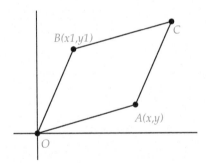

Figure 3.4

Proof: First we have that $AC = \sqrt{(x + x_1 - x)^2 + (y + y_1 - y)^2}$ (refer to Figure 3.4). Thus, $AC = \sqrt{(x_1)^2 + (y_1)^2} = BO$. Similarly, $AO = BC$. A simple triangle congruence argument shows that $OACB$ is a parallelogram and $\overline{OC}$ is the diagonal. □

The choice of a coordinate system will divide the set of non-axes points into four groups, called *quadrants*. Quadrant I consists of points with both coordinates positive. Quadrant II is the set of points where $x < 0$ and $y > 0$. Quandrant III has $x < 0$ and $y < 0$, and quadrant IV has $x > 0$ and $y < 0$. Thus, all points are either on an axis or in one of the four quadrants.

3.2 VECTOR GEOMETRY

Analytic geometry has been incredibly useful in modeling natural systems such as gravity and temperature flow. One reason for analytic geometry's effectiveness is the ease of representing physical properties using coordinates. Many physical systems are governed by variables such as force, velocity, electric charge, and so forth, that are uniquely determined by their size and direction of action. Such variables are called *vectors*.

The concept of a *vector* can be traced back to the work of William Rowan Hamilton (1805–1865) and his efforts to treat complex numbers both as ordered pairs and as algebraic quantities. His investigations into the algebraic properties of vector systems led to one of the greatest discoveries of the nineteenth century, that of *quaternions*. While trying to create an algebra of multiplication for three-dimensional vectors (a task at which he repeatedly failed), he realized that by extending his vector space by one dimension, into four dimensions, he could define an operation of multiplication on vectors.

In 1843, as he was walking along the Royal Canal in Dublin, Hamilton came to a realization:

> And here there dawned on me the notion that we must admit, in some sense, a fourth dimension of space for the purpose of calculating with triples . . . An electric circuit seemed to close, and a spark flashed forth. [39]

This realization had such a profound effect on Hamilton that he stopped by a bridge on the canal and carved into the stone this famous formula:

$$i^2 = j^2 = k^2 = ijk = -1$$

Here i, j, and k are unit length vectors along the y, z, and w axes in four dimensions. Other important figures in the development of vector algebra include Arthur Cayley (1821–1895) and Hermann Grassman (1809–1877). Vector methods have come to dominate fields such as mathematical physics. We will cover the basics of vector methods and then use vectors to study geometric properties.

Definition 3.1. *A vector is a quantity having a length and a direction. Geometrically, we represent a vector in the plane by a directed line segment or arrow. The starting point of the vector is called the* tail *and the point of direction is called the* head. *Two parallel directed segments with the same length and direction will represent the same vector. Thus, a vector is really a set of equivalent directed segments in the plane.*

What is the coordinate representation of a vector? Let the vector be represented by the segment from (h, k) to (x, y) as shown in Figure 3.5.

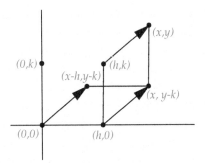

Figure 3.5

If we consider the vector from $(h, 0)$ to $(x, y-k)$, which has its tail on the x-axis, then, the quadrilateral formed by the heads and tails of these two vectors forms a parallelogram, since by the distance formula the sides will be congruent in pairs. These two directed segments are equivalent as vectors since they have the same length and direction. Similarly, the directed segment from $(0, 0)$ to $(x - h, y - k)$ represents the same vector.

Thus, given any vector in the plane, there is a unique way of identifying that vector as a directed segment from the origin. The vector from $A = (h, k)$ to $B = (x, y)$ can be represented as the ordered pair $(x - h, y - k)$ if we assume that the vector's tail is at the origin. This is the standard way to represent vectors.

Definition 3.2. *Let v be a vector represented by the ordered pair (x, y). The norm of v will be defined as the length of the directed segment v and will be denoted by $\|v\|$. Thus,*

$$\|v\| = \sqrt{x^2 + y^2}$$

It will be convenient to define certain algebraic operations with vectors.

Definition 3.3. *Let u, v be vectors and k a positive number. Then*

- *The vector $u + v$ is the vector representing the diagonal of the parallelogram determined by u and v, if u, v are not on the same line through the origin. If u, v are on the same line through the origin, then $u + v$ is the vector representing the sum of the x and y coordinates of the vectors.*

- *The vector $-v$ is the vector having the same length but opposite direction from v.*

- *The vector ku is the vector in the same direction as u whose length is k times the length of u.*

It follows from the definitions that the vector $u - v$, which is $u + (-v)$, is the vector from v to u (Figure 3.6).

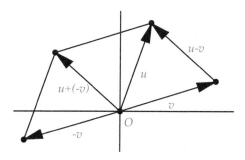

Figure 3.6

> **Theorem 3.5.** *Let $u = (a, b)$, $v = (s, t)$ be two vectors in a coordinate system and k be a number. Then*
>
> - *The vector $u + v$ has coordinates $(a + s, b + t)$.*
>
> - *The vector $-v$ has coordinates $(-s, -t)$.*
>
> - *The vector ku has coordinates (ka, kb).*

Proof: Since vectors can be represented by segments from the origin to points in the plane, then this theorem is basically a restatement of the algebraic-geometric properties discussed earlier in our discussion of coordinate systems. □

Vectors can be used to give some very elegant geometric proofs. For example, consider the following theorem about the medians of a triangle.

> **Theorem 3.6.** *The medians of a triangle intersect at a common point that lies two-thirds of the way along each median.*

Let's see how we can use vectors to prove this theorem (Figure 3.7).

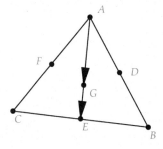

Figure 3.7

Proof: First, it is left as an exercise to show that we can represent the line through A and B by the set of all $\vec{A} + t(\vec{B} - \vec{A})$, where t is real. ($\vec{A}$ stands for the vector from the origin to A.) The midpoint of the segment $\overline{AB}$ will then be $\frac{1}{2}(\vec{A} + \vec{B})$ (the proof of this fact is left as an exercise).

Similarly, the midpoints of $\overline{AB}$, $\overline{AC}$, and $\overline{BC}$ are given by the three vectors

$$\vec{D} = \frac{1}{2}(\vec{A} + \vec{B})$$

$$\vec{E} = \frac{1}{2}(\vec{B} + \vec{C})$$

$$\vec{F} = \frac{1}{2}(\vec{A} + \vec{C})$$

$$(3.1)$$

Consider the median $\overline{AE}$. Let G be the point two-thirds along this segment from A to E. Then

$$\vec{G} - \vec{A} = \vec{AG} = \frac{2}{3}\vec{AE} = \frac{2}{3}(\vec{E} - \vec{A}) = \frac{2}{3}(\frac{1}{2}(\vec{B}+\vec{C}) - \vec{A}) = (\frac{1}{3}\vec{B} + \frac{1}{3}\vec{C} - \frac{2}{3}\vec{A})$$

Adding $\vec{A}$ to both sides yields

$$\vec{G} = \frac{1}{3}(\vec{A} + \vec{B} + \vec{C})$$

A similar argument shows that the point that is two-thirds along the other two medians is also $\frac{1}{3}(\vec{A} + \vec{B} + \vec{C})$. Thus, these three points can be represented by the same vector, and thus the three medians meet at the point represented by this vector. □

Exercise 3.2.1. *Prove Theorem 3.2.*

Exercise 3.2.2. *Prove Theorem 3.3.*

Exercise 3.2.3. *Let l be the line through a point $P = (a, b)$ and parallel to vector $v = (v_1, v_2)$. Show that if $Q = (x, y)$ is another point on l, then $(x, y) = (a, b) + t(v_1, v_2)$ for some t. [Hint: Consider the vector from P to Q.]*

Exercise 3.2.4. *Use the previous exercise to show that given a line l, there are constants A, B, and C such that $Ax + By + C = 0$, for any point (x, y) on l.*

Exercise 3.2.5. *Given a segment $\overline{AB}$, show that the midpoint is represented as the vector $\frac{1}{2}(\vec{A} + \vec{B})$.*

Exercise 3.2.6. *Given a quadrilateral $ABCD$, let W, X, Y, Z be the midpoints of sides $\overline{AB}, \overline{BC}, \overline{CD}, \overline{DA}$, respectively. Use vectors to prove that $WXYZ$ is a parallelogram.*

3.3 PROJECT 5 - BÉZIER CURVES

So far we have considered fairly simple geometric figures such as lines and circles, for which we have correspondingly simple equations in x and y. While these have incredibly useful features for the design of many everyday objects, they are not so useful for describing complex shapes such as the curves one finds in automobile designs or in the outlines of the letters of a font.

For such non-linear and non-circular curves it is very difficult to derive simple polynomial functions of x and y that completely represent the curve. In general, the more a curve oscillates, the higher the degree of polynomial we need to represent the curve. For a designer, having to calculate high-order polynomials in order to sketch curves such as the fender of a car is not a pleasant or efficient prospect.

In the 1950s and 1960s the problem of mathematically describing curves of arbitrary shape was a primary area of research for engineers and mathematicians in the aircraft and automobile industries. As computational tools for design became more and more widely used, the need for efficient mathematical algorithms for modeling such curves became one of the highest priorities for researchers.

To image a curve on a computer screen, one has to plot each individual screen pixel that makes up the curve. If one pre-computes all the (x, y) points needed for accurately representing the curve to the resolution of the screen, then this will consume significant portions of the computer's system memory. A better solution would be to have an algorithm for computing the curve that stored just a few special points and then computed the rest of the points "on the fly."

In this project we will look at a clever method of computing smooth curves by using a simple set of "control" points in the plane. This method is due to two automobile designers: Pierre Bézier (1910–1999), who worked for the French automaker Rénault, and Paul de Casteljau, who also worked for a French automaker—Citroën. The curves which these engineers discovered have become known as *Bézier curves*, whereas the algorithm we will consider for computing them has become known as *de Casteljau's algorithm*. A complete review of Bézier curves, and other curves used in computer-aided design, can be found in textbooks on computer graphics such as [24] or [14].

To start our discussion, let's consider how we would create a curved path in the plane by using a small number of defining points. Clearly, one point cannot define a two-dimensional curve and two points uniquely define a line. Thus, to have any hope of defining a truly curved path in the plane, we need at least three points.

We will be using dynamic geometry software for this project. Notes on how to use particular software packages for this project can be found at http://www.gac.edu/~hvidsten/geom-text.

Start up your geometry software and create three (non-collinear) points A, B, and C joined by two segments as shown.

Let's imagine a curve that passes through points A and C and that is pulled toward B like a magnet. We will define such a curve *parametrically*. That is, the curve will be given as a vector function $\vec{c}(t)$ where t will be a parameter running from 0 to 1. We want $\vec{c}(0) = A$ and $\vec{c}(1) = C$, and for each $0 < t < 1$ we want $\vec{c}(t)$ to be a point on a smooth curve bending toward B. By *smooth* we mean a curve that has no sharp corners, that is, one whose derivative is everywhere defined and continuous.

So, our task is to find a way to associate the parameter t with points $\vec{c}(t)$ on the curve we want to create.

Bézier's solution to this problem was to successively *linearly interpolate* points on the segments to define $\vec{c}(t)$. Attach a point A' on the segment $\overline{AB}$ as shown and measure the ratio of AA' to AB.

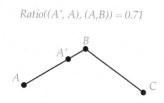

Ratio((A', A), (A,B)) = 0.71

The value of this ratio is the value of t that would appear as the parameter value for the point A' in a parametric equation for $\overline{AB}$. This parametric equation is given by $\vec{l_1}(t) = \vec{A} + t(\vec{B} - \vec{A})$, with $0 \leq t \leq 1$. (Refer to Exercise 3.2.3.) We will use this ratio parameter as the defining parameter for the curve $\vec{c}(t)$ we are trying to create. Note that when $t = 0$ we are at A, which is what we want. But when $t = 1$, we are only at B and not at C.

We will now interpolate a point between B and C that has the same ratio as the point A' did along the segment $\overline{AB}$. To do this, we will use the dilating (or scaling) capability of your geometry software. A dilation is defined by a center point and a numerical ratio. If you are unfamiliar with how to define dilations is your geometry software, consult the notes on this project at at http://www.gac.edu/~hvidsten/geom-text.

To define the dilation that will create the desired point between B and C, set point B as the center of the dilation and use the preceding ratio measurement as the dilation ratio.

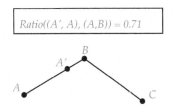

Once the dilation is defined we select point C and dilate it towards B by this dilation. A new point B' will be created on $\overline{BC}$ with the same relative parameter value as A' has on $\overline{AB}$. To convince yourself of this, measure the ratio of $B'B$ to BC as shown.

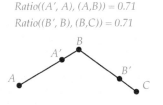

We now have parameterized both segments with two vector functions $\vec{l_1}(t) = \vec{A} + t(\vec{B} - \vec{A})$ and $\vec{l_2}(t) = \vec{B} + t(\vec{C} - \vec{B})$, where the parameter t represents the relative distance points are along their respective segments.

Since linear interpolation has worked nicely so far, we carry it one step further on the new segment $\overline{A'B'}$. Create this segment and set A' as a new center of dilation. Using the same dilation ratio as before, select B' and dilate it towards A' by this ratio.

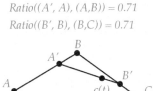

This new point is the point we will use for our definition of $\vec{c}(t)$. Let's calculate the precise form for $\vec{c}(t)$.

Since $\vec{c}(t)$ is the interpolated point along $\overline{A'B'}$, then it will have equation

$$\vec{c}(t) = \vec{A'} + t(\vec{B'} - \vec{A'})$$

We also know that $\vec{A'} = \vec{l}_1(t) = \vec{A} + t(\vec{B} - \vec{A})$ and $\vec{B'} = \vec{l}_2(t) = \vec{B} + t(\vec{C} - \vec{B})$. Substituting these values into $\vec{c}(t)$ we get

$$\begin{aligned} \vec{c}(t) &= \vec{A} + t(\vec{B} - \vec{A}) + t((\vec{B} + t(\vec{C} - \vec{B})) - (\vec{A} + t(\vec{B} - \vec{A}))) \\ &= \vec{A} + t(2\vec{B} - 2\vec{A}) + t^2(\vec{C} - 2\vec{B} + \vec{A}) \end{aligned}$$

We note that $\vec{c}(0) = \vec{A} = A$ and $\vec{c}(1) = \vec{C} = C$, which is exactly what we wanted! Also, this function is a simple quadratic function in t and thus is perfectly smooth. But, what does $\vec{c}(t)$ really look like?

To answer this question, hide the two measurements, and hide the label for $\vec{c}(t)$, to unclutter our figure. Then, select $\vec{c}(t)$ and set your program to *trace* this point. This will cause your program to draw copies of the point as it moves. Drag point A' back and forth along $\overline{AB}$ and see how $\vec{c}(t)$ traces out the curve we have just calculated.

Exercise 3.3.1. *It appears from our picture that the curve $\vec{c}(t)$ is actually tangent to the vector $\vec{B} - \vec{A}$ at $\vec{c}(0)$ and to the vector $\vec{C} - \vec{B}$ at $\vec{c}(1)$. Calculate the derivative of $\vec{c}(t)$ and prove that this is actually the case.*

By controlling the positions of A, B, and C we can create curves that are of varying shapes, but that have only one "bump." As we discussed earlier, for more oscillatory behavior, we will need higher order functions. A simple way to do this is to have more control points.

Clear the screen and draw four points connected with segments as shown. Attach a point A' to $\overline{AB}$ and then measure the ratio of $A'A$ to AB.

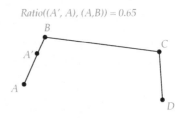

Now, review our methods of the previous construction, and construct B' and C' on $\overline{BC}$ and $\overline{CD}$ with the same parameter as A' has on $\overline{AB}$. (In your project write-up, describe the steps you take to accomplish this step and the next set of steps.)

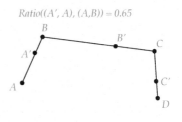

Next, construct $\overline{A'B'}$ and $\overline{B'C'}$. Now, we have just three points, as we did at the start of the previous construction. Construct points A'' and B'' on $\overline{A'B'}$ and $\overline{B'C'}$ using the same parameter.

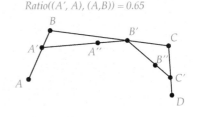

Finally, carry out the interpolation process one more time to get $\vec{c}(t)$.

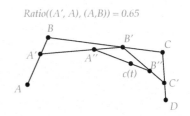

If we now trace $\vec{c}(t)$ we again get a nice smooth curve that passes through the endpoints A and D.

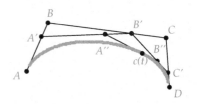

If we move C to the other side of $\overleftrightarrow{AD}$ we get an oscillation in the curve. (Make sure to clear the tracing and then re-start the tracing after you move C.)

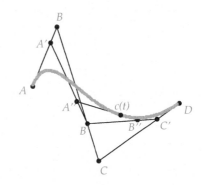

What is the equation for this new four-point Bézier curve? Clearly, the parametric form for $\vec{c}(t)$ will be

$$\vec{c}(t) = \vec{A''} + t(\vec{B''} - \vec{A''})$$

From our previous work we know that $\vec{A''} = \vec{A} + t(2\vec{B} - 2\vec{A}) + t^2(\vec{C} - 2\vec{B} + \vec{A})$ and $\vec{B''} = \vec{B} + t(2\vec{C} - 2\vec{B}) + t^2(\vec{D} - 2\vec{C} + \vec{B})$.

Exercise 3.3.2. *Substitute these values of $\vec{A''}$ and $\vec{B''}$ into the equation of $\vec{c}(t)$ and show that*

$$\vec{c}(t) = \vec{A} + t(3\vec{B} - 3\vec{A}) + t^2(3\vec{C} - 6\vec{B} + 3\vec{A}) + t^3(\vec{D} - 3\vec{C} + 3\vec{B} - \vec{A})$$

Bézier curves generated from four points are called *cubic* Bézier curves due to the t^3 term in $\vec{c}(t)$. Bézier curves generated from three points are called *quadratic* Bézier curves.

Exercise 3.3.3. *Show that $\vec{c}(t)$ has the same tangent properties as the quadratic Bézier curve. That is, show that the tangent at $t = 0$ is in the direction of $\vec{B} - \vec{A}$ and the tangent at $t = 1$ is in the direction of $\vec{D} - \vec{C}$.*

We could continue this type of construction of Bézier curves for five control points, or six control points, or as many control points as we wish, in order to represent more and more complex curves. However, as we add more points, the degree of our parametric Bézier curve increases as well. As the number of operations to compute a point increases, so too does the numerical instability inherent in finite precision computer algebra. Also, the oscillatory behavior of higher order curves is not always easy to control. Altering one control point to achieve a bump in one part of the curve can have unwanted ripple effects on other parts of the curve.

In practice, quadratic and cubic Bézier curves are the most widely used curves for computer graphics. They are easy to calculate, relatively easy to control as to shape, and can model highly complex figures by joining together separately defined curves to form one continuous curve.

As an example, let's look at one of the most important uses of Bézier curves in computer graphics—the representation of fonts. Suppose we want to represent the letter "S".

Here we have defined two four-point (cubic) Bézier curves to approximate the two bumps in the "S" shape. One curve is controlled by A, B, C, and D, with A_1 the initial parameter point and c_1 the curve point. The other curve is controlled by D, E, F, and G with A_2 the parameter point and c_2 the curve point. By "doubling up" on the point D we make the two curve pieces join together, and by making C, D, and E lie along a line, we make the curve look smooth across the joining region (the curves' tangents line up).

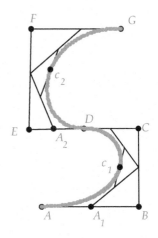

Exercise 3.3.4. *Use cubic (or quadratic) Bézier curves to design your own version of the letter "b." Describe your construction in detail and provide a printout of the curves used, similar to that in the preceding example.*

In the actual design of a font, we would need to define more than a simple curve representing the shape. Fonts are not made of one-dimensional curves, but are made of filled-in areas that are bounded by curves. For example, look closely at the bold letter **G** of this sentence. Notice how the vertical bulge on the left side of the shape is much thicker than the top and bottom parts. When designing a font, what we actually design is the *outline* of each letter shape; that is, the curve that surrounds the shape. Once this is defined, we fill in the outline to form the completed shape.

Project Report

The ability to model complex geometric figures is a critical component of modern computer graphics systems. Computational geometry methods, like the ones explored in this project, are at the heart of computer-aided design and also appear in the computer-generated images we see on television and in film.

For your project report, provide detailed answers to each of the questions, along with an explanation of the steps you took to construct the four-point cubic Bézier curve.

3.4 ANGLES IN COORDINATE GEOMETRY

As mentioned in the previous chapter, angles have a very precise definition in classical Euclidean geometry, a definition that is independent of questions of orientation. In this section we will expand the notion of angle in coordinate geometry to include the idea of orientation. We start off by defining two familiar functions.

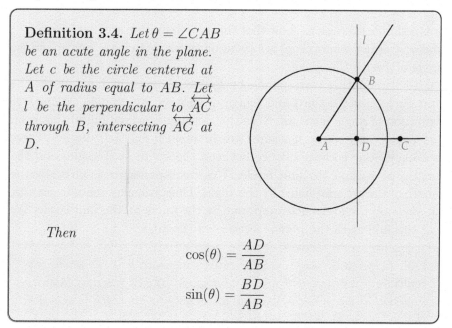

Definition 3.4. *Let* $\theta = \angle CAB$ *be an acute angle in the plane. Let* c *be the circle centered at* A *of radius equal to* AB. *Let* l *be the perpendicular to* $\overleftrightarrow{AC}$ *through* B, *intersecting* $\overleftrightarrow{AC}$ *at* D.

Then

$$\cos(\theta) = \frac{AD}{AB}$$

$$\sin(\theta) = \frac{BD}{AB}$$

These are the standard trig formulas everyone learns to love in high school trigonometry. Note that this definition is consistent with the definition presented in Exercise 2.5.5 of Chapter 2.

Theorem 3.7. *Let (x, y) be a point in the plane making an acute angle of θ with the x-axis. Let r be the distance from (x, y) to the origin. Then*

$$x = r \cos(\theta), \ y = r \sin(\theta)$$

Proof: This is an immediate consequence of the definition of sine and cosine. □

Corollary 3.8. *If (x, y) is a point on the unit circle (in the first quadrant) making an angle of θ with the x-axis, then $x = \cos(\theta)$ and $y = \sin(\theta)$.*

We now want to extend the definition of sine and cosine to *obtuse* angles. If $90 < \theta \leq 180$, we define

$$\cos(\theta) = -\cos(180 - \theta)$$
$$\sin(\theta) = \sin(180 - \theta)$$

It is a simple exercise to show that this new definition still has $x = \cos(\theta)$ and $y = \sin(\theta)$, where (x, y) is a point on the unit circle making an obtuse angle of θ with the x-axis.

To extend this correspondence between angles and points on the unit circle into the third and fourth quadrants, we need to expand our definition of angle measure. Note that for angles whose initial side is on the x-axis, and whose measures are between 0 and 180 degrees, there is a one-to-one correspondence between the set of such angles and the lengths of arcs on the unit circle. This correspondence is given by the length of the arc subtended by the angle. Thus, it makes sense to *identify* an angle by the arclength swept out by the angle on the unit circle. We call this arclength the *radian measure* of the angle.

Definition 3.5. *An angle has* radian *measure θ if the angle subtends an arc of length θ on the unit circle. If θ is positive, then the arclength is swept out in a* counterclockwise *fashion. If θ is negative, the arc is swept out in a* clockwise *fashion.*

Thus, an angle of π radians corresponds to an angle of 180 degrees swept out in the counterclockwise direction from the x-axis along the unit circle. Likewise, an angle of 2π would be the entire circle, or 360 degrees, while an angle of $-\frac{\pi}{2}$ would correspond to sweeping out an angle of 90 degrees *clockwise* from the x-axis. In this definition of angle, the *direction* or *orientation* of the angle is measured by the positive or negative nature of the radian measure.

Consistent with this extension of the definition of angle measure, we extend the definitions of sine and cosine in the usual fashion consistent with the x, y values on the unit circle.

The next result is a generalization of the Pythagorean Theorem.

> **Theorem 3.9.** *(The Law of Cosines) Let ABC be a triangle with sides of lengths a, b, c, with a opposite A, b opposite B, and c opposite C. Then*
> $$c^2 = a^2 + b^2 - 2ab \cos(\angle ACB).$$

Proof: Let $\overline{AD}$ be the altitude to side $\overline{BC}$ at D. There are four cases to consider.

First, suppose that $D = B$ or $D = C$. If $D = B$, then $\cos(\angle ACB) = \frac{a}{b}$ and we would have $c^2 = a^2 + b^2 - 2a^2 = b^2 - a^2$, which is just a restatement of the Pythagorean Theorem. If $D = C$, we again have a restatement of the Pythagorean Theorem.

If D is not B or C, then either D is between B and C, or D is on one or the other of the sides of $\overline{BC}$ on $\overleftrightarrow{BC}$. In all three cases (Figure 3.8) we have

$$DB^2 + AD^2 = c^2$$

and

$$DC^2 + AD^2 = b^2$$

Solving for AD^2, we get

$$b^2 - DC^2 = c^2 - DB^2$$

Suppose that D is between B and C. Then

$$a^2 = DB^2 + 2(DB)(DC) + DC^2$$

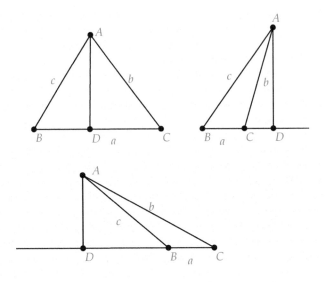

Figure 3.8

Solving for DB^2 and substituting into the equation above, we get

$$c^2 - a^2 + 2(DB)(DC) + DC^2 = b^2 - DC^2$$

or

$$c^2 = a^2 + b^2 - 2(DB + DC)DC = a^2 + b^2 - 2a(DC)$$

Since $\cos(\angle ACB) = \frac{DC}{b}$, we have

$$c^2 = a^2 + b^2 - 2ab \, \cos(\angle ACB)$$

A similar argument can be used to show the result in the other two cases. □

Another very useful result connecting angles and triangles is the Law of Sines.

Theorem 3.10. *(Law of Sines) Let ABC be a triangle with sides of lengths a, b, c, with a opposite A, b opposite B, and c opposite C. Then*

$$\frac{a}{\sin(\angle A)} = \frac{b}{\sin(\angle B)} = \frac{c}{\sin(\angle C)} = d \qquad (3.2)$$

where d is the diameter of the circle that circumscribes the triangle.

Proof: Let σ be the circumscribing circle, which can be constructed using the techniques from Project 2.2 of Chapter 2. Let O be the center of σ (Figure 3.9).

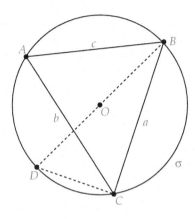

Figure 3.9

Let D be the intersection of $\overrightarrow{BO}$ with σ. If A and D are on the same side of $\overleftrightarrow{BC}$, then $\angle BAC$ and $\angle BDC$ are congruent, as they share the same arc (refer to Corollary 2.32 in Chapter 2). Thus, $\sin(\angle A) = \sin(\angle D)$, and since $\triangle DCB$ is a right triangle (refer to Corollary 2.33), we have $a = \sin(\angle D)(BD) = \sin(\angle A)d$.

If A and D are on opposite sides of $\overleftrightarrow{BC}$, as illustrated in Figure 3.10, it is left as an exercise to show that $\sin(\angle A) = \sin(\angle D)$ and thus $a = \sin(\angle A)d$. Similar arguments can be used to show $b = \sin(\angle B)d$ and $c = \sin(\angle C)d$. $\square$

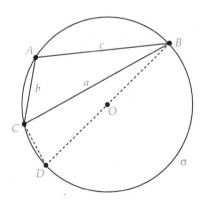

Figure 3.10

We can connect vectors and angles via the following results.

Definition 3.6. *The* dot product *of two vectors* $v = (v_1, v_2)$ *and* $w = (w_1, w_2)$ *is defined as*

$$v \bullet w = v_1 w_1 + v_2 w_2$$

Theorem 3.11. *Let* v, w *be two vectors with tails at the origin in a coordinate system, as illustrated in Figure 3.11. Let* θ *be the angle between these two vectors. Then*

$$v \bullet w = \|v\| \|w\| \cos(\theta)$$

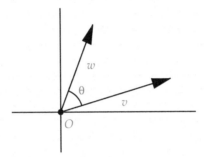

Figure 3.11

Proof: Let $v = (v_1, v_2)$ and $w = (w_1, w_2)$. The vector from v to w will be the vector $w - v$. Let θ be the angle between v and w. By the Law of Cosines, we have that

$$\|v - w\|^2 = \|v\|^2 + \|w\|^2 - 2\|v\| \|w\| \cos(\theta)$$

Also, since $\|v\|^2 = v \bullet v$, we have that

$$\|v - w\|^2 = (v - w) \bullet (v - w) = v \bullet v - 2v \bullet w + w \bullet w$$

and, thus,

$$\|v - w\|^2 = \|v\|^2 + \|w\|^2 - 2v \bullet w$$

Clearly, $v \bullet w = \|v\| \|w\| \cos(\theta)$. □

We list here some of the useful identities from trigonometry. The proof of the first identity is clear from the definition of cosine and sine and the fact that the point $(\cos(\alpha), \sin(\alpha))$ lies on the unit circle. The proofs of the other two identities are left as exercises.

Theorem 3.12. *For any angles $\alpha, \beta,$*

- $\sin^2(\alpha) + \cos^2(\alpha) = 1$

- $\sin(\alpha + \beta) = \sin(\alpha)\cos(\beta) + \sin(\beta)\cos(\alpha)$

- $\cos(\alpha + \beta) = \cos(\alpha)\cos(\beta) - \sin(\beta)\sin(\alpha)$

Exercise 3.4.1. *Let $A = (\cos(\alpha), \sin(\alpha))$ and $B = (\cos(\beta), \sin(\beta))$. Use the definition of the dot product, and the preceding result connecting the dot product to the angle between vectors, to prove that $\cos(\alpha - \beta) = \cos(\alpha)\cos(\beta) + \sin(\beta)\sin(\alpha)$.*

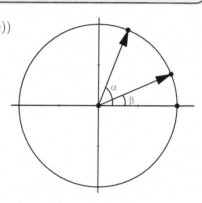

Exercise 3.4.2. *Use the previous exercise to show that*

$$\cos(\alpha + \beta) = \cos(\alpha)\cos(\beta) - \sin(\beta)\sin(\alpha)$$

Exercise 3.4.3. *Show that*

$$\sin(\alpha + \beta) = \cos(\frac{\pi}{2} - (\alpha + \beta))$$

and then use this to prove that

$$\sin(\alpha + \beta) = \sin(\alpha)\cos(\beta) + \sin(\beta)\cos(\alpha)$$

Exercise 3.4.4. *Finish the proof of the Law of Sines in the case where the points A and D are on opposite sides of $\overleftrightarrow{BC}$.*

3.5 THE COMPLEX PLANE

Complex numbers were created to extend the real numbers to a system where the equation $x^2 + 1 = 0$ has a solution. This requires the definition of a new number (symbol), $\sqrt{-1}$. The Swiss mathematician Leonhard Euler (1707–1783) introduced the notation i for this new number in the late 1700s.

By adding this number to the reals, we get a new number system consisting of real numbers and products of real numbers with powers of

i. Since $i^2 = -1$, these combinations can be simplified to numbers of the form

$$\alpha = a + ib$$

where a and b are real. The set of all possible numbers of this form will be called the set of *complex numbers* and the plane containing these numbers will be called the *complex plane.*

Historically, complex numbers were created strictly for solving algebraic equations, such as $x^2 + 1 = 0$. In the early part of the nineteenth century, Carl Friedrich Gauss (1777–1855) and Augustin Louis Cauchy (1789–1857) began working with complex numbers in a *geometric* fashion, treating them as points in the plane.

Thus, a complex number has an interesting dual nature. It can be thought of geometrically as a vector (i.e., an ordered pair of numbers), or it can be thought of algebraically as a single (complex) number having real components.

> **Definition 3.7.** *If $z = x + iy$ is a complex number, then x is called the* real part *of z, denoted $Re(z)$, and y is called the* imaginary part, *denoted $Im(z)$.*

Given a complex number $z = x + iy$, or two complex numbers $z_1 = x_1 + iy_1$ and $z_2 = x_2 + iy_2$, we define basic algebraic operations as follows:

Addition-Subtraction $z_1 \pm z_2 = (x_1 \pm x_2) + i(y_1 \pm y_2)$

Multiplication by Real Scalar $kz = kx + iky$, for k a real number

Multiplication $z_1 z_2 = (x_1 x_2 - y_1 y_2) + i(x_1 y_2 + x_2 y_1)$

Complex Conjugate $\overline{z} = x - iy$

Modulus $|z| = \sqrt{z\overline{z}} = \sqrt{x^2 + y^2}$

Note that complex addition (subtraction) is defined so that this operation satisfies the definition of vector addition (subtraction). In particular, complex addition satisfies the parallelogram property described earlier in this chapter.

The modulus of z is the same as the norm (length) of the vector that z represents. The conjugate of z yields a number that is the reflection of z (considered as a vector) across the x-axis.

3.5.1   Polar Form

The x and y coordinates of a complex number z can be written as

$$x = r \, \cos(\theta), \; y = r \, \sin(\theta)$$

where r is the length of $v = (x, y)$ and θ is the angle that v makes with the x-axis. Since $r = |z|$, then

$$z = x + iy = |z|(\cos(\theta) + i \, \sin(\theta))$$

The term $(\cos(\theta) + i \, \sin(\theta))$ can be written in a simpler form using the following definition.

Definition 3.8. *The complex exponential function e^z is defined as*

$$e^z = e^{x+iy} = e^x(\cos(y) + i \, \sin(y))$$

From this definition we can derive *Euler's Formula*:

$$e^{i\theta} = \cos(\theta) + i \, \sin(\theta)$$

Thus, the polar form for a complex number z can be written as

$$z = |z|e^{i\theta}$$

All of the usual power properties of the real exponential hold for e^z, for example, $e^{z_1+z_2} = e^{z_1}e^{z_2}$. Thus,

$$e^{i\theta}e^{i\phi} = e^{i(\theta+\phi)}$$

We also note that if $z = |z|e^{i\theta}$, then

$$\overline{z} = |z|e^{-i\theta}$$

The angle coordinate for z will be identified as follows:

Definition 3.9. *Given $z = |z|e^{i\theta}$, the argument or arg of z is a value between 0 and 2π defined by*

$$arg(z) = \theta \, (\text{mod} \, 2\pi)$$

We use here the modular arithmetic definition that $a \bmod n$ represents the remainder (in $[0, n)$) left when a is divided by n. For example, 24 mod 10 is 4.

From the definition of arg, and using the properties of the complex exponential, we see that if z and w are complex numbers, then

$$arg(zw) = (arg(z) + arg(w))(\bmod 2\pi)$$

The proof is left as an exercise. Also, if $z = |z|e^{i\theta}$ and $w = |w|e^{i\phi}$, then $wz = |w||z|e^{i(\phi+\theta)}$.

Exercise 3.5.1. *Prove that $e^{i\theta}e^{i\phi} = e^{i(\theta+\phi)}$ using Euler's Formula and the trigonometric properties from section 3.4.*

Exercise 3.5.2. *Use Euler's Formula to prove the remarkable identity $e^{i\pi} + 1 = 0$, relating five of the most important constants in mathematics.*

Exercise 3.5.3. *Show that $arg(zw) = (arg(z) + arg(w))(\bmod 2\pi)$, where z and w are complex.*

Exercise 3.5.4. *Use the polar form of a complex number to show that every non-zero complex number has a multiplicative inverse.*

Exercise 3.5.5. *Express these fractions as complex numbers by rationalizing the denominator, namely, by multiplying the numerator and denominator by the conjugate of the denominator.*

$$\frac{1}{2i}, \quad \frac{1+i}{1-i}, \quad \frac{1}{2+4i}$$

3.6 BIRKHOFF'S AXIOMATIC SYSTEM FOR ANALYTIC GEOMETRY

In this last section of the chapter, we will consider Birkhoff's axiomatic system, a system that is quite different from Euclid's original axiomatic system. Euclid's set of axioms is somewhat cumbersome in developing analytic geometry. One has to *construct* the set of real numbers and also to develop the machinery necessary for computing distance and working with angle measure. Birkhoff's system, on the other hand, *assumes* the existence of the reals and gives four axioms for angles and distances directly.

Birkhoff's system starts with two undefined terms (*point* and *line*) and assumes the existence of two real-valued functions—a *distance* function, $d(A, B)$, which takes two points and returns a non-negative real number, and an *angle* function, $m(\angle A, O, B)$, which takes an ordered triple of points ($\{A, O, B\}$ with $A \neq O$ and $B \neq O$) and returns a number between 0 and 2π.

Birkhoff's Axioms are as follows:

The Ruler Postulate The points of any line can be put into one-to-one correspondence with the real numbers x so that if x_A corresponds to A and x_B corresponds to B, then $|x_A - x_B| = d(A, B)$ for all points A, B on the line.

The Euclidean Postulate One and only one line contains any two distinct points.

The Protractor Postulate Given any point O, the rays emanating from O can be put into one-to-one correspondence with the set of real numbers (mod 2π) so that if a_m corresponds to ray m and a_n corresponds to ray n and if A, B are points (other than O) on m, n, respectively, then $m(\angle AOB) = a_m - a_n \pmod{2\pi}$. Furthermore, if the point B varies continuously along a line n not containing O, then a_n varies continuously also.

The SAS Similarity Postulate If in two triangles ABC and $A'B'C'$, and for some real number $k > 0$, we have $d(A', B') = k\, d(A, B), d(A', C') = k\, d(A, C)$, and $m(\angle BAC) = m(\angle B'A'C')$, then the remaining angles are pair-wise equal in measure and $d(B', C') = k\, d(B, C)$.

With these four axioms and the assumed structure of the real numbers, Birkhoff was able to derive the standard set of results found in planar Euclidean geometry [6].

For example, given three distinct points A, B, and C, Birkhoff defines *betweenness* as follows: B is *between* A and C if $d(A, B) + d(B, C) = d(A, C)$. From this definition, Birkhoff defines a *segment* $\overline{AB}$ as the points A and B along with all points between them. Birkhoff then defines rays, triangles, and so on, and shows the standard set of results concerning betweenness in the plane.

The protractor postulate allows us to define *right angles* as follows: two rays m,n from a point O form a right angle if $m(\angle AOB) = \pm\frac{\pi}{2}$, where A, B are points on m, n. In this case, we say the rays are *perpendicular*. Parallel lines are defined in the usual way—as lines that never meet.

Let's see how analytic geometry can be treated as a *model* for

Birkhoff's system. That is, we will see that Birkhoff's postulates can be satisfied by the properties of analytic geometry.

Define a *point* as an ordered pair (x, y) of numbers and define the distance function as

$$d((x_1, y_1), (x_2, y_2)) = \sqrt{(x_1 - x_2)^2 + (y_1 - y_2)^2}$$

Note that this definition does not suppose any *geometric* properties of points. All we are assuming is that points are ordered pairs of numbers.

Define a *line* as the set of points (x, y) satisfying an equation of the form $ax + by + c = 0$, where a, b, c are real constants, uniquely given up to a common scale factor.

We next define vectors as ordered pairs and define the angle determined by two vectors by

$$\cos(\theta) = \frac{v \bullet w}{\|v\|\|w\|}$$

where the dot product and norm of vectors is defined algebraically as before, and the cosine function can be defined as a Taylor series, thus avoiding any geometric interpretation. The angle determined by two rays emanating from a point will be defined as the angle determined by two vectors along these rays.

Does this definition of angles satisfy Birkhoff's Protractor Postulate? First of all, given any point O we can identify a ray from O with a direction vector $w = (x, y)$, with $w \neq (0, 0)$. To create a correspondence between angles and numbers, we will fix a particular direction given by $v = (1, 0)$ and measure the angle determined by w and v:

$$\cos(\theta) = \frac{v \bullet w}{\|v\|\|w\|} = \frac{x}{\|w\|}$$

We then define the sine function by

$$\sin(\theta) = \frac{y}{\|w\|}$$

If we divide w by its length, we still have a vector in the same direction as w and thus we can assume $w = (x, \sqrt{1 - x^2})$, or $w = (x, -\sqrt{1 - x^2})$ with $-1 \leq x \leq 1$. If $w = (x, \sqrt{1 - x^2})$ we get from the above equation that $-1 \leq \cos(\theta) \leq 1$, and thus $0 \leq \theta \leq \pi$. We can extend this correspondence to angles between π and 2π in the usual way by identifying those angles between π and 2π with values of w, where $w = (x, -\sqrt{1 - x^2})$.

Thus, we have created a correspondence between rays from O (and thus angles) and real numbers between 0 and 2π. Now, to check whether the second part of the Protractor Postulate holds, we suppose m, n are two rays from O, with direction vectors $w_1 = (x_1, y_1)$ and $w_2 = (x_2, y_2)$. If θ_1, θ_2 are the angles made by w_1, w_2 under the correspondence described, then

$$\cos(\theta_1) = \frac{x_1}{\|w_1\|}$$
$$\cos(\theta_2) = \frac{x_2}{\|w_2\|}$$

Also, using the definitions of sine and cosine, we can show algebraically that

$$\cos(\theta_1 - \theta_2) = \cos(\theta_1)\cos(\theta_2) + \sin(\theta_1)\sin(\theta_2)$$

(Review the exercises at the end of section 3.3 to convince yourself that this formula can be proved using only the dot product and cosine expression above and the algebraic properties of sine and cosine.)

Now, by definition, if θ is the angle formed by m,n, then

$$\begin{aligned}
\cos(\theta) &= \frac{w_1 \bullet w_2}{\|w_1\|\|w_2\|} = \frac{x_1 x_2 + y_1 y_2}{\|w_1\|\|w_2\|} \\
&= \frac{x_1}{\|w_1\|}\frac{x_2}{\|w_2\|} + \frac{y_1}{\|w_1\|}\frac{y_2}{\|w_2\|} \\
&= \cos(\theta_1)\cos(\theta_2) + \sin(\theta_1)\sin(\theta_2) \\
&= \cos(\theta_1 - \theta_2)
\end{aligned}$$

Thus, under the correspondence described above between rays and numbers, $a_m = \theta_1$ and $a_n = \theta_2$ are the "angles" assigned to m and n, and then $m(\angle AOB) = a_n - a_m(\text{mod } 2\pi)$, where A, B are points on n and m.

The analytic geometry model also satisfies Birkhoff's Euclidean Postulate. Let (x_1, y_1) and (x_2, y_2) be two distinct points. To find the "line" these points are on, we need to find the constants a, b, c in the line equation $ax + by + c = 0$.

Since (x_1, y_1) and (x_2, y_2) are on this line, we have

$$ax_1 + by_1 + c = 0$$

$$ax_2 + by_2 + c = 0$$

Subtracting, we get $a(x_1 - x_2) + b(y_1 - y_2) = 0$. Since (x_1, y_1) and (x_2, y_2) are distinct, we know that one of $x_1 - x_2$ or $y_1 - y_2$ are non-zero. Suppose $y_1 - y_2 \neq 0$. Then

$$b = -a\frac{x_1 - x_2}{y_1 - y_2}$$

and

$$c = -ax_1 - by_1$$

Since the constants need only be defined up to a constant multiple, we can choose $a = 1$, and the line through the two points is now well defined and, in fact, uniquely determined (up to the constant scale factor).

The Ruler Postulate follows from first choosing a point $O = (x_0, y_0)$ on a given line l to serve as the origin. Let (x_1, y_1) be a second point on l. Let $\Delta x = x_1 - x_0$ and $\Delta y = y_1 - y_0$. Then

$$a(\Delta x) = -b(\Delta y)$$

If (x, y) is any other point on l, then

$$a(x - x_0) = -b(y - y_0)$$

If $x - x_0 = t(\Delta x)$, then clearly, $y - y_0 = t(\Delta y)$, in order for the preceding equations to hold. Thus, we can write the line's equation in vector form as

$$(x, y) = (x_0, y_0) + t(\Delta x, \Delta y)$$

For the correspondence required in the Ruler Postulate, simply associate the point (x, y) with the value of $t\sqrt{(\Delta x)^2 + (\Delta y)^2}$. It is left as an exercise to check that $|x_A - x_B| = d(A, B)$ for all points A, B on l.

Finally, does the analytic geometry model satisfy the last postulate, the SAS condition for similar triangles?

Consider the Law of Cosines, described earlier in this chapter. Given a triangle ABC with side lengths $a = BC$, $b = AC$, and $c = AB$ and angles $\alpha = m(\angle BAC)$, $\beta = m(\angle ABC)$, and $\gamma = m(\angle ACB)$, then

$$c^2 = a^2 + b^2 - 2a\,b\,\cos(\gamma)$$

It is clear that if two pairs of sides in two triangles are proportional and if the included angles are equal, then by the Law of Cosines the third pair of sides must have the same constant of proportionality. Then, using the Law of Cosines again with all three sides proportional, it is clear that the other angles must be equal.

It is left as an exercise to show that the Law of Cosines can be proven from within the Birkhoff model.

We have thus proved that Birkhoff's Postulates are satisfied within the model of analytic geometry. Since Birkhoff showed that classical Euclidean geometry can be derived from his set of postulates, we see that Euclidean geometry can be derived from within an analytic model, just by using the properties of ordered pairs of numbers and analytic equations of lines and angles. Birkhoff's system does not suffer the foundational problems of Euclid's original axiomatic system. For example, there is no need to axiomatize the idea of "betweeness" since it can be analytically defined. However, the system derives its power and elegance at a price—the assumption of the existence of the real numbers. To be complete, we would have to develop the real numbers *before* using Birkhoff's system, by axiomatically creating the reals from within another system.

Exercise 3.6.1. *Finish the verification that the Ruler Postulate is satisfied by the analytic geometry model. That is, show that $|x_A - x_B| = d(A, B)$ for all points A, B on l, where l is represented as $(x, y) = (x_0, y_0) + t(dx, dy)$.*

Exercise 3.6.2. *We have to take care when using a result, such as the Law of Cosines, to prove that Birkhoff's SAS Similarity Postulate holds for the analytic geometry model. We have to ensure that the Law of Cosines can be proved solely within the model itself. Derive the Law of Cosines by using the vector equation for the cosine of an angle and by using general properties of vectors. [Hint: $c^2 = \|\vec{B} - \vec{A}\|^2 = \|(\vec{B} - \vec{C}) - (\vec{A} - \vec{C})\|^2.$]*

Exercise 3.6.3. *Using the vector definition of sine and cosine, prove that $\sin^2(\theta) + \cos^2(\theta) = 1$.*

Exercise 3.6.4. *We have seen in this chapter two approaches to constructing analytic geometry. In the first approach, we construct analytic geometry from a basis of synthetic Euclidean geometry, building it from prior work on triangles, parallels, perpendiculars, angles, and so on, in the style of Euclid and Hilbert. In the second approach, we start with Birkhoff's axioms for geometry and then consider analytic geometry as a model for this axiom set. Which approach do you think is better pedagogically?*

Exercise 3.6.5. *The invention of analytic geometry has often been described as the "arithmetization" of geometry. What is meant by this?*

CHAPTER 4

Constructions

Geometry is the science of correct reasoning on incorrect figures. – George Pólya (1887–1985) (from [35, page 208])

4.1 EUCLIDEAN CONSTRUCTIONS

The quote by Pólya is somewhat tongue-in-cheek, but contains an important nugget of wisdom that is at the heart of how the Greeks viewed geometric constructions. For Euclid a geometric figure drawn on paper was only an *approximate* representation of the abstract, *exact* geometric relationship described by the figure and established through the use of axioms, definitions, and theorems.

When we think of drawing a geometric figure, we typically imagine using some kind of straightedge (perhaps a ruler) to draw segments and a compass to draw circles. Euclid, in his first three axioms for planar Euclidean geometry, stipulates that there are *exact, ideal* versions of these two tools that can be used to construct *perfect* segments and circles. Euclid is making an abstraction of the concrete process of drawing geometric figures so that he can provide logically rigorous proofs of geometric results.

But Euclid was paradoxically quite *concrete* in his notions of what constituted a proof of a geometric result. It was not enough to show that a figure or result *could* be constructed—the *actual* construction had to be demonstrated, using an ideal straightedge and compass or other constructions that had already been proved valid.

In this section we will follow Euclid and assume the ability to construct a segment connecting two given points and the ability to construct a circle with a given point as the center and a constructed segment as the radius. From these two basic constructions, we will develop some of the more useful constructions that appear in Euclidean geometry.

Many of these constructions are well known from high school geometry and proofs of validity will be developed in the exercises.

Construction 4.1. *(Copying an Angle) To copy the angle defined by two rays $\overrightarrow{AB}$ and $\overrightarrow{AC}$ to ray $\overrightarrow{DE}$, we*

- *Construct the circle at A of radius AB. This will intersect $\overrightarrow{AC}$ at point F.*

- *Copy segment $\overline{AB}$ to $\overrightarrow{DE}$, getting segment $\overline{DG}$.*

- *Copy a circle centered at B of radius BF to point G. Let H be an intersection of this circle with the arc used in part 2. Then $\angle HDG$ will have the same angle measure as $\angle BAC$ (Figure 4.1).*

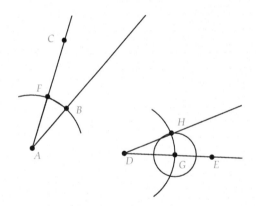

Figure 4.1 Copying an angle

Proof: The proof is left as an exercise. □

Note in this construction the implied ability to copy a segment from one place to another. This assumes the existence of a *non-collapsing* compass, that is, a compass that perfectly holds the relative position of its dividers as you move it from one place to another.

Euclid, in the second proposition of Book I of *The Elements*, proves that one can copy the length of a given segment to another point, using only the two axiomatic constructions and Proposition 1 (the construction of an equilateral triangle). It can be inferred that Euclid does not assume that a compass preserves its position as it is moved, for if he did, then there would be no need for the second proposition.

Why would Euclid make the assumption of a *collapsing* compass? Just as he strove to make his set of axioms as simple and economical as possible, so too he sought to make the *ideal* tools of construction as simple and free of ambiguity as possible. Thus, if one could not guarantee that the dividers of a compass would stay perfectly fixed when lifted from a drawing, then Euclid did not want to assume this property for his ideal compass.

It is interesting to note, however, that Proposition 2 implies that one can assume either a collapsing compass or a non-collapsing compass with equivalent results. For Proposition 2 implies that a collapsing compass and straightedge allow one to copy segment lengths from one position to another. Because of this equivalence, we will assume compasses are non-collapsing for the rest of this chapter.

Two quite elegant and simple constructions are those of bisecting a segment and bisecting an angle.

> **Construction 4.2.** *(Perpendicular Bisector of a Segment) To bisect segment $\overline{AB}$ we set the compass center at A and construct a circle of radius AB. Likewise we set the compass center at B and draw a circle of radius AB. These two circles will intersect at two points C, D. The line through C, D will be a perpendicular bisector of $\overline{AB}$ and will intersect $\overline{AB}$ at the midpoint M of $\overline{AB}$ (Figure 4.2).*

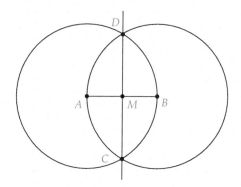

Figure 4.2 Perpendicular bisector of a segment

Proof: The proof is left as an exercise. □

Construction 4.3. *(Bisecting an Angle) To bisect the angle defined by rays $\overrightarrow{AB}$ and $\overrightarrow{AC}$:*

- *At A construct a circle of radius AB, intersecting $\overrightarrow{AC}$ at D.*

- *At B construct a circle of radius BD and at D construct a circle of radius BD. These will intersect at E.*

- *Construct ray $\overrightarrow{AE}$. This will be the bisector of the angle (Figure 4.3).*

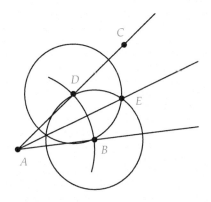

Figure 4.3 Bisector of an angle

Proof: The proof is left as an exercise. □

To construct a line perpendicular to a given line l at a point A, there are two possible constructions, depending on whether A is on l or off l.

Construction 4.4. *(Perpendicular to a Line through a Point on the Line) To construct a perpendicular to line l at a point A on l, we first set our compass center at A and draw a circle of some positive radius. This creates a segment $\overline{BC}$ where the circle intersects l. Then do the construction for the perpendicular bisector (Figure 4.4).*

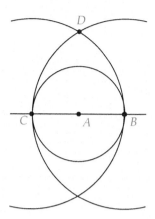

Figure 4.4 Perpendicular to a line through a point on the line

Proof: The proof is similar to that of the perpendicular bisector. □

Construction 4.5. *(Perpendicular to a Line through a Point Not on the Line) To construct a perpendicular to line l at a point A that is not on l, we first select a point B on l and set our compass center at A and draw a circle of some radius AB. Either l will be tangent to the circle at B (in which case $\overline{AB}$ will be perpendicular to l) or the circle will intersect l at two points B, C. If the circle intersects at two points, construct the angle bisector to ∠BAC. This will intersect $\overline{BC}$ at right angles (Figure 4.5).*

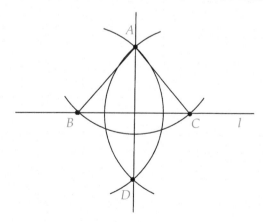

Figure 4.5 Perpendicular to a line through a point not on the line

Proof: The proof is left as an exercise. □

We have already mentioned Euclid's construction of an equilateral triangle. We include it here for completeness.

Construction 4.6. *(Equilateral Triangle)*

To construct an equilateral tri-angle on a segment $\overline{AB}$, we construct two circles, one with center A and one with cen-ter B, with both having radius AB. Let C be an intersection of these circles. Then $\triangle ABC$ is equilateral.

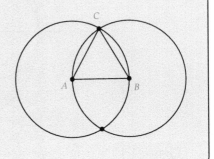

Proof: The proof is left as an exercise. □

For the next set of constructions, we will not include all of the helping marks from earlier constructions but will instead just outline the major construction steps. The proofs of the correctness of these constructions can be found in Project 2.2 from Chapter 2.

Construction 4.7. *(Circumcenter) To construct the circumcenter P of $\triangle ABC$, we find the intersection of two perpendicular bisectors of the triangle's sides (Figure 4.6).*

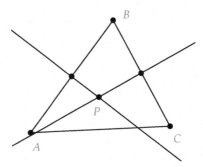

Figure 4.6 Triangle circumcenter

Construction 4.8. *(Orthocenter) To construct the orthocenter O of △ABC, we find the intersection of two altitudes of the triangle. An altitude is a perpendicular to a side of the triangle that passes through the opposite vertex (Figure 4.7).*

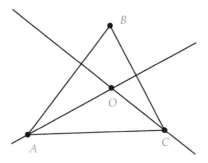

Figure 4.7 Triangle orthocenter

Construction 4.9. *(Incenter) To construct the incenter I of △ABC, we find the intersection of two angle bisectors of the angles of the triangle (Figure 4.8).*

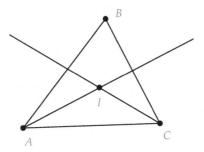

Figure 4.8 Triangle incenter

Construction 4.10. *(Centroid) To construct the centroid Q of △ABC, we find the intersection of two medians of the triangle. A median is a segment from the midpoint of a side to the opposite vertex (Figure 4.9).*

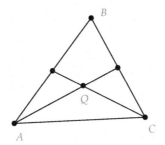

Figure 4.9 Triangle centroid

We will now look at some circle constructions. Again we will not put in all of the helping construction marks that are based on constructions covered earlier. The proofs of the correctness of these constructions can be found in the section on circle geometry in Chapter 2.

Construction 4.11. *(Tangent to a Circle at a Point on Circle) To construct the tangent to a circle c at a point P on the circle, we construct the line through the center O of the circle and P and then find the perpendicular to this line at P (Figure 4.10).*

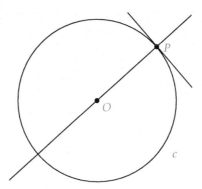

Figure 4.10 Tangent to a circle at a point on circle

Construction 4.12. *(Tangent to a Circle at a Point Not on Circle)*
To construct the two tangents to a circle c, with center O, at a point P not on the circle, we

- *Construct segment $\overline{OP}$ and find the midpoint M of $\overline{OP}$.*

- *Construct the circle at M of radius OM.*

- *Let T_1 and T_2 be the two intersections of this new circle with c.*

- *Construct two lines: one through P and T_1 and the other through P and T_2. These will be tangents to c through P (Figure 4.11).*

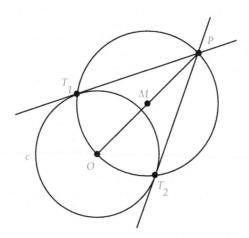

Figure 4.11 Tangent to a circle at a point not on circle

We note here that all of the preceding constructions are independent of the parallel postulate; that is, the proof of their validity does not depend on any result derived from Euclid's fifth postulate.

Here are three Euclidean circle constructions that do rely on the parallel property of Euclidean geometry.

Construction 4.13. *(Circle through Three Points) To construct the circle through three non-collinear points A, B, C, we*

- *Construct the segments $\overline{AB}$ and $\overline{BC}$.*

- *Construct the perpendicular bisectors of these segments. Let O be the intersection of these bisectors.*

- *Construct the circle with center O and radius OA. This is the desired circle (Figure 4.12).*

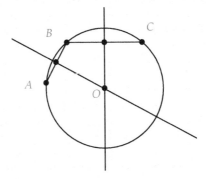

Figure 4.12 Circle through three points

Construction 4.14. *(Inversion of a Point through a Circle) To construct the inverse of a point P inside a circle c, we*

- *Construct the ray $\overrightarrow{OP}$, where O is the center of c.*

- *Construct the perpendicular to this ray at P. Let T be an intersection of this ray with the circle.*

- *At T, construct the tangent to the circle. The intersection P' of this tangent with $\overrightarrow{OP}$ will be the inverse point to P through c (Figure 4.13).*

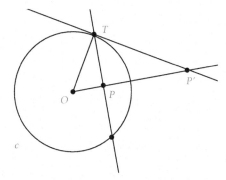

Figure 4.13 Inversion of a point through a circle

Construction 4.15. *(Orthogonal Circles) To construct a circle orthogonal to a circle c through two points P, Q inside c, we*

- *Construct the inverse point P′ to P through c.*

- *Construct the circle through the three points P, P′, and Q (Figure 4.14).*

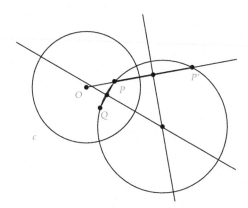

Figure 4.14 Orthogonal circles

This concludes our review of some of the most basic Euclidean constructions. Other basic constructions involving parallels and triangles are covered in the exercises of this section.

Exercise 4.1.1. *Prove that Construction 4.1 creates a new angle congruent to the original angle.*

Exercise 4.1.2. *Prove that Construction 4.2 produces the perpendicular bisector of a segment.*

Exercise 4.1.3. *Prove that Construction 4.3 produces the angle bisector of an angle.*

Exercise 4.1.4. *Prove that Construction 4.5 produces the perpendicular to a line through a point not on the line.*

Exercise 4.1.5. *Prove that Construction 4.6 produces an equilateral triangle.*

Exercise 4.1.6. *In this exercise we will investigate a construction that will allow us to copy a circle with a* collapsing *compass. Given a circle c with center O and radius point A and another point B, we wish to construct a circle centered at B of radius OA. It suffices to prove the result for the case where B is outside c. (Why?) First, construct a circle centered at O of radius OB. Then construct a circle at B of radius OB. Let C and D be the intersection points of these circles. Let E be an intersection of the circle centered at B with the original circle c.*

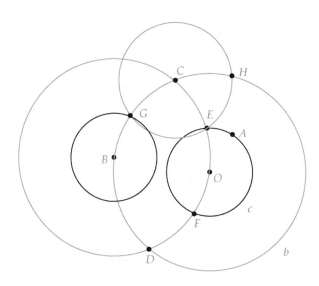

Figure 4.15 Copying a circle

At intersection point C, construct a circle of radius CE. This circle will intersect the circle at B of radius OB at a point G. Show that the circle with center B and radius point G is the desired circle. [Hint: Show that △CGO and △CEB are congruent, and then consider △GBO and △EOB.] Why is this construction valid for a collapsing compass?

Exercise 4.1.7. *Using the perpendicular construction, show how one can construct a line parallel to a given line through a point not on the line.*

Exercise 4.1.8. *Draw a segment $\overline{AB}$ and devise a construction of the isosceles right triangle with $\overline{AB}$ as a base.*

Exercise 4.1.9. *Show how to construct a triangle given two segments a and b and an angle $\angle ABC$ that will be the included angle of the triangle. To what triangle congruence result is this construction related?*

Exercise 4.1.10. *Given a segment $\overline{AB}$ and a positive integer n, devise a construction for dividing $\overline{AB}$ into n congruent sub-segments.*

4.2 PROJECT 6 - EUCLIDEAN EGGS

So far we have looked at constructions of fairly traditional geometric figures—lines, circles, parallels, perpendiculars, tangents, and so on. In this project we will look at a way of joining circular arcs together in a smooth fashion to make interesting shapes.

The idea of joining curves together so that they look "smooth" across the point of attachment can be traced back to at least the time of the ancient Romans and their construction of arches and oval tracks. The fifteenth-century artist Albrecht Dürer makes great use of this technique in his design of alphabets for the printing press.

In his book *Mathographics* [11], Robert Dixon describes how to make a variety of curves and oval shapes using a simple method of smoothly joining circular arcs.

To see how this method works, we'll try a little experiment. (Notes on how to use particular dynamic geometry software packages for this experiment can be found at http://www.gac.edu/~hvidsten/geom-text.)

Start up your geometry software and create a segment $\overline{AB}$. Attach a point C to the segment and create a circle c_1 with center A and radius point C, and a circle c_2 with center B and radius point C.

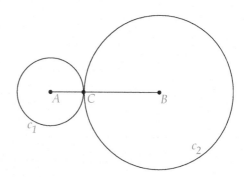

Now attach two points D and E on circle c_1 and two points F and G on c_2 as shown.

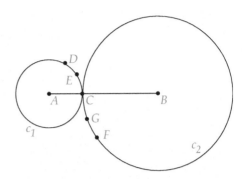

Hide the two circles and construct an arc through the three points C, D, and E. Similarly, construct the arc on C, F, and G.

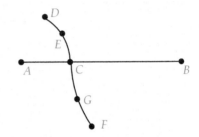

Note how smoothly these two different arcs join together at C, even though they are constructed from circles of different radii.

Exercise 4.2.1. *Show that two circular arcs joined together in this fashion will always be smooth across the join point (that is, their tangents will coincide) if the following condition holds: a straight line drawn through the centers of the two arcs passes through the point where they are joined. (In the preceding figure, this would refer to $\overleftrightarrow{AB}$ passing through point C.)*

We will use this idea to construct an oval, or more precisely an *egg*.

Clear the screen and create segment $\overline{AB}$. Construct the midpoint M of $\overline{AB}$ and construct a perpendicular to $\overline{AB}$ at M (refer to Figure 4.16). Next create three circles: c_1 with center M and radius point A, c_2 with center A and radius point B, and c_3 with center B and radius point A.

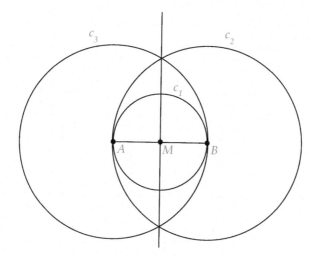

Figure 4.16

Next construct the intersection points I_1 and I_2 of the perpendicular with c_1, and then create two rays, one from B through I_1, and one from A through I_1 (refer to Figure 4.17). Construct the intersection points C and D of these rays with circles c_2 and c_3. Then create a circle with center I_1 and radius point C. Can you see an egg emerging from the figure?

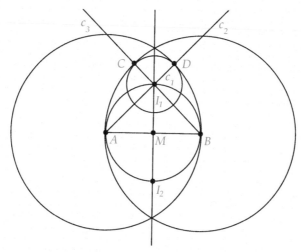

Figure 4.17

To finish off the construction, construct the points I_3 and I_4 of the intersection of the circle with center I_1 and the perpendicular to $\overline{AB}$

(refer to Figure 4.18). Construct two rays: one through A and I_4 and the other though B and I_4. Then construct the intersection points E and F of these rays with c_2 and c_3.

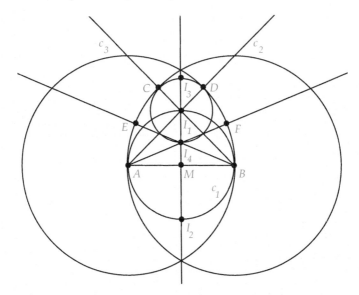

Figure 4.18

The four arcs $\{A, E, C\}$, $\{C, I_3, D\}$, $\{D, F, B\}$, and $\{B, I_2, A\}$, will define our egg. Construct these arcs and hide all helper circles and lines.

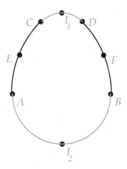

Exercise 4.2.2. *Verify that these four arcs meet smoothly across the join points, using the smoothness definition discussed in the last exercise (coincident tangent lines).*

Exercise 4.2.3. *Here is another interesting oval shape from Dixon's book. It is made of four circular arcs. Study Figure 4.19 and describe the construction steps necessary to construct this oval.*

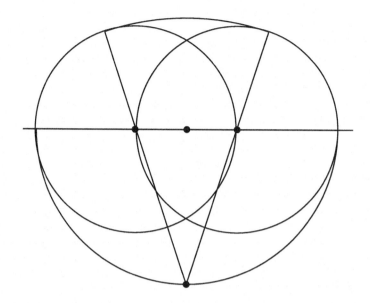

Figure 4.19

The ability to construct complex figures from simple curves, such as circular arcs, is one of the most significant practical uses for geometry. Everything from the fonts used on a computer screen to the shapes of airframes depends on the ability to smoothly join together simple curves. In Project 3.2 we saw how quadratic and cubic Bézier curves could be used to model curved shapes. Interestingly enough, it is not possible to perfectly model a circle using Bézier curves, although we can construct approximations that will fool the eye. Thus, for shapes that require the properties of a circle (at least in small sections), we can use the joining method above to create such shapes.

In your project report you should include a discussion of the construction steps you used to create the preceding egg shapes and also include complete answers to each of the exercises.

4.3 CONSTRUCTIBILITY

In the last two sections we looked at some of the basic construction techniques of Euclidean geometry. Constructions were fundamental to Euclid's axiomatic approach to geometry, as is evident by the fact that the first two Propositions of Book I of *Elements* deal with the construction of equilateral triangles and the transferring of segment lengths (the question of collapsing versus non-collapsing compasses).

The notion of constructibility was also fundamental to the way Euclid viewed numbers. To Euclid a number existed only in relation to a particular geometric figure—the length of a *constructed* segment. A number was the end result of a series of straightedge and compass constructions, starting with a given segment that was assumed to be of unit length.

Euclid did allow the independent existence of some numbers—the positive integers. The number theory found in the *Elements* can be traced back to the mystical beliefs of the Pythagoreans. Integers could be "perfect" or "prime." Euclid also allowed the consideration of the *ratio* of integers independent of any geometric context.

However, any other number not expressible as a ratio of positive integers (that is, *irrational*) could only be discussed as a purely geometric quantity—the end product of a sequence of geometric constructions. For example, the irrational number $\sqrt{2}$ is constructible, as it is possible to construct a right triangle with two base sides of unit length, given an existing segment of unit length. The hypotenuse of this triangle would be a segment of length $\sqrt{2}$.

> **Definition 4.1.** *A number α is* constructible *if a segment of length α can be constructed by a finite sequence of straightedge and compass constructions, starting with a given segment of unit length.*

Given Euclid's insistence on the construction of numbers, it is not surprising that constructibility puzzles—riddles asking whether certain numbers could be constructed—came to be a celebrated, almost mythic, aspect of Euclidean geometry.

Three of these puzzles have occupied the attention of mathematicians from the time of Euclid (300 BC) until the work of Niels Henrik Abel (1802–1829) and Evariste Galois (1811–1832) in the early 1800s:

Doubling the Cube   Given an already constructed cube, construct another cube with twice the volume of the given cube.

Angle Trisection   Given an arbitrary constructed angle, construct an angle that divides the given angle into three congruent parts.

Squaring the Circle Given an already constructed circle, construct a square of area equal to the area of the circle.

The solution of these puzzles involves a deeper understanding of how constructibility relates to solutions of *algebraic* equations. The connection with algebra comes from embedding the construction of geometric figures within the context of analytic geometry.

As was mentioned previously, all constructions start with a segment assumed to be of unit length. Let $\overline{OI}$ be this segment. We know by the constructions of the first section of this chapter that we can construct a perpendicular to the line $\overleftrightarrow{OI}$ at O, and thus we can construct a Cartesian coordinate system with origin O and with $x = 1$ at the point I (Figure 4.20).

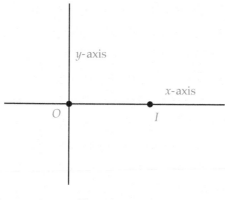

Figure 4.20

One of the great insights of Descartes was that straightedge and compass constructions were equivalent to the solution of linear and quadratic equations.

Theorem 4.1. *Let $P_1 = (x_1, y_1)$, $P_2 = (x_2, y_2)$, ..., $P_n = (x_n, y_n)$ be a set of points that have been constructed from the initial segment $\overline{OI}$. Then a point $Q = (\alpha, \beta)$ can be constructed from these points if and only if the numbers α and β can be obtained from x_1, x_2, ..., x_n and y_1, y_2, ..., y_n by simple arithmetic operations of $+$, $-$, $\cdot$, and $\div$ and the solution of a finite set of linear and quadratic equations.*

Proof: Since straightedge and compass constructions involve the construction of segments (and lines by extension) and circles, then a point in the plane is constructible if and only if it is either the intersection of two lines, a line and a circle, or two circles. (Or, it could be one of the two initial points O or I.)

From our work in Chapter 3, we know that a line through two constructed points, say P_1 and P_2, has the equation

$$(y - b_1)(a_2 - a_1) = (x - a_1)(b_2 - b_1)$$

This can be transformed into the form

$$Ay + Bx + C = 0$$

where A, B, and C are simple arithmetic combinations of the numbers a_1, a_2, b_1, and b_2 of the type described in the statement of the theorem.

A circle through a constructed point, say P_1, of radius equal to the length of a constructible segment, say $\overline{P_1 P_2}$, will have the equation

$$(x - a_1)^2 + (y - b_1)^2 = ((a_2 - a_1)^2 + (b_2 - b_1)^2)$$

where $((a_2 - a_1)^2 + (b_2 - b_1)^2) = r^2$, r being the radius of the circle.

Again, this can be written as

$$x^2 + Bx + y^2 + Cy + D = 0$$

where the coefficients are again simple arithmetic combinations of the x and y coordinates of the constructed points.

If we are given two lines, say $A_1 y + B_1 x + C_1 = 0$ and $A_2 y + B_2 x + C_2 = 0$, then the intersection of these two lines will be given by a simple formula involving the addition, subtraction, multiplication, and division of the coefficients (proved as an exercise).

If we are given a line $Ay + Bx + C = 0$ and a circle $x^2 + Dx + y^2 + Ey + F = 0$, then if the lines intersect, we know from elementary algebra that the solution will involve solving the quadratic formula for an expression involving simple arithmetic combinations of the coefficients. However, a new operation, the square root, will be introduced in the solution.

To find the point of intersection of two circles, say $x^2 + B_1 x + y^2 + C_1 y + D_1 = 0$ and $x^2 + B_2 x + y^2 + C_2 y + D_2 = 0$, we first subtract the two equations yielding $(B_2 - B_1)x + (C_2 - C_1)y + (D_2 - D_1) = 0$. We solve this for x or y (depending on which terms are non-zero) and substitute into one of the original equations to get a quadratic equation in a single variable. The solution then follows from the quadratic formula.

Thus, we have proved that if a point can be constructed by a sequence of straightedge and compass constructions on a given set of already constructed points, then the coordinates of the new point can be obtained from simple arithmetic operations of $+$, $-$, $\cdot$, $\div$, and $\sqrt{}$ used in combination on the coordinates of the existing points.

The converse is also true. That is, any combination of using $+$, $-$, $\cdot$, $\div$, and $\sqrt{}$ on the coordinates of a set of already constructed points, that results in a pair of numbers α and β, can be realized as the coordinates of a point that arises from a sequence of straightedge and compass constructions on the given points.

To show this, it is enough to show that if $a \neq 0$ and $b \neq 0$ are given constructible numbers, then the numbers

$$a + b, \; a - b, \; ab, \; \frac{a}{b}, \; \sqrt{a}$$

are also constructible.

It is fairly trivial to show that the numbers $a + b$ and $a - b$ are constructible. For $a + b$ we lay out a segment $\overline{OA}$ of length a on the x-axis and extend $\overline{OA}$ to a point B such that $AB = b$ (this is possible since we can transfer segment lengths). Then $\overline{OB}$ has length $a + b$. For $a - b$ we just lay off a segment for b in the negative direction from A on the x-axis.

For the product of two numbers, ab, consider the following construction:

On the x-axis, lay off the length a so that $OA = a$. Likewise, on the y-axis, lay off the unit length and b to get points I_y and B, with $OB = b$. Construct the segment from I_y to A and construct a parallel to this line through B, cutting the x-axis at C. Let $c = OC$.

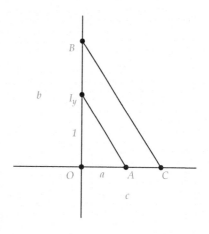

By similar triangles we have that

$$\frac{a}{1} = \frac{c}{b}$$

Clearly, $c = ab$ and we have constructed the product of a and b.

The construction of the ratio of two numbers is left as an exercise, as is the construction of the square root of a positive number. □

Starting with a unit length segment, what kinds of numbers are constructible? Clearly, all non-zero integers can be constructed. Using the construction of the ratio of two numbers, we see that all rational numbers are constructible, a rational number being a fraction of two integers. In fact, if we restrict our constructions to those that involve only the intersection of lines, it is clear that the set of rational numbers contains all such constructed numbers (if we include 0).

The set K of all constructible numbers (together with 0) forms a special type of algebraic structure called a *field*. A field is a set of elements having two operations (like $+$ and $\cdot$) that satisfies a set of properties such as associativity and commutativity for the operations. A field also satisfies the distributive property for the two operations and is *closed* under the operations and their inverses.

The prime example of a field is perhaps the set of real numbers with ordinary addition and multiplication. Given two (non-zero) real numbers, a and b, the sum $a+b$ is again a real number, as is the additive inverse $-a$, the product ab, and the multiplicative inverse $\frac{1}{a}$. This shows that the real numbers are closed under each of these operations.

By Theorem 4.1 we know that the set of constructible numbers K is a field, as this set is closed under the algebraic operations $+$, $-$, $\cdot$, and $\div$.

What about the taking of square roots? How does this operation affect the algebraic structure of a given set of constructible numbers? Let F be the field of rational numbers. Suppose we construct the number $\sqrt{2}$ and then consider all possible algebraic combinations of rational numbers with this new number, for example, $\sqrt{2}+1$, $\frac{1+\sqrt{2}}{10-\sqrt{2}}$, and so on.

It turns out that all such combinations can be written in a simpler form, as we will see next.

Definition 4.2. *Let F be a field contained in the real numbers and let k be a positive number in F. Suppose that $\sqrt{k}$ does not belong to F. Then the set*

$$F(k) = \{x + y\sqrt{k} \,|\, x, y \in F\}$$

is called a quadratic extension of F.

Theorem 4.2. *Let F be a field contained in the real numbers, let k be an element of F, and suppose $\sqrt{k}$ does not belong to F. Then the set F' of all possible simple algebraic combinations (ones using $+$, $-$, $\cdot$, and $\div$) of elements of F and the number $\sqrt{k}$ is a field and is equal to $F(k)$.*

Proof: We first note that F' contains $F(k)$ by definition. Let z be an element of F'. Then z is obtained by algebraic operations using elements of F and the number $\sqrt{k}$. If we can show that each of these operations can be represented in $F(k)$, then we will have shown that $F(k)$ contains F' (which implies $F(k) = F'$), and we will also have shown that $F(k)$ is a field, as it will be closed under the basic operations and their inverses.

Clearly, any sum or difference of elements of F or $\sqrt{k}$ can be represented in $F(k)$. What about products? Any product of elements in F is again in F, as F is a field. Also, if a, b, c are in F, then $a(b + c\sqrt{k}) = ab + ac\sqrt{k}$ is an element in $F(k)$, as is $a(c\sqrt{k})$.

In fact, the only algebraic operation that is not obviously represented in $F(k)$ is division. Suppose a and b are in F and suppose $a + b\sqrt{k} \neq 0$. Then

$$
\begin{aligned}
\frac{1}{a + b\sqrt{k}} &= \frac{1}{a + b\sqrt{k}} \left(\frac{a - b\sqrt{k}}{a - b\sqrt{k}} \right) \\
&= \frac{a - b\sqrt{k}}{a^2 - b^2 k}
\end{aligned}
$$

which is again an element in $F(k)$. □

As an application of this notion of field extensions, we will consider cubic polynomials.

Theorem 4.3. *Given a cubic polynomial*

$$p(z) = z^3 + az^2 + bz + c = 0$$

with coefficients in a field F (contained in the reals), if $p(w) = 0$, where w is an element of a quadratic extension $F(k)$ but not an element of F, then the polynomial has another root in F.

Proof: We know that every polynomial with real coefficients can be factored into $p(z) = (z - z_1)(z - z_2)(z - z_3)$ with possibly some of the z_i being complex numbers. Equivalently, we have

$$p(z) = z^3 - (z_1 + z_2 + z_3)z^2 + (z_1 z_2 + z_1 z_3 + z_2 z_3)z - z_1 z_2 z_3 = 0$$

The coefficients for a polynomial can only be represented one way, and so

$$-(z_1 + z_2 + z_3) = a$$

Now suppose the root w of $p(z)$ is represented by $z_1 = x + y\sqrt{k}$, with x and y in F. Then $x - y\sqrt{k}$ is also a root. This can be proved by showing that $x - y\sqrt{k}$ acts like the complex conjugate when compared to $x + y\sqrt{k}$. Then we use the fact that roots come in conjugate pairs. We can assume that $z_2 = x - y\sqrt{k}$. Then

$$-((x + y\sqrt{k}) + (x - y\sqrt{k}) + z_3) = a$$

which implies that $z_3 = -(a + 2x)$ is in F. □

Corollary 4.4. *Let F be a field contained in the reals. If a cubic polynomial has a root w that is in a field F_n that is the result of a series of quadratic extensions*

$$F = F_0, F_1, F_2, \ldots, F_n$$

where each F_i is a quadratic extension of the previous F_{i-1}, then the polynomial must have a root in F.

Proof: The proof of this result is just a repeated application of the preceding theorem. □

We are now in a position to tackle two of the classic constructibility puzzles.

Duplication of the Cube

Given a cube constructed from a segment $\overline{AB}$, is it possible to construct another segment $\overline{CD}$ such that the cube on $\overline{CD}$ has volume double that of the cube on $\overline{AB}$?

If it is possible, then we have the algebraic relationship

$$(CD)^3 = 2(AB)^3$$

Or

$$\left(\frac{CD}{AB}\right)^3 = 2$$

This implies that if it is possible to double the original volume of the cube, then it must be possible to construct $\overline{CD}$ and thus it must be possible to construct the fraction $\frac{CD}{AB}$. This fraction is then a root of the cubic polynomial $z^3 - 2 = 0$. Thus, if we can carry out a sequence of straightedge and compass constructions that yield $\frac{CD}{AB}$, then this construction sequence would be mirrored in an algebraic sequence of larger and larger quadratic extensions of the rational numbers. By Corollary 4.4, letting F be the rationals, we would have that $z^3 - 2 = 0$ has a root in the rationals. But from algebra we know that the only possible rational roots of this polynomial are 1, -1, 2, and -2. Since none of these are actually roots, then no element in a quadratic extension can be a root, and the construction is impossible.

Trisection of an Angle

Given an arbitrary angle, is it always possible to construct an angle that is $\frac{1}{3}$ the measure of the given angle?

Note that a solution to this problem would imply that *all* angles can be trisected and, in particular, a 60 degree angle. We will show that it is impossible to construct a 20 degree angle and thus that the trisection puzzle has no solution.

If it is possible to construct a 20 degree angle, then it must be possible to construct the number $\cos(20°)$, as this will be the base of a right triangle with angle of 20 degrees and hypotenuse equal to 1.

We make use of several of the trigonometric formulas from Chapter 3 to consider the formula for the cosine of 3θ for a given angle θ:

$$
\begin{aligned}
\cos(3\theta) &= \cos(2\theta + \theta) \\
&= \cos(2\theta)\cos(\theta) - \sin(2\theta)\sin(\theta) \\
&= (\cos^2(\theta) - \sin^2(\theta))\cos(\theta) - 2\sin(\theta)\cos(\theta)\sin(\theta) \\
&= (2\cos^2(\theta) - 1)\cos(\theta) - 2(1 - \cos^2(\theta))\cos(\theta) \\
&= 4\cos^3(\theta) - 3\cos(\theta)
\end{aligned}
$$

Letting $\theta = 20°$ and using the fact that $\cos(60°) = \frac{1}{2}$, we have

$$\frac{1}{2} = 4\cos^3(20°) - 3\cos(20°)$$

which is equivalent to $8\cos^3(20°) - 6\cos(20°) - 1 = 0$.

We conclude that if we can construct $\cos(20°)$, then it must be a root of the polynomial $8z^3 - 6z - 1 = 0$. We will simplify the analysis of this polynomial by using the observation that a construction of $\cos(20°)$ would imply the construction of $2\cos(20°)$. Making the substitution $x = 2z$ in the polynomial $8z^3 - 6z - 1$, we get that $2\cos(20°)$ is a root of $x^3 - 3x - 1 = 0$. The only possible rational roots of this polynomial are 1 and -1, and neither are actually roots. So the number $2\cos(20°)$ is not constructible, and general angle trisection is impossible.

Squaring the Circle

Given a circle of radius r, can we construct a square whose area is equal to the area of the circle? Since the area of the circle is πr^2, then we are looking for a segment whose length is $\sqrt{\pi}r$. Since r is assumed constructible, then if $\sqrt{\pi}r$ is constructible, we would be able to divide by r and have a construction for $\sqrt{\pi}$. Multiplying this number by itself, we would have a construction for π.

The proof of the impossibility of the construction of π is beyond the level of this text. While the previous two puzzles could be resolved by considering roots of polynomials with rational coefficients, no such analysis will prove the impossibility of squaring the circle. This is because the number π is *transcendental*; that is, it is not the root of a polynomial with rational coefficients. This was first proved by Carl Louis Ferdinand von Lindemann (1852–1939) in 1882. An interesting historical note about Lindemann is that David Hilbert was one of his doctoral students in Germany.

In this section we have just scratched the surface as to the connection between constructibility and the algebraic theory of fields. For a more detailed review of this connection, see Chapter 19 of Moise's text [32] or the excellent book by Robin Hartshorne [19].

Exercise 4.3.1. *Prove that if two lines $A_1y + B_1x + C_1 = 0$ and $A_2y + B_2x + C_2 = 0$ intersect, then the coordinates of the intersection point will be given by a simple formula involving the addition, subtraction, multiplication, and division of the coefficients of the lines.*

Exercise 4.3.2. *Show that the number $\sin(22\frac{1}{2}°)$ is constructible. [Hint: Use a trigonometric formula for $45°$.]*

Exercise 4.3.3. *Devise a construction for the ratio of two numbers.*

Exercise 4.3.4. *In this exercise we see how to construct square roots of* *positive numbers. Let $\overline{AB}$ be a segment representing a length a (Figure 4.21). Extend $\overrightarrow{AB}$ beyond B by the unit length to a point C. Let M be the midpoint of $\overline{AC}$ and construct a circle centered at M of radius AM. Construct the perpendicular to $\overline{AC}$ at B and let D be an intersection point of this perpendicular with the circle. Show that $BD = \sqrt{a}$. [Hint: Use right triangles.]*

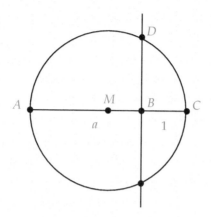

Figure 4.21 Square root

Exercise 4.3.5. *Devise constructions for the numbers $\sqrt{3}$ and $\sqrt{5}$ that are different than the one given in the previous exercise.*

Exercise 4.3.6. *Show that there are an infinite number of non-constructible numbers.*

Exercise 4.3.7. *Show that there is at least one non-constructible number in every interval $[0, a]$ for $a > 0$.*

Exercise 4.3.8. *Show that every circle centered at the origin contains at least two points which are not constructible. [Hint: Use the preceding exercise.]*

Exercise 4.3.9. *Show that every circle contains at least two points which are not constructible. [Hint: Use the preceding exercise.]*

Exercise 4.3.10. *Galois was one of the more colorful figures in the history of mathematics. His work on the solvability of equations revolutionized algebra. Research this area and prepare a short report on the significance of Galois's work.*

4.3.1   Mini-Project - Origami Construction

So far we have looked at constructions where the constructing tools are primarily a straightedge and a (collapsing) compass. We have seen that a non-collapsing compass can be substituted for a collapsing one, with equivalent capabilities for Euclidean construction. We have also discussed the issue of *constructibility* of numbers and of certain geometric figures such as the trisection of an angle. We showed that the trisection of a general angle was impossible with straightedge and compass. Interestingly enough, the trisection of a general angle is possible with a ruler (*marked* straightedge) and compass. (For the proof see [19, page 260].)

The question of constructibility is thus dependent on the tool set that one is permitted to use. Over the years, mathematicians have experimented with using other types of tools; for example, rusty compasses (ones where the divider length is permanently fixed).

One method of construction that has become popular in recent years is that of paper folding, or *origami*. The seemingly simple practice of folding paper can produce quite complex geometric configurations. In this project we will investigate the geometric constructions possible in origami by setting up a set of axioms for "perfect" paper folding. This is similar to the first few axioms of Euclid's geometry, where he postulates the ability to do perfect straightedge and compass constructions.

The axioms we will use for paper folding were first formulated by Humiaki Huzita in 1992 [26]. Huzita postulates six axioms for paper folding:

(Axiom O1) Given two constructed points P and Q, we can construct (fold) a line through them.

(Axiom O2) Given two constructed points P and Q, we can fold P onto Q.

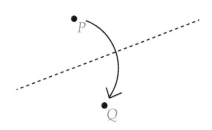

(Axiom O3) Given two constructed lines l_1 and l_2, we can fold line l_1 onto l_2.

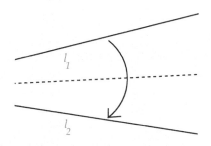

(Axiom O4) Given a constructed point P and a constructed line l, we can construct a perpendicular to l passing through P.

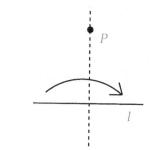

(Axiom O5) Given two constructed points P and Q and a constructed line l, then whenever possible, the line through Q, which reflects P onto l, can be constructed.

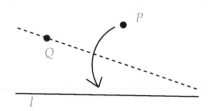

(Axiom O6) Given two constructed points P and Q and two constructed lines l_1 and l_2, then whenever possible, a line that reflects P onto l_1 and also reflects Q onto l_2 can be constructed.

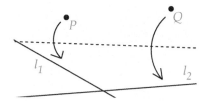

In the illustrations for the axioms, a fold line created by the axiom is indicated by a dotted line, and the direction of the fold is indicated by a curved arrow.

We are assuming in the statement of these axioms that a constructed line is either one of the original four edge lines of a square piece of origami paper or is a line created by using one or more of the folding axioms. A constructed point is either one of the original four vertices of the square paper or a point created by using one or more of the folding axioms. We will also assume that folds take lines to lines and preserve segment lengths and angles. We will prove this fact carefully in Chapter 5, where we will see that a fold is essentially a Euclidean *reflection* across the fold crease line. For now we will take this fact as a rule of reasoning, assumed without proof—thus the references to reflections in the axioms will be assumed without explanation or proof.

Another rather strange property of Axioms O5 and O6 is the phrase "whenever possible." All of the axiomatic systems we have studied up to this point have been quite definitive. For example, Euclid's third postulate states that circles are *always* constructible.

Before continuing on with this project, practice each of the six axiomatic foldings using a square sheet of paper, preferably origami paper or waxed paper.

Exercise 4.3.11. *Using the six axiomatic foldings, devise a construction for the perpendicular bisector of a segment.*

Exercise 4.3.12. *Devise a folding construction for the parallel to a given line through a point not on the line.*

Many of the axioms for paper folding are quite similar to the constructions of Euclidean geometry. Axiom O1 is essentially equivalent to Euclid's first axiom on the construction of a line joining two points. Axiom O2 mimics the construction of the perpendicular bisector of a segment. For the third axiom, if the given pair of lines intersect, then the axiom gives the angle bisector construction for the angle formed by the lines. If the lines are parallel, the construction is slightly more complicated but still possible (convince yourself of this fact). Axiom O4 is equivalent to the perpendicular to a line through a point.

What straightedge and compass construction has the equivalent effect of Axiom O5? After carrying out this construction, it must be the case that $PQ = P'Q$, where P' is the folding (reflection) of P onto l.. Thus, we are looking for an intersection point of a circle, centered at Q of radius PQ, with the line l, as depicted in Figure 4.22.

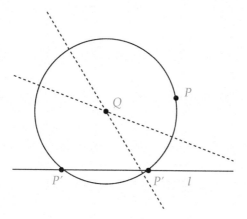

Figure 4.22

It is clear that there are three possibilities: either the circle has no intersections with the line (in which case the construction is impossible), or there is one intersection (at a point of tangency), or there are two different intersections. (The third case is illustrated in Figure 4.22, with the two different lines of reflection shown as dotted lines.)

In fact, Axiom O5 allows us to construct a parabola with focus P and directrix l. Recall that a parabola is the set of points that are equidistant from a given point (the focus) and a given line (the directrix).

Exercise 4.3.13. *Show that Axiom O5 can be used to construct a parabola with focus P and directrix l by referring to Figure 4.23. In this figure $\overleftrightarrow{P'R}$ is the perpendicular to l at the constructed point P' and R is the intersection of this perpendicular with the line of reflection t taking P to P'. [Hint: Use the distance-preserving properties of a folding (reflection).]*

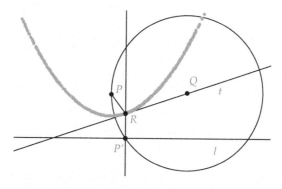

Figure 4.23

The creation of point R involves the solution of a quadratic equation and thus allows for the construction of square roots. In fact, the set of constructible numbers using Axioms O1–O5 is the same as the set of constructible numbers using straightedge and compass (allowing for arbitrary large initial squares of paper). A proof of this can be found in [2].

We see, then, that almost all of the folding axioms can be carried out by simple straightedge and compass constructions. How about the last axiom? It turns out that the last axiom is not constructible using a straightedge and a compass. In fact, using the last axiom we can actually *trisect* a general angle.

Let's see how this is done.

Let the given angle ($\angle A$) be defined in the lower left corner of the paper square by line m as shown. Construct two lines l_1 and l_2 that are parallel to the bottom edge l_b with the property that l_1 is equidistant from l_2 and l_b. (What is an easy way to construct l_2 given just the initial four lines and four points?)

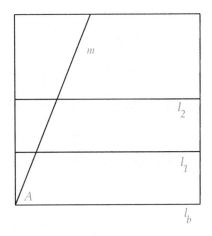

Let P be the lower left corner vertex of the square, and let Q be the intersection of l_2 with the left edge of the square. Then carry out the fold in Axiom O6 to place P on l_1 at P' and Q on m at Q'.

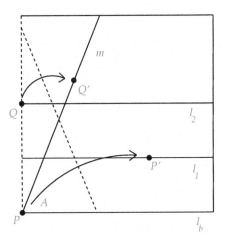

Leaving the paper folded, fold the paper once again along the folded-over portion of l_1. This will create line l_3. Unfold the paper. The claim is that line l_3 will make an angle with l_b of $\frac{2}{3}$ the angle A.

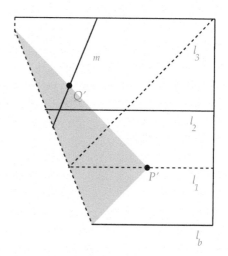

Exercise 4.3.14. *Prove that the preceding construction actually does give an angle that is $\frac{2}{3}$ the angle A. [Hint: Prove that the three triangles $\triangle PQ'R$, $\triangle PP'R$, and $\triangle PP'S$ are congruent in Figure 4.24.]*

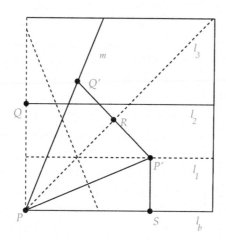

Figure 4.24 Origami trisection

Since we have constructed an angle that is $\frac{2}{3}$ of angle A, we can easily bisect this new angle to get the trisection of angle A. It is clear that the new folding axiom construction (Axiom O6) *cannot* be equivalent to a series of straightedge and compass constructions since we know it is impossible to trisect a general angle with straightedge and compass alone.

How can Axiom O6 solve the trisection problem? Since Axiom O6 is essentially a simultaneous solution to two Axiom O5 constructions, then what we are really looking for is a reflection line that is simultaneously tangent to two parabolas (see the "Hint" at the end of Exercise 4.3.13). The solution of this simultaneous tangent problem leads to a cubic equation of the form developed in the last section when we looked at the trisection puzzle in detail. Thus, Axiom O6 guarantees that such cubic equations are solvable, and therefore angle trisection is possible. For complete details on the connection between Axiom O6 and cubic polynomials, see [2].

CHAPTER 5

Transformational Geometry

Geometry is the study of those properties of a set which are preserved under a group of transformations on that set.

– Felix Klein (1849–1925)

Classical Euclidean geometry, such as the material covered in Chapter 2, is primarily concerned with *static* properties of objects. To expand this static geometry to a more *dynamic* geometry, we need to explore what it means to *transform* objects.

A transformation will be some *function* on points in the plane. That is, it will be some process whereby points are transformed to other points. This process could be the simple movement of points or could be a more complex alteration of the points.

Transformations are basic to both a practical and theoretical understanding of geometry. Object permanence, the idea that we can move an object to a different position, but the object itself remains the same, is one of the first ideas that we learn as infants.

Felix Klein, one of the great geometers of the late nineteenth century, gave an address at Erlanger, Germany, in 1872, in which he proposed that geometry should be *defined* as the study of transformations and of the objects that transformations leave unchanged, or *invariant*. This view has come to be known as the *Erlanger Program*.

If we apply the Erlanger Program to Euclidean geometry, what kinds of transformations characterize this geometry? That is, what are the transformations that leave basic Euclidean figures, such as lines, segments, triangles, and circles, invariant? Since segments are the basic building blocks of many geometric figures, Euclidean transformations must, at least, preserve the "size" of segments; that is, they must preserve *length*.

5.1 EUCLIDEAN ISOMETRIES

Definition 5.1. *A function f on the plane is called a* Euclidean isometry *(or a* Euclidean motion*) if f has the property that for all points A and B the segment* $\overline{AB}$ *and the transformed segment* $\overline{f(A)f(B)}$ *have the same length.*

This simple definition has important implications.

Theorem 5.1. *Let f be a Euclidean isometry. Then*

(i) f is one-to-one. That is, if $f(A) = f(B)$, then $A = B$.

(ii) If $f(A) = A'$ and $f(B) = B'$, then f maps all points between A and B to points between A' and B'. That is, $f(\overline{AB}) = \overline{A'B'}$.

(iii) f maps lines to lines.

(iv) f preserves angles.

(v) f is onto the plane. That is, for all points P', there is a point P such that $f(P) = P'$.

(vi) f preserves parallel lines.

Proof: (i) Suppose $f(A) = f(B)$. Then, $f(A)f(B) = 0$ and by the definition of an isometry, $AB = 0$. Then, $A = B$ and f is one-to-one.

(ii) Let C be a point between A and B and let $C' = f(C)$. We need to show that C' is on the line through A', B' and that C' is between A' and B'. Since f is one-to-one, C' cannot be A' or B'. Now, $AB = AC + CB$. Since f is an isometry we have that

$$A'B' = A'C' + C'B'$$

This implies that C' is on the line through A', B'. For if it were not on this line, then the triangle inequality would imply that $A'B' < A'C' + C'B'$.

Now either A' is between B' and C', or B' is between A' and C', or C' is between A' and B'. In the first case, we would get

$$B'C' = B'A' + A'C'$$

If we subtract this from the equation above, we would get

$$A'B' - B'C' = C'B' - B'A'$$

and

$$2A'B' - 2B'C' = 0$$

So $A'B' = B'C'$, which would contradict the fact that $A'B' < B'C'$ if A' is between B', C'. Likewise, we cannot have B' between A', C', and so C' must be between A', B'.

(iii) Let A, B be points on a line l. By part (ii) of this theorem, we know that segment $\overline{AB}$ gets mapped to segment $\overline{f(A)f(B)}$. Let D be a point on the ray $\overrightarrow{AB}$ not on segment $\overline{AB}$. Then B is between A, D on $\overrightarrow{AB}$ and since f preserves betweenness, $f(B)$ will be between $f(A)$ and $f(D)$ and so will be on the ray $\overrightarrow{f(A)f(B)}$. Thus, we have that ray $\overrightarrow{AB}$ gets mapped to ray $\overrightarrow{f(A)f(B)}$ and similarly ray $\overrightarrow{BA}$ gets mapped to ray $\overrightarrow{f(B)f(A)}$. This implies that the line through A, B gets mapped to the line through $f(A), f(B)$.

(iv) Let $\angle ABC$ be an angle with vertex B. Since f preserves length, by SSS triangle congruence, $\triangle ABC$ and $\triangle f(A)f(B)f(C)$ will be congruent and their angles will be congruent.

(v) We know that f is one-to-one. Thus, given P' we can find two points A, B such that $f(A) \neq f(B) \neq P'$. Let $f(A) = A'$ and $f(B) = B'$. There are two cases for A', B', P': either they lie on the same line or not.

If A', B', P' are collinear, then P' is either on the ray $\overrightarrow{A'B'}$ or on the opposite ray. Suppose P' is on $\overrightarrow{A'B'}$. Let P be a point on $\overrightarrow{AB}$ such that $AP = A'P'$. Since $AP = f(A)f(P) = A'f(P)$ and since P' and $f(P)$ are on the same ray $\overrightarrow{A'B'}$, then $P' = f(P)$. If P' is on the opposite ray to $\overrightarrow{A'B'}$, we would get a similar result.

If A', B', P' are not collinear, then consider $\angle P'A'B'$. On either side of the ray through A, B, we can find two points P, Q such that $\angle P'A'B' \cong \angle PAB \cong \angle QAB$ (Figure 5.1).

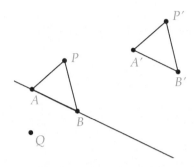

Figure 5.1

We can also choose these points such that $AP = AQ = A'P'$. Since f preserves betweenness (proved in statement (ii)) we know that one of $\overline{A'f(P)}$ or $\overline{A'f(Q)}$ will be on the same side as $\overline{A'P'}$. We can assume that $\overline{A'f(P)}$ is on this same side. Then, since f preserves angles, we have that $\angle P'A'B' \cong \angle f(P)A'B'$ and thus $\overrightarrow{A'P'} \cong \overrightarrow{A'f(P)}$. Since f preserves lengths, we have that $A'P' = AP = f(A)f(P) = A'f(P)$ and thus $P' = f(P)$.

(vi) Let l, m be parallel lines. Suppose that $f(l), f(m)$ were not parallel. Then, for some P on l and Q on m, we would have $f(P) = f(Q)$. But, we know that $PQ \neq 0$ as l and m are parallel. As f is an isometry, we then have that $f(P)f(Q) \neq 0$. Thus, it cannot be true that $f(P) = f(Q)$, and $f(l), f(m)$ must be parallel. $\square$

We have shown that isometries are

- Length-preserving

- One-to-one

- Onto

Isometries are a special type of transformation, but we have not yet explicitly defined what we mean by a "transformation." We have said that a transformation is a function on points in the plane, but this definition is too general. In the spirit of Klein's Erlanger Program, we want to consider "reasonable" functions that leave Euclidean figures invariant. Functions that map lines to points, or areas to segments, are not reasonable geometric equivalences. However, functions that are one-to-one and onto do not have such pathological behavior. Thus, we will define transformations as follows:

Definition 5.2. *A function f on the plane is a* transformation *of the plane if f is a one-to-one function that is also onto the plane.*

An isometry is then a length-preserving transformation. One important property of any transformation is that it is *invertible*.

Definition 5.3. *Let f, g be functions on a set S. We say that g is the* inverse *of f if $f(g(s)) = s$ and $g(f(s)) = s$ for all s in S. That is, the composition of g and f (f and g) is the identity function on S. We denote the inverse by f^{-1}.*

It is left as an exercise to show that a function that is one-to-one and onto must have a unique inverse. Thus, all transformations have unique inverses.

A nice way to classify transformations (isometries) is by the nature of their fixed points.

Definition 5.4. *Let f be a transformation. P is a* fixed point *of f if $f(P) = P$.*

How many fixed points can an isometry have?

Theorem 5.2. *If points A, B are fixed by an isometry f, then the line through A, B is also fixed by f.*

Proof: We know that f will map the line $\overleftrightarrow{AB}$ to the line $\overleftrightarrow{f(A)f(B)}$. Since A, B are fixed points, then $\overleftrightarrow{AB}$ gets mapped back to itself.

Suppose that P is between A and B. Then, since f preserves betweenness, we know that $f(P)$ will be between A and B. Also

$$AP = f(A)f(P) = Af(P)$$

This implies that $P = f(P)$.

A similar argument can be used in the case where P lies elsewhere on $\overleftrightarrow{AB}$. □

Definition 5.5. *The isometry that fixes all points in the plane will be called the* identity *and will be denoted as* id.

Theorem 5.3. *An isometry* f *having three non-collinear fixed points must be the identity.*

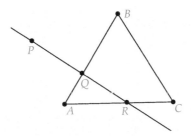

Figure 5.2

Proof: Let A, B, C be the three non-collinear fixed points. From the previous theorem we know that f will fix lines $\overleftrightarrow{AB}$, $\overleftrightarrow{AC}$, and $\overleftrightarrow{BC}$.

Let P be a point not on one of these lines. Let Q be a point between A, B (Figure 5.2). Consider the line through P, Q. By Pasch's axiom, this line will intersect one of $\overline{AC}$ or $\overline{BC}$ at some point R. By the previous theorem, f fixes the line $\overleftrightarrow{QR}$ and thus fixes P. Since P was chosen arbitrarily, then f fixes all points in the plane and is the identity. $\square$

Corollary 5.4. *If two isometries* f, g *agree on any three non-collinear points, then the two isometries must agree everywhere, that is,* $f = g$.

The proof of this result is left as an exercise.

It is clear from this theorem that we can classify isometries into three non-trivial (non-identity) types: those with two fixed points, those with one fixed point, and those with no fixed points. In the following sections we will study the properties of isometries with two, one, or zero fixed points. We will make extensive use of techniques from both synthetic and analytic geometry in our proofs and development.

It is interesting to note that the preceding results on isometries do not depend on Euclid's fifth postulate, the parallel postulate. They are part of *neutral geometry* (or *absolute geometry*). We will make use of this fact in Chapter 7, where we explore a *non-Euclidean* geometry called Hyperbolic geometry.

For future reference, we will also note that many of the results in this section depend on the notion of *betweenness*. As was mentioned in Chapter 2, the notion of betweenness is one that must be axiomatically developed, and this is done in the on-line chapter covering Hilbert's axioms. The axioms of betweenness guarantee that given three distinct points on a line, exactly one of the points is between the other two. This is reasonable for Euclidean and Hyperbolic geometries, but is not true in another non-Euclidean geometry – Elliptic geometry – where lines have finite length and circle back on themselves.

Elliptic geometry is developed in Chapter 8. In the on-line chapter on Hilbert's axioms, we develop a new axiomatic basis for betweenness in Elliptic geometry. We show that Euclid's first 15 propositions (up to the Exterior Angle Theorem) hold with this new definition of betweenness. Thus, the results of Theorem 5.1 that do not ultimately rely on the Exterior Angle Theorem will hold in Elliptic geometry. The proof of statement (ii) of the Theorem relies on the Triangle Inequality, which is Euclid's Proposition 20, and thus is based on the Exterior Angle Theorem. In Chapter 8 we will resolve this problem with the proof of statement (ii) by assuming that elliptic isometries are *defined* to be transformations that preserve elliptic betweenness.

Exercise 5.1.1. *Prove that every function f on a set S that is one-to-one and onto has a unique inverse. [Hint: First, define f^{-1} using f and show that it is a valid function. Then, show that $f \circ f^{-1} = id$ and $f^{-1} \circ f = id$, where id is the identity on S. Finally, show that the inverse is unique.]*

Exercise 5.1.2. *Prove that the inverse of an isometry is again an isometry. (This implies that the set of isometries is closed under the inverse operation.)*

Exercise 5.1.3. *Let f, g be two invertible functions from a set S to itself. Let $h = f \circ g$; that is, h is the composition of f and g. Show that $h^{-1} = g^{-1} \circ f^{-1}$.*

Exercise 5.1.4. *Let f, g be two isometries. Show that the composition $f \circ g$ is again an isometry. (This says the set of isometries is closed under composition.)*

Exercise 5.1.5. *Show that isometries map circles of radius r to circles of radius r. That is, isometries preserve circles.*

Exercise 5.1.6. *Prove that the image of a triangle under an isometry is a new triangle congruent to the original.*

Exercise 5.1.7. *Given an equilateral triangle ABC, show that there are exactly six isometries that map the triangle back to itself. [Hint: Consider how the isometry acts on the vertices of the triangle.]*

Exercise 5.1.8. *Prove Corollary 5.4.*

Exercise 5.1.9. *Consider points in the plane as ordered pairs (x, y) and consider the function f on the plane defined by $f(x, y) = (kx + a, ky + b)$, where k, a, b are real constants, and $k \neq 0$. Is f a transformation? Is f an isometry?*

Exercise 5.1.10. *Define a* similarity *to be a transformation on the plane that preserves the betweenness property of points and preserves angle measure. Prove that under a similarity, a triangle is mapped to a* similar *triangle.*

Exercise 5.1.11. *Use the previous exercise to show that if f is a similarity, then there is a positive constant k such that*

$$f(A)f(B) = k\, AB$$

for all segments $\overline{AB}$.

Exercise 5.1.12. *Consider points in the plane as ordered pairs (x, y) and consider the function f on the plane defined by $f(x, y) = (kx, ky)$, where k is a non-zero constant. Show that f is a similarity.*

5.2 REFLECTIONS

Definition 5.6. *An isometry with two different fixed points, and that is not the identity, is called a* reflection.

What can we say about a reflection? By Theorem 5.2 if A, B are the fixed points of a reflection, then the reflection also fixes the line through A, B. This line will turn out to be the equivalent of a "mirror" through which the isometry reflects points.

Theorem 5.5. *Let r be a reflection fixing A and B. If P is not collinear with A, B, then the line through A and B will be a perpendicular bisector of the segment connecting P and $r(P)$.*

Proof: Drop a perpendicular from P to $\overleftrightarrow{AB}$, intersecting at Q (Figure 5.3). At least one of A or B will not be coincident with Q; suppose B is not. Consider $\triangle PQB$ and $\triangle r(P)QB$.

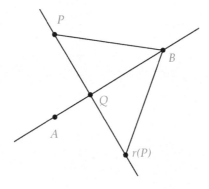

Figure 5.3

Since we know that Q and B are fixed points of r, then $PQ = r(P)Q$, $BP = Br(P)$, and the two triangles are congruent by SSS. Since the two congruent angles at Q make up a straight line, $\angle r(P)QB$ will be a right angle and $\overleftrightarrow{AB}$ will be a perpendicular bisector of the segment $\overline{Pr(P)}$. □

We call the line through A, B the *line of reflection* for r.

Theorem 5.6. *Let P, P' be two points. Then there is a unique reflection taking P to P'. The line of reflection will be the perpendicular bisector of $\overline{PP'}$.*

Proof: Let $\overleftrightarrow{AB}$ be the perpendicular bisector of PP' (Figure 5.4). Define a function r on the plane as follows: If a point C is on $\overleftrightarrow{AB}$, let $r(C) = C$. If C is not on this line, drop a perpendicular from C to $\overleftrightarrow{AB}$ intersecting at Q, and let $r(C)$ be the unique point on this perpendicular such that $r(C) \neq C$, Q is between C and $r(C)$, and $\overline{r(C)Q} \cong \overline{CQ}$.

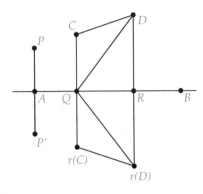

Figure 5.4

Will r be an isometry? We need to show that for all $C \neq D$, $r(C)r(D) = CD$. Let C be a point not on $\overleftrightarrow{AB}$ and D a point on the same side of $\overleftrightarrow{AB}$ as C. Consider Figure 5.4. By SAS, $\triangle QRD \cong \triangle QR\,r(D)$. Again using SAS congruence, we have $\triangle CQD \cong \triangle r(C)\,Q\,r(D)$. Thus, $CD = r(C)\,r(D)$. Similar arguments using congruent triangles can be used if D is on $\overleftrightarrow{AB}$ or on the other side of $\overleftrightarrow{AB}$ as C. (The proof is left as an exercise.)

If C is a point on $\overleftrightarrow{AB}$ and if D is also on $\overleftrightarrow{AB}$, then clearly $CD = r(C)\,r(D)$. If D is not on $\overleftrightarrow{AB}$, then a simple SAS argument will show that $CD = r(C)\,r(D)$.

Thus, r is an isometry. Is the reflection r unique? Suppose there was another reflection r' taking P to P'. By the previous theorem we know that the fixed points of r' are on the perpendicular bisector of $\overline{PP'}$. Since the perpendicular bisector is unique, we have that the fixed points of r' are on $\overleftrightarrow{AB}$. Thus, r and r' have the same values on three non-collinear points P, P', and B and so $r = r'$. □

5.2.1 Mini-Project - Isometries through Reflection

In the first part of this chapter, we discussed Felix Klein's idea of looking at geometry as the study of figures that are invariant under sets of transformations. In the case of transformations that are isometries, invariance means that lengths and angles are preserved, and lines get mapped to lines. Thus, isometries must not only map triangles to triangles, but must map triangles to *congruent* triangles.

Recall that $\triangle ABC \cong \triangle PQR$ if and only if

$$\overline{AB} \cong \overline{PQ}, \overline{AC} \cong \overline{PR}, \overline{BC} \cong \overline{QR}$$

and

$$\angle BAC \cong \angle QPR, \angle CBA \cong \angle RQP, \angle ACB \cong \angle PRQ$$

In other words, there is an *ordering* to the vertex listing for two congruent triangles. It will be important to keep this in mind during the rest of this project.

We know that an isometry f maps a triangle $\triangle ABC$ to a triangle $\triangle PQR$, with $\triangle ABC \cong \triangle PQR$. We can conversely ask whether, given two congruent triangles, there is an isometry that maps one to the other.

We will start with an easy case. Clearly, if two triangles are identical, then the identity isometry will map the triangle to itself. What if the triangles are not identical?

Exercise 5.2.1. *Suppose you have two congruent triangles $\triangle ABC$ and $\triangle PQR$ with $A = P$ and $B = Q$. Show that either the triangles are the same or that there is a reflection that takes $\triangle ABC$ to $\triangle PQR$.*

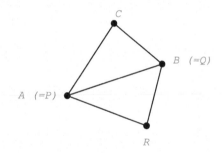

Exercise 5.2.2. *Now suppose you have two congruent triangles that share only one point in common. Let $\triangle ABC \cong \triangle PQR$ with $A = P$. Let l_1 be the angle bisector of $\angle BAQ$ and r_1 the reflection across l_1.*

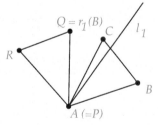

Show that $r_1(B) = Q$. Then, use the preceding exercise to argue that there is a sequence of at most two reflections that will take $\triangle ABC$ to $\triangle PQR$, if the two triangles share one point in common.

Finally, suppose two congruent triangles share no point in common.

Exercise 5.2.3. *Show that there is a sequence of at most three reflections that will take $\triangle ABC$ to $\triangle PQR$ in this case.*

We have now proved the following theorem:

> **Theorem 5.7.** *Let* $\triangle ABC \cong \triangle PQR$. *Then there is an isometry composed of at most three reflections that takes* $\triangle ABC$ *to* $\triangle PQR$.

This theorem has the following amazing corollary:

> **Corollary 5.8.** *Every isometry can be written as the product of at most three reflections.*

Proof: Let f be an isometry and consider any triangle $\triangle ABC$. Then $\triangle f(A) f(B) f(C)$ is a triangle congruent to $\triangle ABC$ and, by the preceding theorem, there is an isometry g composed of at most three reflections taking $\triangle ABC$ to $\triangle f(A) f(B) f(C)$. Since two isometries that agree on three non-collinear points must agree everywhere, then f must be equal to g. $\square$

Exercise 5.2.4. *Let two triangles be defined by coordinates as follows: $\triangle ABC$ with $A = (-3, 2)$, $B = (-3, 6)$, $C = (-6, 2)$ and $\triangle DEF$ with $D = (1, -4)$, $E = (4, -4)$, $F = (1, -8)$. Verify that these two triangles are congruent and then find a sequence of three (or fewer) reflection lines such that $\triangle ABC$ can be transformed to $\triangle DEF$. [Hint: The reflections need not be those defined by the first three exercises of this section. Try to find simple reflections to accomplish the transformation.]*

We note here that the results in this project, and in the preceding section on reflections, are *neutral*—they do not depend on Euclid's fifth postulate, the parallel postulate. We will make use of this fact in Chapter 7.

Also, all of the results of this section, except for the uniqueness part of Theorem 5.6 will hold in Elliptic geometry. The uniqueness of the reflection in Theorem 5.6 depends on the uniqueness of the perpendicular bisector to a segment. In Elliptic geometry, there are possibly two segments defined by two points, and thus possibly two perpendicular bisectors.

5.2.2   Reflection and Symmetry

The word *symmetry* is usually used to refer to objects that are in *balance*. Symmetric objects have the property that parts of the object look similar to other parts. The symmetric parts can be interchanged, thus creating a visual balance to the entire figure. How can we use transformations to mathematically describe symmetry?

Perhaps the simplest definition of mathematical symmetry is the one that most dictionaries give: an arrangement of parts equally on either side of a dividing line. While this type of symmetry is not the only one possible, it is perhaps the most basic in that such symmetry pervades the natural world. We will call this kind of symmetry *bilateral* symmetry.

Definition 5.7. *A figure F in the plane is said to have a* line of symmetry *or* bilateral symmetry *if there is a reflection r that maps the figure back to itself having the line as the line of reflection. For example, in this figure line l is a line of symmetry for* $\triangle ABC$ *since if we reflect the triangle across this line, we get the exact same triangle back again.*

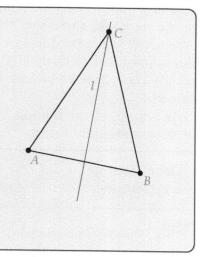

Where is bilateral symmetry found in nature? Consider the insect body types in Figure 5.5.

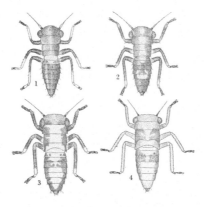

Figure 5.5 Insect symmetry, Manual of Vegetable-Garden Insects, Cyrus R. Crosby and Mortimer D. Leonard, Macmillan, New York, 1918

All of the body types exhibit bilateral symmetry. In fact, most animals, insects, and plants have bilateral symmetry. Why is this the case? Living creatures need bilateral symmetry for *stability*. Consider an animal that needs to be mobile, that needs to move forward and backward. To move with the least expenditure of energy, it is necessary that a body shape be balanced from side to side so that the creature does not waste energy keeping itself upright. Likewise, an immobile living creature, such as a tall pine tree, needs to be bilaterally symmetric in order to keep itself in a vertical equilibrium position.

We know that a reflection will map its line of symmetry back to the same line of symmetry. Lines perpendicular to the line of symmetry are also mapped back to themselves. (The proof is one of the exercises that follow.) These lines are perhaps the simplest figures that are bilaterally symmetric.

In general, we can ask which lines are preserved under the action of an arbitrary transformation.

Definition 5.8. *Lines that are mapped back to themselves by a transformation f are called* invariant lines *of f.*

Note that we do not require that *points* on the line get mapped back to themselves, only that the line as a set of points gets mapped back to itself. Thus, a line may be invariant under f, but the points on the line need not be fixed by f.

In the next chapter we will use invariant lines extensively to classify different sets of symmetries in the plane.

Exercise 5.2.5. *Find examples of five objects in nature that have two or more lines of bilateral symmetry. Draw sketches of these along with their lines of symmetry.*

Definition 5.9. *A polygon is a* regular polygon *if it has all sides congruent and all interior angles congruent.*

Exercise 5.2.6. *Show that the angle bisectors of a regular pentagon are lines of symmetry. Would your proof be extendable to show that the angle bisectors of any regular polygon are lines of symmetry?*

Exercise 5.2.7. *Show that the perpendicular bisector of a side of a regular pentagon is a line of symmetry. Would your proof be extendable to show that* the perpendicular bisectors of the sides of any regular polygon are lines of symmetry?

Exercise 5.2.8. *Show that if a parallelogram has a diagonal as a line of symmetry, then the parallelogram must be a rhombus (i.e., have all sides congruent).*

Exercise 5.2.9. *Show that if a parallelogram has a line of symmetry parallel to a side, then the parallelogram must be a rectangle.*

Exercise 5.2.10. *Finish the proof of Theorem 5.6. That is, prove that the function r defined in the proof is length-preserving for the case where C is not on $\overleftrightarrow{AB}$ and D is either on $\overleftrightarrow{AB}$ or on the other side of $\overleftrightarrow{AB}$.*

Exercise 5.2.11. *Prove that the composition of a reflection with itself is always the identity. Thus, a reflection is its own inverse.*

Exercise 5.2.12. *Show that the lines invariant under a reflection r_m, where m is the line of reflection for r_m, consist of the line m and all lines perpendicular to m.*

Exercise 5.2.13. *Let r_l and r_m be two reflections with lines of reflection l and m, respectively. Show that the composition $r_m \circ r_l \circ r_m = r_{l'}$, where l' is the reflection of l across m. [Hint: Let A, B be distinct points on l. Show that $r_m(A)$ and $r_m(B)$ are fixed points of $r_m \circ r_l \circ r_m$.]*

Exercise 5.2.14. *An object at point O is visible in a mirror from a viewer at point V. What path will the light take from O to the mirror to V? Light always travels through a homogeneous medium to minimize total travel distance. Consider a possible light ray path from the object that hits the mirror at P and then travels to V. Show that the total length of this path is the same as the path from O' to P to V, where O' is the reflection of O across the mirror. Use this to find the shortest path for the light ray from O to the mirror to V. Describe this path in terms of the two angles made at P by the light ray.*

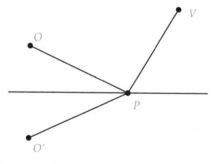

5.3 TRANSLATIONS

> **Definition 5.10.** *An isometry that is made up of two reflections, where the lines of reflection are parallel, or identical, is called a* translation.

What can we say about a translation?

> **Theorem 5.9.** *Let T be a translation that is not the identity. Then, for all points $A \neq B$, if A, B, $T(A)$, and $T(B)$ form a quadrilateral, then that quadrilateral is a parallelogram.*

Proof: Let r_1, r_2 be the two reflections comprising T and let l_1, l_2 be the two lines of reflection. Since T is not the identity, we know that l_1 is parallel to l_2. If we set up a coordinate system where l_1 is the x-axis, then $r_1(x, y) = (x, -y)$.

Since l_1 is parallel to l_2, we can assume l_2 is the line at $y = -K$, $K \neq 0$ (Figure 5.6).

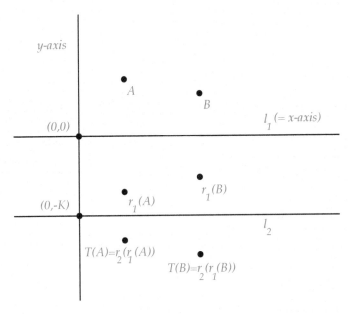

Figure 5.6

A point (x, y) reflected across l_2 must be transformed to a point at a distance $y + K$ below l_2. Thus, the y-coordinate must be $-K - (y + K)$ and $r_2(x, y) = (x, -2K - y)$. Note that K is the distance between l_1, l_2.

Now, if $A = (x, y)$, then $A\,T(A) = A\,r_2(r_1(A))$ will be just the difference in the y values of A and $r_2(r_1(A))$; that is, $y - (-2K - (-y)) = 2K$.

So, $A\,T(A) = 2K$ and $B\,T(B) = 2K$. Since T is an isometry, we also have that $T(A)\,T(B) = AB \neq 0$, as $A \neq B$. Thus, the sides of quadrilateral $A\,B\,T(B)\,T(A)$ are pair-wise congruent. Also, since $A\,\overleftrightarrow{T}(A)$ is perpendicular to l_1 and $B\,\overleftrightarrow{T}(B)$ is perpendicular to l_1, then these two sides of the quadrilateral are parallel. By constructing the diagonal of the quadrilateral, and using a triangle congruence argument, we see that the other pair of sides in the quadrilateral are also parallel, and thus the quadrilateral must be a parallelogram. □

From this theorem we see that for every point A in the plane, a translation T (not equal to the identity) will map A to $T(A)$ in such a way that the length of the segment from A to $T(A)$ will be constant and the direction of this segment will also be constant. Thus, a translation is determined by a *vector*, the vector from A to $T(A)$ for any A. This will be called the *displacement vector* of the translation.

Since the displacement vector is the vector from A to $T(A)$, this vector is given by $T(A) - A$ (considering the points as vectors from the origin). Thus, if $A = (x, y)$, then a translation T with displacement vector $v = T(A) - A = (v_1, v_2)$ will have the coordinate equation

$$T(x, y) = (x, y) + (v_1, v_2)$$

On the other hand, suppose that we are given a function defined by the coordinate equation $T(x, y) = (x, y) + (v_1, v_2)$.

Let l_1 be a line through the origin that is perpendicular to the vector v and l_2 a line parallel to l_1, but passing through the midpoint of the segment along vector v.

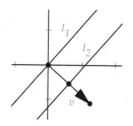

Then the translation T' defined by successive reflection across l_1 and l_2 will map the origin to the head of v and thus will have a translation vector equal to v. The coordinate equation for T' will then be $T'(x, y) = (x, y) + (v_1, v_2)$. Thus, the translation T' and the function T must be the same function.

> **Corollary 5.10.** *Every translation T can be expressed in rectangular coordinates as a function $T(x,y) = (x,y) + (v_1, v_2)$ and, conversely, every coordinate function of this form represents a translation.*

We also have the following result that expresses the translation vector in terms of the original lines of reflection.

> **Corollary 5.11.** *Let a translation be defined by reflection across two parallel lines l_1, l_2 (in that order). Let m be a line perpendicular to both lines at A on l_1 and B on l_2. Then the displacement vector is given by $2\overrightarrow{AB}$.*

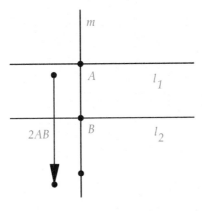

Figure 5.7

Proof: Set up a coordinate system with the x-axis on l_1 (Figure 5.7). Then, as in the proof of the last theorem, the translation will move a point a distance of twice the distance between the lines l_1, l_2. The direction will be perpendicular to the lines. Thus, the vector will be $2\overrightarrow{AB}$, as this vector has the right length and direction. $\square$

How many fixed points will a translation have? If the two lines of reflection defining the translation are not coincident, then the translation always moves points a finite distance and thus there are *no* fixed points. On the other hand, if the two lines are the same, then the translation is the identity. Thus, a non-trivial translation (one that is not the identity) has no fixed points.

5.3.1 Translational Symmetry

When we discussed bilateral symmetry in the last section, we saw that many objects in nature exhibited bilateral symmetry. Such objects, when reflected across a line of symmetry, remain unchanged or invariant. Bilateral symmetry is a property that a single, *finite* object can exhibit.

What can we say about an object that is invariant under translation, that is, an object that has translational symmetry? To remain unchanged under translation, an object must repeat its form at regular intervals, defined by the translation vector. When we take a portion of the object and translate it, we must overlap the exact same shape at the new position. Thus, an object that is invariant under translation is necessarily *infinite* in extent. As Hermann Weyl stated in his foundational work on symmetry:

> A figure which is invariant under a translation t shows what in the art of ornament is called "infinite rapport," i.e., repetition in a regular spatial rhythm [41, page 47].

Translational symmetry, being infinite in extent, cannot be exhibited by finite living creatures. However, we can find evidence of a limited form of translation symmetry in some animals and plants. For example, the millipede has leg sections that are essentially invariant under translation (Figure 5.8).

Figure 5.8 Millipede, National Park Service, Petroglyph National Monument

Also, many plants have trunks (stems) and branching systems that are translation invariant.

Whereas translational symmetry is hard to find in nature, it is extremely common in human ornamentation. For example, wallpaper must have translational symmetry in the horizontal and vertical directions so that when you hang two sections of wallpaper next to each other the seam is not noticeable. Trim patterns called *friezes*, which often run horizontally along tops of walls, also have translational invariance (Figure 5.9).

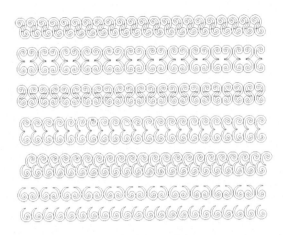

Figure 5.9 Frieze patterns

Exercise 5.3.1. *Find and sketch two examples of translational symmetry in nature.*

Exercise 5.3.2. *Find and sketch three examples of wallpaper or frieze patterns that have translational invariance. Indicate on your sketch the translation vector(s) for each pattern.*

Exercise 5.3.3. *Given a translation $T = r_1 \circ r_2$ defined by two reflections r_1, r_2 and with displacement vector v, show that the inverse of T is the translation $T^{-1} = r_2 \circ r_1$ having displacement vector $-v$.*

Exercise 5.3.4. *Show that the composition of two translations T_1 and T_2 is again a translation. Find the translation vector for $T_1 \circ T_2$.*

Exercise 5.3.5. *Show that the composition of translations is a commutative operation. That is, if T_1 and T_2 are translations, then $T_1 \circ T_2 = T_2 \circ T_1$.*

Exercise 5.3.6. *Given a reflection r across a line l and a translation T in the same direction as l, show that $r \circ T = T \circ r$. [Hint: Choose a "nice" setting in which to analyze r and T.] Will this result hold if T is not in the direction of l?*

Exercise 5.3.7. *In the section on reflections, we saw that a simple reflection across the x-axis, which we will denote by r_x, could be expressed as $r_x(x,y) = (x, -y)$. Let r be the reflection across the line $y = K$. Let T be the translation with displacement vector of $v = (0, -K)$. Show that the function $T^{-1} \circ r_x \circ T$ is equal to r and find the coordinate equation for r. [Hint: Show that $T^{-1} \circ r_x \circ T$ has the right set of fixed points.]*

Exercise 5.3.8. *Let T be a (non-identity) translation. Show that the set of invariant lines for T are all pair-wise parallel and that each is parallel to the displacement vector of T. [Hint: Use Theorem 5.9.]*

Exercise 5.3.9. *Let T be a translation with (non-zero) displacement vector parallel to a line l. Let m be any line perpendicular to l. Show that there is a line n perpendicular to l such that $T = r_n \circ r_m$. [Hint: Suppose m intersects l at P. Choose n to be the perpendicular bisector of $P\,T(P)$. Show that $r_n \circ T$ fixes m.]*

5.4 ROTATIONS

Definition 5.11. *An isometry that is made up of two reflections where the lines of reflection are **not** parallel will be called a rotation.*

To analyze rotations we will make use of the following lemma.

Lemma 5.12. *If two non-coincident lines l_1, l_2 intersect at O, and if m is the bisector of $\angle Q_1 O Q_2$, with Q_1 on l_1 and Q_2 on l_2, then $r_m(l_1) = l_2$ and $r_m \circ r_{l_1} = r_{r_m(l_1)} \circ r_m = r_{l_2} \circ r_m$. $(r_m(l_1) = \{r_m(A)| \ A \ is \ on \ l_1\})$*

Proof: We can assume that Q_1, Q_2 were chosen such that $OQ_1 = OQ_2$ (Fig 5.10). Let P be the intersection of the bisector m with $\overline{Q_1 Q_2}$.

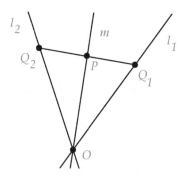

Figure 5.10

Then, by SAS we know that P is the midpoint of $\overline{Q_1 Q_2}$ and m is the perpendicular bisector of $\overline{Q_1 Q_2}$. Thus, $r_m(Q_1) = Q_2$. Since $r_m(O) = O$, the line l_1 (defined by O and Q_1) must get mapped to l_2 (defined by O and Q_2), or $r_m(l_1) = l_2$.

For the second part of the lemma, we know by Exercise 5.2.13 that $r_m \circ r_{l_1} \circ r_m = r_{r_m(l_1)} = r_{l_2}$. Thus, $r_m \circ r_{l_1} = r_{l_2} \circ r_m$. $\square$

Rotations are characterized by the fact that they have a single fixed point.

Theorem 5.13. *An isometry $R \neq id$ is a rotation iff R has exactly one fixed point.*

Proof: Suppose R has a single fixed point, call it O (Figure 5.11). Let A be another point with $A \neq O$.

Let l be the line through O and A and let m be the bisector of $\angle AOR(A)$ (or the perpendicular bisector of $\overline{A\,R(A)}$ if the three points are collinear).

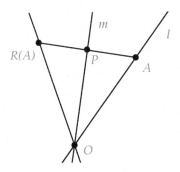

Figure 5.11

If $A, O, R(A)$ are non-collinear, we know from the previous lemma that $r_m \circ R$ will fix point A. If they are collinear, $r_m \circ R$ will also fix A, as m is the perpendicular bisector of $\overline{A\,R(A)}$. Since $r_m \circ R$ also fixes O, $r_m \circ R$ is either the identity or a reflection. If it is the identity, then $R = r_m$. But, then R would have more than one fixed point. Thus, $r_m \circ R$ must be a reflection fixing O and A, and thus $r_m \circ R = r_l$ and $R = r_m \circ r_l$.

Conversely, let $R = r_m \circ r_l$ for lines l, m intersecting at O. If R had a second fixed point, say $B \neq O$, then $r_m(B) = r_l(B)$. Clearly, B cannot be on m or l. But, then the segment joining B to $r_m(B)$ is perpendicular to m, and this same segment would also be perpendicular to l, as $r_m(B) = r_l(B)$. This is impossible. □

The next lemma describes triples of reflections about coincident lines.

Lemma 5.14. *Let l, m, n be three lines intersecting at a point O. Then, $r_l \circ r_m \circ r_n$ is a reflection about a line p passing through O. Also, $r_l \circ r_m \circ r_n = r_n \circ r_m \circ r_l$.*

Proof: Let $f = r_l \circ r_m \circ r_n$, and let A be a point on n not equal to O (Figure 5.12). For $A' = f(A)$, either $A' \neq A$ or $A' = A$.

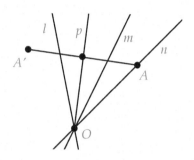

Figure 5.12

If $A' \neq A$, let line p be the angle bisector of $\angle AOA'$. A simple triangle argument shows that p is also the perpendicular bisector of $\overline{AA'}$ and thus r_p maps A to A'. If $A' = A$ choose $p = n$.

In either case, $r_p \circ f$ will fix both A and O and thus fixes n. Thus, $r_p \circ f$ is either equal to r_n or is the identity. If $r_p \circ f = r_p \circ r_l \circ r_m \circ r_n = r_n$, then $r_p \circ r_l \circ r_m = id$ and $r_p = r_m \circ r_l$. Since m and l intersect, then r_p would be either a rotation or the identity. Clearly, r_p cannot be a rotation or the identity, and thus $r_p \circ f = id$; that is, $r_p = f$.

We conclude that f is a reflection about a line p passing through O, and $r_p = f = r_l \circ r_m \circ r_n$.

For the second part of the theorem, we note that $(r_p)^{-1} = (r_n)^{-1} \circ (r_m)^{-1} \circ (r_l)^{-1}$. Since reflections are their own inverses, we have $r_p = r_n \circ r_m \circ r_l$.

□

The next theorem tells us how rotations transform points through a fixed angle.

> **Theorem 5.15.** *Let $R \neq id$ be a rotation about a fixed point O. For any point $A \neq O$, there is a unique line m passing through O such that $R = r_m \circ r_l$, where l is the line through O and A. Also, if $m\angle AOR(A)$ is θ degrees, then for any point $B \neq O$, we have $m\angle BOR(B)$ is also θ degrees.*

Proof: Let m be the angle bisector of $\angle AOR(A)$. By the first part of the proof of Theorem 5.13, we know that $R = r_m \circ r_l$.

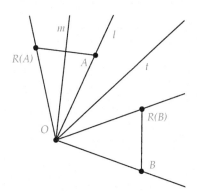

Figure 5.13

For the second part of the theorem, note that if $m\angle BOR(B)$ is θ, then $m\angle B'OR(B')$ is θ for all points B' on $\overleftrightarrow{OB}$. So, we can assume B has the property that $\overline{OA} \cong \overline{OB} \cong \overline{O\,R(B)}$

Let t be the angle bisector of $\angle R(B)\,OA$. Then, $r_t(A) = R(B) = r_m(r_l(B))$. Thus, $A = r_t(r_m(r_l(B)))$ and $B = r_l(r_m(r_t(A)))$. By the previous lemma we also have $B = r_t(r_m(r_l(A)))$. Since $r_m(r_l(A)) = R(A)$, then $B = r_t(R(A))$.

Since $R(B) = r_t(A)$ and $B = r_t(R(A))$, then $\overline{A\,R(A)} \cong \overline{B\,R(B)}$. If A, O, and $R(A)$ are collinear, then A and $R(A)$ are on opposite sides of O on $\overleftrightarrow{OA}$. Also, B, O, and $R(B)$ must be collinear because r_t maps A to $R(B)$, O to O and $R(A)$ to B, and reflections map lines to lines. Also, B and $R(B)$ are on opposite sides of O as reflections preserve betweenness. If A, O and $R(A)$ are non-collinear, then by SSS congruence $\triangle AOR(A) \cong \triangle BOR(B)$. In either case, $\angle AOR(A) \cong \angle BOR(B)$. Thus, the two angles have the same measure. □

From this theorem we can see that the construction of a rotation about O of a specific angle θ requires the choice of two lines that meet at O and make an angle of $\frac{\theta}{2}$.

We note here that the preceding theorems on rotations do not depend on Euclid's fifth postulate and are thus part of *neutral geometry*. We will use this fact to consider non-Euclidean rotations in Chapter 7.

Also, all of the results of this section hold in Elliptic geometry, except for the converse implication of Theorem 5.13 (That a rotation has exactly one fixed point). The proof that a rotation has exactly one fixed point depends on the Neutral geometry fact that two distinct, but intersecting, lines cannot have a common perpendicular. In Elliptic geometry, pairs of intersecting lines can have a common perpendicular and there are rotations having more than one fixed point.

Definition 5.12. *The point O of intersection of the reflection lines of a rotation R is called the* center of rotation. *The angle ϕ defined by $\angle AO\,R(A)$ for $A \neq O$ is called the* angle of rotation.

Now let's consider the coordinate form of a rotation R.

Given a point (x, y) in a coordinate system, we know that we can represent the point as $(r\,\cos(\theta), r\,\sin(\theta))$. A rotation of (x, y) through an angle of ϕ about the origin $O = (0, 0)$ is given by

$$Rot_\phi(x, y) = (r\,\cos(\theta + \phi), r\,\sin(\theta + \phi))$$

From the trigonometric formulas covered in Chapter 3, we know that the right side of this equation can be written as

$$(r\,\cos(\theta)\cos(\phi) - r\,\sin(\theta)\sin(\phi), r\,\sin(\theta)\cos(\phi) + r\,\cos(\theta)\sin(\phi))$$

Therefore,

$$Rot_\phi(x, y) = (x\,\cos(\phi) - y\,\sin(\phi), x\,\sin(\phi) + y\,\cos(\phi))$$

This is the coordinate form for a rotation about the origin by an angle of ϕ.

We note for future reference that a rotation about a point O of 180 degrees is called a *half-turn* about O.

5.4.1 Rotational Symmetry

> **Definition 5.13.** *A figure is said to have* rotational symmetry *(or* cyclic symmetry*) of angle ϕ if the figure is preserved under a rotation about some center of rotation with angle ϕ.*

Rotational symmetry is perhaps the most widespread symmetry in nature.

Rotational symmetry can be found in the very small, such as this radiolarian illustrated by Ernst Haeckel in his book *Art Forms in Nature* [17], to the very large, as exhibited by the rotationally symmetric shapes of stars and planets.

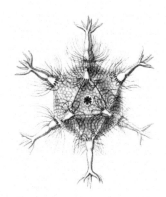

Many flowers exhibit five-fold symmetry, the rotational symmetry of the regular pentagon. Let's prove that the regular pentagon has five-fold symmetry.

> **Theorem 5.16.** *The regular pentagon has rotational symmetry of 72 degrees.*

Proof: In earlier exercises it was shown that the angle bisectors of the pentagon are lines of symmetry and the perpendicular bisectors of the sides are also lines of symmetry. Let l_1 be the angle bisector of $\angle CAB$ and l_2 be the perpendicular bisector of side $\overline{AB}$ at M in the pentagon shown in Figure 5.14.

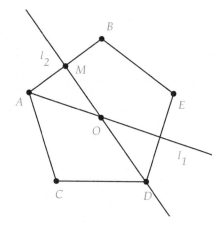

Figure 5.14

Then l_1 and l_2 must intersect. For, suppose that they were parallel. Then $\angle CAB$ must be a straight angle, that is, C, A, and B are collinear, which is clearly impossible.

Let O be the intersection of l_1, l_2 and ϕ be the measure of $\angle AOB$. The composition R of the reflections through l_1, l_2 is a rotation through an angle of $2\angle AOM = \phi$. We know that the pentagon is invariant under R, since it is invariant under the component reflections. If C, D, E are the other vertices, we have that iterated application of the rotation R on A will cycle through these other vertices, and thus $OC = OA = OD = OE$, making all of the interior triangles $\triangle AOB$, $\triangle COA$, and so on, congruent. Since there are five angles at O of these triangles, then $\phi = \frac{360}{5} = 72$. $\square$

In a similar fashion, we could show that the regular n-gon has rotational symmetry of $\frac{360}{n}$ degrees.

Exercise 5.4.1. *Suppose we are given a coordinate system centered at a point O. Let Rot_ϕ be a rotation about O of angle ϕ. Let $C = (x, y)$ be a point not equal to O and let T be the translation with vector $v = (-x, -y)$. Show that $T^{-1} \circ Rot_\phi \circ T$ is a rotation about C of angle ϕ.*

Exercise 5.4.2. *Given a coordinate system centered at a point O and a line l that does not pass through O, find an expression for reflection across l, by using translations, rotations, and a reflection across the x-axis.*

Exercise 5.4.3. *Find three examples in nature of each of the following rotational symmetries: 90 degrees (square), 72 degrees (pentagon), and 60 degrees (hexagon). Sketch your examples and label the symmetries of each.*

Exercise 5.4.4. *Show that if R is a rotation of θ degrees about O, with $R \neq id$, and l is a line not passing through O, then $R(l) \neq l$. That is, if a* *(non-identity) rotation has an invariant line, it must pass through the center of rotation. [Hint: Drop a perpendicular from O to l at A. If θ < 180 consider the triangle AO R(A). If θ = 180 use a different argument.]*

Exercise 5.4.5. *Show that if a rotation R ≠ id has an invariant line, then it must be a rotation of 180 degrees. Also, the invariant lines for such a rotation are all lines passing through the center of rotation O. [Hint: Use the preceding exercise.]*

Exercise 5.4.6. *Let R be a rotation about a point O and m be a line. Show that if R(m) ∥ m, then the rotation angle for R is 180 degrees. [Hint: Suppose the angle is less than 180 and consider △AO R(A) where A is a point on m.]*

Exercise 5.4.7. *Let R be a rotation about a point O that is not the identity or a half-turn. Let m be a line. Show that R(m) and m must intersect and that if P is the point of intersection, then the angle made by m and R(m) at P is equal to the rotation angle of R. [Hint: Show that you get a contradiction if m and R(M) intersect and then drop perpendiculars to the two lines from P. Consider the quadrilateral thus created.]*

Exercise 5.4.8. *Show, using reflections, that the inverse to a rotation about a point of ϕ degrees is a rotation about the same point of −ϕ degrees.*

Exercise 5.4.9. *Show that the composition of two rotations centered at the same point is again a rotation centered at that point.*

Exercise 5.4.10. *Suppose that two rotations R, R′ centered at O have the same effect on a point A ≠ O. Show that R = R′.*

Exercise 5.4.11. *Suppose that the composition of a rotation R ≠ id with a reflection r_1 is again a reflection r_2. That is, suppose that $r_1 \circ R = r_2$. Show that r_1 and r_2 must pass through the center of rotation for R.*

Exercise 5.4.12. *Let l be perpendicular to m at a point A on l (or m). Let $H = r_l \circ r_m$. Show that H is a rotation about A of 180 degrees; that is, a half-turn about A.*

Exercise 5.4.13. *Let A, B be distinct points. Let H_A, H_B be half-turns about A, B, respectively. Show that $H_B \circ H_A$ is a translation in the direction of the vector from A to B.*

Exercise 5.4.14. *Let f be any isometry and H_A a half-turn about a point A. Show that $f \circ H_A \circ f^{-1}$ is a half-turn. [Hint: Show it is a rotation and that it maps lines through a point back to themselves.]*

5.5 PROJECT 7 - QUILTS AND TRANSFORMATIONS

Before we look at the last type of isometries, those composed of three reflections, we will take a break to have a little fun with our current toolbox of isometries (reflections, translations, and rotations).

One of the uniquely American craft forms is that of quilting. In making a quilt, we start with a basic block, usually a square, made up of various pieces of cloth. This basic block is then copied to form a set of identical blocks that are sewn together to make a quilt.

For example, here is a quilt block we'll call "square-in-square" that is made up of a piece of white cloth on top of a black background.

If we translate this block up and down (equivalently, sew copies of this block together), we will get the quilt shown here.

Note that the square-in-square block can be built up from a simpler shape, that of the triangular-divided square found in the bottom left corner of the square-in-square block. The square-in-square block is built of four copies of this basic shape, using various reflections.

In the first part of this project, we will see how we can use the transformation capability of dynamic geometry software to construct the triangular-divided square. We will then use this square to build the square-in-square block, which will then be used to construct a quilt. (Notes on how to use transformations in particular dynamic geometry software packages can be found at http://www.gac.edu/~hvidsten/geom-text.)

We note here that when we refer to a "block" of a quilt, we are referring to a square region of the quilt that can cover the entire quilt when repeatedly transformed via reflections, rotations, and translations. Thus, in the quilt shown in the middle figure in the preceding series of figures, either the square-in-square shape, or the triangular-divided square, or the entire quilt itself, could be considered basic quilting blocks.

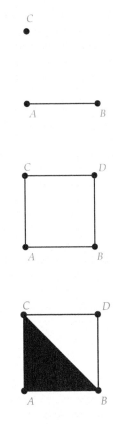

Start up your geometry software and create a segment $\overline{AB}$. Set point A to be a center of rotation. Then, rotate point B 90 degrees about point A.

Now, set C as the center of rotation and rotate point A by 90 degrees to get point D. Then, connect the vertices with segments to form the square shown.

To construct the triangular-divided square, we will construct the figure shown. Select points A, B, and C, and construct a filled-polygon area.

We can finish the creation of the square-in-square block either by reflection or rotation about D. Carry out whichever transformations you wish to get the block shown. (We will ignore the extra lines and points for now.)

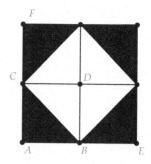

To create a larger quilt from this block, we will translate the block vertically and horizontally. For example, to translate horizontally, we will want to shift the block to the right by the vector $\overrightarrow{AE}$.

Define the translation from A to E and then translate the entire block to get the image shown.

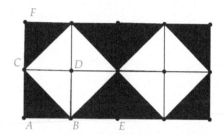

We can likewise set $\overrightarrow{AF}$ as a vector and translate the previous figure in a vertical direction multiple times.

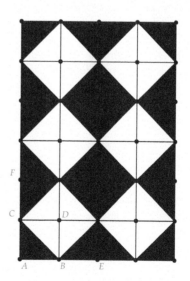

Finally, let's hide all the segments and points used in the construction.

Exercise 5.5.1. *Using the same triangular-divided square and appropriate transformations, construct the Yankee Puzzle shown here. Feel free to use different colors for the various shapes in the quilt. Describe the steps (i.e., sequence of transformations) that you took to build the quilt.*

Exercise 5.5.2. *Design your own pattern, based on triangular-divided squares or simple squares, and use it to build a quilt. Below are a few quilts you can use for ideas.*

Star Puzzle

Dutch Man's Puzzle

25-Patch Star

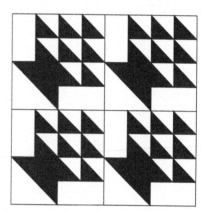

Flower Basket

Exercise 5.5.3. *Which of the four quilt patterns in the previous exercise have*

- *Bilateral symmetry (Specify the lines of reflection.)*
- *Rotational symmetry (Specify the rotation angle and center of rotation.)*
- *Both rotational and bilateral symmetry*

Exercise 5.5.4. *Why must a quilt having two perpendicular lines of reflection have a rotational symmetry? What is the rotation angle?*

The quilt patterns in this project, and many other intriguing quilt patterns, can be found in [5, pages 305–311] and also in [36].

5.6 GLIDE REFLECTIONS

Now we are ready to look at the last class of isometries—those made up of three reflections. Such isometries will turn out to be equivalent to either a reflection or a *glide reflection.*

> **Definition 5.14.** *An isometry that is made up of a reflection and a translation parallel to the line of the reflection is called a* glide reflection.

A glide reflection is essentially a flip across a line and then a glide (or translate) along that line. If $\overrightarrow{AB}$ is a vector with T_{AB} translation by this vector and if l is a line parallel to $\overrightarrow{AB}$ with r_l reflection across l, then the glide reflection defined by these isometries is

$$G_{l,AB} = T_{AB} \circ r_l$$

Our first theorem about glide reflections says that it doesn't matter if you glide and then reflect or reflect and then glide. You always end up at the same place.

> **Theorem 5.17.** *Let l be a line and $\overrightarrow{AB}$ a vector parallel to l.*
>
> - $G_{l,AB} = T_{AB} \circ r_l = r_l \circ T_{AB}$
>
> - $G_{l,AB}^{-1} = T_{BA} \circ r_l$

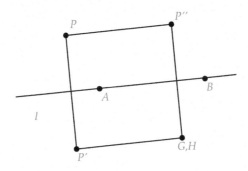

Figure 5.15

Proof: For the first statement of the theorem, let P be a point not on l. Let $P' = r_l(P)$ and $P'' = T_{AB}(P)$ (Figure 5.15). Let $G = T_{AB}(r_l(P))$ and $H = r_l(T_{AB}(P))$. We know that $\overline{PP'}$ is perpendicular to l. We also know that $APP''B$ and $AP'GB$ will be parallelograms, by Theorem 5.9. Thus, $\angle P'PP''$ and $\angle PP'G$ are right angles, as $\overline{PP''}$ and $\overline{P'G}$ are both parallel to l and $\overline{PP'}$ crosses l at right angles. The angles at P'' and G in quadrilateral $PP''GP'$ are also right angles, as translation preserves angles. Thus, $PP''GP'$ is a rectangle.

A similar argument will show that $PP''HP'$ is also a rectangle and thus $G = H$, or $T_{AB}(r_l(P)) = r_l(T_{AB}(P))$.

If P lies on l, then since translation of $T_{AB}(P)$ will still lie on l, we have that $r_l(T_{AB}(P)) = T_{AB}(P) = T_{AB}(r_l(P))$.

For the second statement we reference one of the earlier exercises of the chapter, which said that if a function h was the composition of f and g ($h = f \circ g$), then $h^{-1} = g^{-1} \circ f^{-1}$. So

$$G_{l,AB}^{-1} = r_l^{-1} \circ T_{AB}^{-1}$$

Since the inverse of a reflection is the reflection itself and the inverse of a translation from A to B is the reverse translation from B to A, we get that

$$G_{l,AB}^{-1} = r_l \circ T_{BA} = T_{BA} \circ r_l$$

□

Now we are ready to begin the classification of isometries that consist of three reflections.

Theorem 5.18. *Let l_1, l_2, l_3 be three lines such that exactly two of them meet at a single point. Then the composition of reflections across these three lines ($r_3 \circ r_2 \circ r_1$) is a glide reflection.*

Proof: Let r_1, r_2, r_3 be the reflections across l_1, l_2, l_3. There are two cases to consider. Either the first two lines l_1, l_2 intersect or they are parallel.

Suppose that l_1, l_2 intersect at P (Figure 5.16). Drop a perpendicular from P to l_3 intersecting at Q.

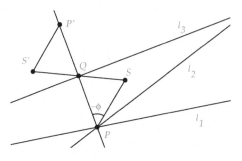

Figure 5.16

We know that the composition of r_1 and r_2 will be a rotation about P of some non-zero angle, say ϕ degrees. Let S be the rotation of Q about P of $-\phi$ degrees. Since $r_2(r_1(S)) = Q$ and $r_3(Q) = Q$, we have

$$Q = r_3(r_2(r_1(S)))$$

Now, let P' be the reflection of P across l_3. Let $S' = r_3(r_2(r_1(Q)))$. We know that $S' \neq S$ as S' must be Q rotated about P by ϕ and then reflected across l_3 and thus must be on the other side of $\overleftrightarrow{QP}$ from S. Since $r_3 \circ r_2 \circ r_1$ is an isometry and since this composition maps ΔPSQ to $\Delta P'QS'$, we know that these two triangles are congruent. Thus, $\angle PQS \cong \angle P'S'Q$. But, both triangles must also be isosceles ($\overline{PS} \cong \overline{PQ}$). Thus, $\angle PQS \cong \angle P'QS'$. This means that S, Q, S' must lie on a line. Since isometries take lines to lines then $r_3 \circ r_2 \circ r_1$ must map $\overleftrightarrow{QS}$ back to itself, with a shift via the vector from S to Q.

Let $G = T_{SQ} \circ r_{SQ}$, where r_{SQ} is reflection across $\overleftrightarrow{QS}$ and T_{SQ} is translation from S to Q. It is left as an exercise to show that $G(P) = P'$. Then, G and $r_3 \circ r_2 \circ r_1$ match on three non-collinear points P, S, Q and so $G = r_3 \circ r_2 \circ r_1$, and thus the composition $r_3 \circ r_2 \circ r_1$ is a glide reflection.

Now, what about the second case, where l_1, l_2 are parallel? Then it must be the case that l_2 and l_3 intersect at a single point. Then, by the argument above, $r_1 \circ r_2 \circ r_3$ is a glide reflection. But, $r_1 \circ r_2 \circ r_3 = (r_3 \circ r_2 \circ r_1)^{-1}$. We know that the inverse of a glide reflection is again a glide reflection by the previous theorem. Thus, $((r_3 \circ r_2 \circ r_1)^{-1})^{-1} = r_3 \circ r_2 \circ r_1$ is a glide reflection. □

We can now give a complete classification of isometries that consist of three reflections.

Theorem 5.19. *The composition of three different reflections is either a reflection or a glide reflection.*

Proof: There are three cases.

First, suppose that the three lines l_1, l_2, l_3 of reflection are parallel. We can suppose that there is a coordinate system set up so that each line is parallel to the x-axis. Then, as was discussed in the section on reflections, we know that the three reflections r_1, r_2, r_3 associated with l_1, l_2, l_3 can be given by

$$r_1(x,y) = (x, -y-2K_1), r_2(x,y) = (x, -y-2K_2), r_3(x,y) = (x, -y-2K_3)$$

for some non-negative constants K_1, K_2, K_3. Then

$$r_3(r_2(r_1(x,y))) = (x, -y - 2K_1 + 2K_2 - 2K_3)$$

This clearly fixes the line at $y = -K_1 + K_2 - K_3$ and thus is a reflection.

Secondly, suppose that only two of the lines meet at exactly one point. By the previous theorem we have that $r_3 \circ r_2 \circ r_1$ is a glide reflection.

Finally, suppose that all three meet at a single point P. Then, by Lemma 5.14 we have that $r_3 \circ r_2 \circ r_1$ is a reflection about some line through the common intersection point. □

5.6.1 Glide Reflection Symmetry

Definition 5.15. *A figure is said to have* glide symmetry *if the figure is preserved under a glide reflection.*

Where does glide symmetry appear in nature? You may be surprised to discover that your feet are creators of glide symmetric patterns! For example, if you walk in a straight line on a sandy beach, your footprints will create a pattern that is invariant under glide reflection (footsteps created by Preston Nichols).

Many plants also exhibit glide symmetry in the alternating structure of leaves or branches on a stem.

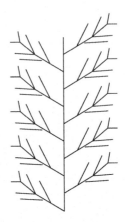

Exercise 5.6.1. *Find examples of two objects in nature that exhibit symmetries of a glide reflection (other than the ones we have given). Draw sketches of these and illustrate the glide reflection for each object on your sketch.*

Exercise 5.6.2. *Finish the proof of Theorem 5.18; that is, show in the proof that $G(P) = P'$. [Hint: Let P'' be the reflection of P across $\overleftrightarrow{QS}$. Show that $\triangle P'P''Q \cong \triangle QSP$ and use this to show the result.]*

Exercise 5.6.3. *Show that the only invariant line under a glide reflection $T_{AB} \circ r_l$ (with $\overrightarrow{AB} \neq (0,0)$) is the line of reflection l. [Hint: If m is invariant, then it is also invariant under the glide reflection squared.]*

Exercise 5.6.4. *Show that if a glide reflection has a fixed point, then it is a pure reflection—it is composed of a reflection and a translation by the vector $v = (0,0)$. [Hint: Use a coordinate argument.]*

Exercise 5.6.5. *Show that the composition of a glide reflection with itself is a translation and find the translation vector in terms of the original glide reflection.*

A *group* of symmetries is a set of Euclidean isometries that have the following properties:

1. Given any two elements of the set, the composition of the two elements is again a member of the set.

2. The composition of elements is an associative operation.

3. The identity is a member of the set.

4. Given any element of the set, its inverse exists and is an element of the set.

Note: Since function composition is associative, the second condition is true for all collections of isometries.

Exercise 5.6.6. *Does the set of glide reflections form a group of symmetries? Why or why not?*

Exercise 5.6.7. *Show that the set of all reflections does not form a group of symmetries.*

Exercise 5.6.8. *Show that the set of all rotations does not form a group of symmetries. [Hint: Use half-turns and exercise 5.4.13.]*

Exercise 5.6.9. *Show that the set of rotations about a fixed center of rotation O does form a group of symmetries.*

Exercise 5.6.10. *Show that the set of translations forms a group of symmetries.*

Definition 5.16. *An isometry is called* direct *(or* orientation-preserving*) if it is a product of two reflections or is the identity. It is called* opposite *(or* orientation-reversing*) if it is a reflection or a glide.*

Exercise 5.6.11. *Given △ABC, the ordering A, B, C is called a "clockwise" ordering of the vertices, whereas A,C,B would be a "counterclockwise" ordering. Describe, by example, how direct isometries preserve such orderings (i.e., map clockwise to clockwise and counterclockwise to counterclockwise) and how opposite isometries switch this ordering.*

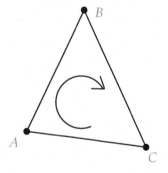

Exercise 5.6.12. *Show that every isometry is either direct or opposite, but not both.*

> **Definition 5.17.** *An isometry f will be called* even *if it can be written as the product of an even number of reflections. An isometry will be called* odd *if it can be written as the product of an odd number of isometries.*

Exercise 5.6.13. *Show that an isometry is even if and only if it is direct, and that it is odd if and only if it is opposite.*

5.7 STRUCTURE AND REPRESENTATION OF ISOMETRIES

We have now completely classified the possible isometries of the Euclidean plane. We can classify isometries in two ways—by the number of fixed points they have or by the number of reflections that make up isometries.

In Table 5.1 we classify isometries by the number of fixed points they have.

TABLE 5.1 Isometry Classification by Fixed Points

Fixed Points	Isometry
0	translation, glide reflection
1	rotation
2	reflection
3 (not collinear)	identity

In Table 5.2 we classify an isometry by the number of reflections that comprise the isometry.

TABLE 5.2 Isometry Classification by Reflections

Number of Reflections	Isometry
1	reflection
2	identity, rotation, translation
3	glide reflection, reflection

We know how an individual isometry behaves, but how do isometries work when combined together? From Exercise 5.4.9 we know that the composition of two rotations about the same center is again a rotation about that center. But, what can we say about the composition of two rotations about *different* centers? We know that the composition of such rotations will again be an isometry, but what type? To answer such questions we will need a uniform way to represent all Euclidean isometries. This will be done through the use of matrices.

5.7.1   Matrix Form of Isometries

Recall that a rotation R_ϕ of a point (x, y) through an angle of ϕ degrees about the origin can be written as

$$R_\phi(x, y) = (x \, \cos(\phi) - y \, \sin(\phi), x \, \sin(\phi) + y \, \cos(\phi))$$

Note that this is equivalent to the matrix equation

$$R_\phi(x, y) = \begin{bmatrix} \cos(\phi) & -\sin(\phi) \\ \sin(\phi) & \cos(\phi) \end{bmatrix} \begin{pmatrix} x \\ y \end{pmatrix}$$

where $\begin{pmatrix} x \\ y \end{pmatrix}$ is the column vector representing the point (x, y). Thus, the action of R_ϕ on points in the plane is the same as matrix multiplication (on the left) of the vector representing the point by the matrix $\begin{bmatrix} \cos(\phi) & -\sin(\phi) \\ \sin(\phi) & \cos(\phi) \end{bmatrix}$. We see, then, that any rotation about the origin can be identified with a 2×2 matrix of the form above, and conversely any matrix of this form will give rise to a rotation about the origin.

Can we find a matrix form for a translation? Let T be a translation with vector $v = (v_1, v_2)$. Then

$$T(x, y) = (x, y) + (v_1, v_2) = (x + v_1, y + v_2)$$

Unfortunately, this cannot be written as the product of a 2x2 matrix with $\begin{pmatrix} x \\ y \end{pmatrix}$. However, consider the following matrix equation:

$$\begin{bmatrix} 1 & 0 & v_1 \\ 0 & 1 & v_2 \\ 0 & 0 & 1 \end{bmatrix} \begin{pmatrix} x \\ y \\ 1 \end{pmatrix} = \begin{pmatrix} x + v_1 \\ y + v_2 \\ 1 \end{pmatrix}$$

The vector $(x, y, 1)$ gets mapped to the vector $(x + v_1, v + v_2, 1)$. In the x and y coordinates, this is exactly what translation by v would produce. But, how can we use this fact about *three-dimensional* vectors in planar geometry?

Recall the discussion on *models* in the first chapter. If we can find a setting where the axioms of a geometry hold, then that setting will serve as a model for the geometry, and all theorems will hold equally well in that model. The normal model for planar geometry is the Euclidean (x, y) plane, which we can interpret as the plane at height 0 in three dimensions. But, the axioms of planar geometry work just as well for the set of points at height $z = 1$, that is, the set of points $(x, y, 1)$. Whatever operation we do on points in this new plane will be equal in effect to the corresponding operation in the abstract Euclidean plane, as long as we ensure that the third coordinate stays equal to 1.

For the rest of this section we will switch to this new model of planar geometry.

> **Definition 5.18.** *A* point *will be a vector* $(x, y, 1)$. *Distance and angle will be defined as the standard distance and angle of the* x *and* y *components.*

In our new model the third coordinate does not carry any geometric significance. It is there just to make our representation of isometries easier.

We will define vector addition of two points $(x_1, y_1, 1)$ and $(x_2, y_3, 1)$ as $(x_1, y_1, 1) + (x_2, y_3, 1) = (x_1 + x_2, y_1 + y_2, 1)$. With this definition, vector addition carries the same geometric meaning as standard vector addition for points in the (x, y) plane. Also, if A is a matrix and v_1, v_2 are two vectors, then $A(v_1 + v_2) = Av_1 + Av_2$. This property of linearity will come in handy.

We have the following classification of "elementary" isometries.

- A rotation about the origin by an angle of ϕ is represented by

$$\begin{bmatrix} \cos(\phi) & -\sin(\phi) & 0 \\ \sin(\phi) & \cos(\phi) & 0 \\ 0 & 0 & 1 \end{bmatrix}$$

- A translation by a vector of $v = (v_1, v_2)$ is represented by

$$\begin{bmatrix} 1 & 0 & v_1 \\ 0 & 1 & v_2 \\ 0 & 0 & 1 \end{bmatrix}$$

- A reflection about the x-axis is represented by

$$\begin{bmatrix} 1 & 0 & 0 \\ 0 & -1 & 0 \\ 0 & 0 & 1 \end{bmatrix}$$

- A reflection about the y-axis is represented by

$$\begin{bmatrix} -1 & 0 & 0 \\ 0 & 1 & 0 \\ 0 & 0 & 1 \end{bmatrix}$$

Any other reflection, translation, rotation, or glide reflection can be built from these elementary isometries, as was shown in the exercises at the end of the last few sections. Thus, any isometry is equivalent to the product of 3×3 matrices of the form above.

But, how exactly do compositions of isometries form new isometries? We know that the composition of two translations is again a translation, that two reflections form either a translation or rotation, and that two rotations about the same center form another rotation about that center. What about the composition of two rotations with different centers? Or the composition of a rotation and translation?

5.7.2 Compositions of Rotations and Translations

In the next two theorems, we will assume that any angle mentioned has been normalized to lie between 0 and 360.

Theorem 5.20. *Let $R_{a,\alpha}$ be rotation about point a by an angle $\alpha \neq 0$. Let $R_{b,\beta}$ be rotation about point $b \neq a$ by $\beta \neq 0$. Let T_v be translation by the vector $v = (v_1, v_2)$. Then*

(i) $R_{a,\alpha} \circ T_v = T_{R_{a,\alpha}(v)} \circ R_{a,\alpha}$.

(ii) $R_{a,\alpha} \circ R_{b,\beta}$ is a translation iff $\alpha + \beta = 0 (\text{mod } 360)$.

(iii) $R_{a,\alpha} \circ T_v$ (or $T_v \circ R_{a,\alpha}$) is a rotation of angle α.

(iv) $R_{a,\alpha} \circ R_{b,\beta}$ is a rotation of angle $\alpha + \beta$ iff $\alpha + \beta \neq 0 (\text{mod } 360)$.

Proof: For the first statement of the theorem, we can set our coordinate system so that a is the origin. Using the matrix forms for $R_{a,\alpha}$ and T_v, we get

$$
R_{a,\alpha} \circ T_v = \begin{bmatrix} \cos(\alpha) & -\sin(\alpha) & 0 \\ \sin(\alpha) & \cos(\alpha) & 0 \\ 0 & 0 & 1 \end{bmatrix} \begin{bmatrix} 1 & 0 & v_1 \\ 0 & 1 & v_2 \\ 0 & 0 & 1 \end{bmatrix}
$$

$$
= \begin{bmatrix} \cos(\alpha) & -\sin(\alpha) & \cos(\alpha)v_1 - \sin(\alpha)v_2 \\ \sin(\alpha) & \cos(\alpha) & \sin(\alpha)v_1 + \cos(\alpha)v_2 \\ 0 & 0 & 1 \end{bmatrix}
$$

The x and y components of the third column of this product are precisely the x and y components of $R_{a,\alpha}(v)$. Let $R_{a,\alpha}(v) = (c, d)$. Then

$$
T_{R_{a,\alpha}(v)} \circ R_{a,\alpha} = \begin{bmatrix} 1 & 0 & c \\ 0 & 1 & d \\ 0 & 0 & 1 \end{bmatrix} \begin{bmatrix} \cos(\alpha) & -\sin(\alpha) & 0 \\ \sin(\alpha) & \cos(\alpha) & 0 \\ 0 & 0 & 1 \end{bmatrix}
$$

$$
= \begin{bmatrix} \cos(\alpha) & -\sin(\alpha) & c \\ \sin(\alpha) & \cos(\alpha) & d \\ 0 & 0 & 1 \end{bmatrix}
$$

This finishes the proof of the first part of the theorem.

For the second part of the theorem, suppose that $R_{a,\alpha} \circ R_{b,\beta}$ is a translation T_v. We can assume that b is the origin. Let R_θ represent rotation about the origin by θ. We know that $R_{a,\alpha} = T_a \circ R_\alpha \circ T_{-a}$. Thus, using statement (i), we get that

$$
\begin{aligned}
T_v &= R_{a,\alpha} \circ R_{b,\beta} \\
&= T_a \circ R_\alpha \circ T_{-a} \circ R_\beta \\
&= T_a \circ T_{R_\alpha(-a)} \circ R_\alpha \circ R_\beta \\
&= T_{a+R_\alpha(-a)} \circ R_{\alpha+\beta}
\end{aligned}
$$

Thus, $T_{-a-R_\alpha(-a)} \circ T_v$ is a rotation ($R_{\alpha+\beta}$) and must have a fixed point. But, the only translation with a fixed point is the identity, and thus $R_{\alpha+\beta}$ must be the identity, and $\alpha + \beta$ must be a multiple of 360 degrees.

Conversely, if $\alpha + \beta$ is a multiple of 360, then $R_{a,\alpha} \circ R_{b,\beta} = T_{a+R_\alpha(-a)}$.

For the third statement of the theorem, we can assume that a is the origin and thus $R_{a,\alpha} = R_\alpha$. Consider the fixed points P of $R_\alpha \circ T_v$. Using the matrix form of these isometries, we get that if P is a fixed point, then

$$\begin{bmatrix} \cos(\alpha) & -\sin(\alpha) & 0 \\ \sin(\alpha) & \cos(\alpha) & 0 \\ 0 & 0 & 1 \end{bmatrix} \begin{bmatrix} 1 & 0 & v_1 \\ 0 & 1 & v_2 \\ 0 & 0 & 1 \end{bmatrix} P = P = \begin{bmatrix} 1 & 0 & 0 \\ 0 & 1 & 0 \\ 0 & 0 & 1 \end{bmatrix} P$$

Set $R_\alpha(v) = (e, f)$. After multiplying out the left side of the previous equation, we get

$$\begin{bmatrix} \cos(\alpha) & -\sin(\alpha) & e \\ \sin(\alpha) & \cos(\alpha) & f \\ 0 & 0 & 1 \end{bmatrix} P = \begin{bmatrix} 1 & 0 & 0 \\ 0 & 1 & 0 \\ 0 & 0 & 1 \end{bmatrix} P$$

If we subtract the term on the right from both sides, we get

$$\begin{bmatrix} \cos(\alpha) - 1 & -\sin(\alpha) & e \\ \sin(\alpha) & \cos(\alpha) - 1 & f \\ 0 & 0 & 0 \end{bmatrix} P = O$$

where O is the origin $(0, 0, 0)$. This equation has a unique solution iff the determinant of the 2×2 matrix in the upper left corner is non-zero. (Remember that the third component is not significant.) This determinant is

$$(\cos(\alpha) - 1)^2 + \sin(\alpha)^2 = \cos(\alpha)^2 - 2\cos(\alpha) + 1 + \sin(\alpha)^2 = 2(1 - \cos(\alpha))$$

Since α is not a multiple of 360, then $\cos(\alpha) \neq 1$ and there is a unique fixed point. Thus, the composition is a rotation, say $R_{g,\gamma}$. Then $R_{-\alpha} \circ R_{g,\gamma}$ is a translation and by statement (ii) we have that $\gamma + (-\alpha) = 0 \pmod{360}$ and thus $\gamma = \alpha \pmod{360}$. A similar argument shows that $T_v \circ R_\alpha$ is a rotation of angle α.

For the fourth statement of the theorem, we know from the work above that $R_{a,\alpha} \circ R_\beta = T_{a + R_\alpha(-a)} \circ R_{\alpha+\beta}$. Since $\alpha + \beta \neq 0 \pmod{360}$, by statement (iii) of the theorem, we know that the composition on the right is a rotation by an angle of $\alpha + \beta$. $\square$

The only compositions left to consider are those involving reflections or glide reflections.

5.7.3 Compositions of Reflections and Glide Reflections

Theorem 5.21. *Let r_l be a reflection with line of symmetry l, $G_{m,v}$ a glide reflection along a line m with the vector v parallel to m, $R_{a,\alpha}$ a rotation with center a and angle $\alpha \neq 0$, and T_w translation along a non-zero vector w. Then*

(i) $r_l \circ R_{a,\alpha}$ (or $R_{a,\alpha} \circ r_l$) is a reflection iff l passes through a. If l does not pass through a, then the composition is a glide reflection.

(ii) $r_l \circ T_w$ (or $T_w \circ r_l$) is a reflection iff the vector w is perpendicular to l. If w is not perpendicular to l, then the composition is a glide reflection.

(iii) $G_{m,v} \circ R_{a,\alpha}$ (or $R_{a,\alpha} \circ G_{m,v}$) is a reflection iff m passes through the center of the rotation defined by $T_v \circ R_{a,\alpha}$ (or $R_{a,\alpha} \circ T_v$). If m does not pass through this center, then the composition is a glide reflection.

(iv) $G_{m,v} \circ T_w$ (or $T_w \circ G_{m,v}$) is a reflection iff the vector $v + w$ is perpendicular to m. If $v + w$ is not perpendicular to m, then the composition is a glide reflection.

(v) $r_l \circ G_{m,v}$ (or $G_{m,v} \circ r_l$) is a translation iff l is parallel to m. The composition is a rotation otherwise.

(vi) The composition of two different glide reflections is a translation iff the reflection lines for both are parallel. The composition is a rotation otherwise.

Proof: For the first two statements we note that the composition of a reflection and either a rotation or a translation will be equivalent to the composition of three reflections, and thus must be either a reflection or a glide reflection. Statements (i) and (ii) then follow immediately from Theorems 5.18 and 5.19 and their proofs.

For statement (iii) we note that $G_{m,v} = r_m \circ T_v$ and so $G_{m,v} \circ R_{a,\alpha} = r_m \circ T_v \circ R_{a,\alpha}$. The result follows from statement (iii) of the previous theorem and statement (i) of this theorem.

For statement (iv) we note that $G_{m,v} \circ T_w = r_m \circ T_v \circ T_w = r_m \circ T_{v+w}$. The result follows from statement (ii).

For statement (v) we note that $r_l \circ G_{m,v} = r_l \circ r_m \circ T_v$. If $r_l \circ G_{m,v}$ is a translation, say T_u, then $r_l \circ r_m = T_{u-v}$ and $l \parallel m$. Conversely, if $l \parallel m$, then $r_l \circ r_m$ is a translation and thus $r_l \circ G_{m,v}$ is a translation. If $l \not\parallel m$, then $r_l \circ r_m$ is a rotation and $r_l \circ G_{m,v}$ is also a rotation.

The last statement of the theorem follows immediately from looking at the structure of two glide reflections and is left as an exercise. □

We now have a complete characterization of how pairs of isometries act to form new isometries. In principle, any complex motion involving rotations, translations, and reflections can be broken down into a sequence of compositions of basic isometries. This fact is used to great effect by computer animators.

5.7.4   Isometries in Computer Graphics

Just as we can represent a point (x, y) as a point $(x, y, 1)$ on the plane at height 1 in three dimensions, we can represent a point in space (x, y, z) as a point $(x, y, z, 1)$ at "hyper" height 1 in four dimensions. Then, translations, rotations, and reflections in three dimensions can be represented as 4×4 matrices acting on these points.

For example, to translate a point (x, y, z) by the vector $v = (a, b, c)$, we use the translation matrix

$$\begin{bmatrix} 1 & 0 & 0 & a \\ 0 & 1 & 0 & b \\ 0 & 0 & 1 & c \\ 0 & 0 & 0 & 1 \end{bmatrix}$$

and multiply this matrix (on the left) by the vector

$$\begin{bmatrix} x \\ y \\ z \\ 1 \end{bmatrix}$$

To animate an object on the computer screen, we need to do two things: represent the object in the computer and then carry out transformations on this representation. We can represent an object as a collection of polygonal patches that are defined by sets of vertices. To move an object we need only move the vertices defining the object and then redraw the polygons making up the object.

To realize movement of an object on the screen, we need to repeatedly carry out a sequence of rotations, reflections, and translations on the points defining the object. All of these isometries can be implemented using 4×4 matrices as described above. Thus, a computer graphics system basically consists of point sets (objects) and sequences of 4×4 matrices (motions) that can be applied to point sets.

One of the most popular graphics systems in use today is the OpenGL system [33]. OpenGL uses the notion of a *graphics pipeline* to organize how motions are carried out. For example, to rotate and then translate an object using OpenGL, you would define the rotation and then the translation and put these two 4×4 matrices into a virtual pipeline. The graphics system then multiplies all the matrices in the pipeline together (in order) and applies the resulting matrix to any vertices that define the object. In a very real sense, computer graphics comes down to being able to quickly multiply 4×4 matrices, that is, to quickly compose three-dimensional transformations.

5.7.5 Summary of Isometry Compositions

We summarize the theorems on compositions of isometries in Table 5.3 for future reference. (The table lists only non-trivial compositions; that is, ones where $l \neq m, a \neq b, v \neq w$.)

TABLE 5.3 Isometry Composition

$\circ$	r_l	$R_{a,\alpha}$	T_v	$G_{l,v}$
r_m	T $(l\|m)$ R $(l\not\|m)$	r $(a\in m)$ G $(a\notin m)$	r $(v\perp m)$ G $(v\not\perp m)$	T $(l\|m)$ R $(l\not\|m)$
$R_{b,\beta}$	r $(b\in l)$ G $(b\notin l)$	$R_{c,\alpha+\beta}$ $(\alpha+\beta\neq0 \pmod{360})$ T $(\alpha+\beta=0 \pmod{360})$	$R_{c,\beta}$	r $(c_1\in l)$ G $(c_1\notin l)$
T_w	r $(w\perp l)$ G $(w\not\perp l)$	$R_{c,\alpha}$	T_{w+v}	r $((w+v)\perp l)$ G $((w+v)\not\perp l)$
$G_{m,w}$	T $(l\|m)$ R $(l\not\|m)$	r $(c_2\in m)$ G $(c_2\notin m)$	r $((w+v)\perp m)$ G $((w+v)\not\perp m)$	T $(l\|m)$ R $(l\not\|m)$

In the table, c stands for an arbitrary center of rotation, c_1 is the center of rotation for $R_{b,\beta}\circ T_v$, and c_2 is the center of rotation for $T_w\circ R_{a,\alpha}$.

Exercise 5.7.1. *Prove statement (vi) of Theorem 5.21. That is, show that the composition of two different glide reflections is a translation if the reflection lines for both are parallel and is a rotation otherwise.*

An important algebraic operation on invertible functions is the idea of the *conjugate* of a function by another function. Given two invertible functions f, g, we construct the conjugate of g by f as $f \circ g \circ f^{-1}$. In the next five exercises, we look at how conjugation acts on types of isometries.

Exercise 5.7.2. *Let f be an isometry and H_O a half-turn about point O (rotation by 180 degrees). Show that conjugation of H_O by f, that is $f \circ H_O \circ f^{-1}$, is equal to $H_{f(O)}$, a half-turn about $f(O)$.*

Exercise 5.7.3. *Let f be an isometry and r_m a reflection about line m. Show that $f \circ r_m \circ f^{-1} = r_{f(m)}$ (reflection about $f(m)$).*

Exercise 5.7.4. *Let f be an isometry and T_{AB} a translation with vector $\overrightarrow{AB}$. Show that $f \circ T_{AB} \circ f^{-1} = T_{f(A)f(B)}$ (translation with vector $\overrightarrow{f(A)f(B)}$).*

Exercise 5.7.5. *Let f be an isometry and $g = r_m \circ T_{AB}$ a glide reflection along m with translation vector $\overrightarrow{AB}$. Show that $f \circ g \circ f^{-1} = r_{f(m)} \circ T_{f(A)f(B)}$ (glide reflection along $f(m)$ with vector $\overrightarrow{f(A)f(B)}$).*

Exercise 5.7.6. *Let f be an isometry and $R_{A,\alpha}$ a rotation about point A by an angle of α. Show that $f \circ R_{A,\alpha} \circ f^{-1} = R_{f(A),\alpha}$ if f is a direct isometry (rotation or translation) and $f \circ R_{A,\alpha} \circ f^{-1} = R_{f(A),-\alpha}$ if f is an indirect isometry (reflection or glide reflection).*

Exercise 5.7.7. *Another model we can use for the Euclidean plane is the complex plane, where a point (x, y) is represented by a complex number $z = x + iy$ with $i = \sqrt{-1}$. Show that a rotation of a point (x, y) about the origin by an angle ϕ is equivalent to multiplication of $x + iy$ by $e^{i\phi}$. [Hint: Recall that $e^{i\phi} = \cos(\phi) + i\sin(\phi)$.] Show that translation of (x, y) by a vector $v = (v_1, v_2)$ is equivalent to adding $v = v_1 + iv_2$ to $x + iy$. Finally, show that reflection across the x-axis is the same as complex conjugation.*

Exercise 5.7.8. *Let $R_{a,\alpha}$ be rotation about a of angle α and R_β be rotation about the origin of β. Show that the center of $R_{a,\alpha} \circ R_\beta$ is the complex number $c = a\frac{1-e^{i\alpha}}{1-e^{i(\alpha+\beta)}}$.*

Exercise 5.7.9. *Let T_v be translation by v and R_β rotation about the origin by an angle of β. Show that the center of $T_v \circ R_\beta$ is the complex number $c = \frac{v}{1-e^{i\beta}}$.*

Exercise 5.7.10. *You are a computer game designer who is designing a two-dimensional space battle on the screen. You have set up your coordinate system so that the screen is a virtual world with visible coordinates running from -5 to 5 in the x direction and likewise in the y direction. Suppose that you want to have your ship start at the origin, move to the position $(2, 3)$, rotate 45 degrees there, and then move to the position $(-2, 3)$. Find the 3×3 matrices that will realize this motion, and then describe the order in which you would put these matrices into a graphics pipeline to carry out the movement.*

5.8 PROJECT 8 - CONSTRUCTING COMPOSITIONS

At first glance the title of this project may seem a bit strange. We have been talking about compositions of isometries, which are essentially compositions of functions. How can you *construct* the composition of two functions? Function composition would seem to be primarily an *algebraic* concept. But composition of *geometric* transformations should have some geometric interpretation as well. In this project we will explore the construction of the composition of rotations and discover some beautiful geometry along the way.

Notes on how to use particular dynamic geometry software packages for this project can be found at http://www.gac.edu/~hvidsten/geom-text.

Start up your geometry software and create a segment $\overline{AB}$. Then, create two rays $\overrightarrow{AC}$ and $\overrightarrow{BD}$ from A and B and construct the intersection point E of these two rays. Create a circle with center F and radius point G.

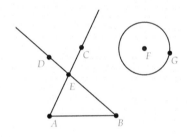

The two rotations we will compose are the rotation about A by $\angle EAB$, which we will denote by $R_{A,\angle EAB}$, and the rotation about B by $\angle ABE$, denoted by $R_{B,\angle ABE}$. Both angles will be *oriented* angles. That is, $\angle EAB$ will be directed clockwise, moving $\overrightarrow{AE}$ to $\overrightarrow{AB}$. $\angle ABE$ will also be directed clockwise.

Set A as a center of rotation and define a rotation of $\angle EAB$. Use this rotation to rotate the circle to a new circle, shown at right with center H.

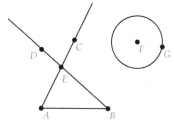

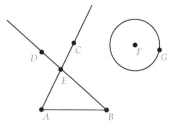

Now we will rotate the circle with center H by the rotation $R_{B,\angle ABE}$. As you did above, set $\angle ABE$ as a new angle of rotation and set B as the new center of rotation. Then, rotate the circle centered at H to get a new circle centered at J.

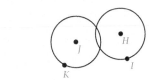

The circle at J is the result of applying $R_{B,\angle ABE} \circ R_{A,\angle EAB}$ to the original circle at F. To better analyze this composition, construct the angle bisector for $\angle BAE$ b Likewise, construct the angle bisector of $\angle EBA$. Label these bisectors "m" and "n."

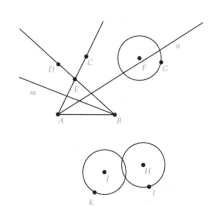

Exercise 5.8.1. *Show that rotation $R_{A,\angle EAB}$ is equivalent to $r_{\overleftrightarrow{AB}} \circ r_n$ and $R_{B,\angle ABE}$ is equivalent to $r_m \circ r_{\overleftrightarrow{AB}}$.*

Find the intersection point O of m and n and note that O appears to be equidistant from the original circle at F and the twice transformed circle at J. Move point D around, thus changing the angles of rotation, and observe how O always seems to have this property. To convince ourselves that this holds true, let's measure the distance from O to F and from O to J.

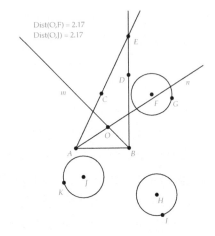

Exercise 5.8.2. *From what you have discovered so far, prove that $R_{B,\angle ABE} \circ R_{A,\angle EAB} = r_m \circ r_n$ and thus $R_{B,\angle ABE} \circ R_{A,\angle EAB}$ is a rotation of some angle γ, with center O.*

Exercise 5.8.3. *What theorem in section 5.7 can be used to show that $\gamma = (\angle EAB + \angle ABE)(\mathrm{mod}\ 360)$? (as long as $(\angle EAB + \angle ABE) \neq 0(\mathrm{mod}\ 360)$.)*

We have shown that the composition of two rotations about different centers is again a rotation, with new rotation angle the sum of the component angles of rotation (mod 360). This was also shown in the last section, but that proof was *algebraic* in nature. Here we have shown that the composition can be completely described by a geometric *construction*. Transformational geometry has this interesting two-sided nature in that one can almost always explain a result either by algebraic manipulation or by geometric construction. This dual nature is what gives transformational geometry its great utility and power.

For what rotations will the preceding construction be valid? From Exercise 5.8.3 the construction gives a rotation if $(\angle EAB + \angle ABE) \neq 0(\mathrm{mod}\ 360)$.

What happens if $(\angle EAB + \angle ABE) = 0(\mathrm{mod}\ 360)$?

Clear the screen and create $\overline{AB}$ and $\overrightarrow{AC}$ at some angle as shown. Create a circle centered at some point D.

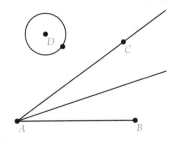

Define a rotation through $\angle CAB$, with center at A, and rotate the circle, yielding a new circle at F.

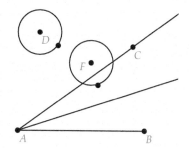

Now, we will rotate the circle at F through an angle β such that $\angle CAB + \beta$ is a multiple of 360. The simplest way to do this is merely to reverse the angle just defined. Define a rotation through $\angle BAC$ with center at B. (Note that the two angles of rotation together must be 360.) Rotate the circle at F to get a new circle at H.

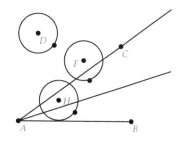

Move the original circle at D about the screen and note that the twice transformed circle at H is always the same distance and direction from the original circle. That is, the circle at H seems to be a *translation* of the circle at D.

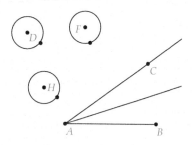

We proved in the last section that this composition must be a translation if the angles of rotation added up to 360. But can we give a more geometric argument for this?

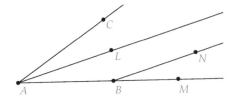

Undo your construction back to the original angle and bisector. Place a point L on the bisector to $\angle BAC$ and extend $\overline{AB}$ to M. Then, rotate point M about B by an angle equal to $\angle BAL$, yielding point N. Construct ray $\overrightarrow{BN}$.

Exercise 5.8.4. *Show that the original rotation about A can be written in terms of the reflections $r_{\overleftrightarrow{AL}}, r_{\overleftrightarrow{AB}}$ and that the rotation about B can be written in terms of $r_{\overleftrightarrow{BM}}, r_{\overleftrightarrow{BN}}$. Use this to prove geometrically that the composition of the two rotations about A and B is a translation by twice the vector between $\overleftrightarrow{AL}$ and $\overleftrightarrow{BN}$.*

For the project report, provide detailed analysis of the constructions used in this project and complete answers to the exercises.

CHAPTER 6

Symmetry

Tyger, Tyger, burning bright,
In the forests of the night;
What immortal hand or eye,
Could frame thy fearful symmetry?

– "The Tyger" by William Blake

In Chapter 1 we saw how our axiomatic and abstract understanding of geometry had its origins with the Greeks' desire for perfection of reasoning. This goal of the "ideal" permeated all aspects of Greek culture, as evidenced by their love of the golden ratio—the perfect harmony of proportion. Western cultures have embraced this love of harmony and balance, as evidenced by a focus on *symmetry* in art, music, architecture, and design.

In the preceding chapter, isometries were used to define several types of *symmetry*. Symmetry is a common property of both natural and man-made objects.

In general, given a geometric figure F, we will call a transformation f a *symmetry* of F if f maps F back to itself. That is, $f(F) = F$. We say that F is *invariant* (or unchanged) under f. Note that this does not mean that every point of F remains unchanged, only that the total *set* of points making up F is unchanged.

Given a figure such as a flower, or snail shell, or triangle, the set of all symmetries of that figure is not just a random collection of functions. The set of symmetries of an object has a very nice *algebraic* structure.

Theorem 6.1. *The set of symmetries of a figure F forms a* group.

Proof: Recall that a group is a set of elements satisfying four properties (refer to the discussion preceding Exercise 1.4.6). For a set of functions, these properties would be:

1. Given any two functions in the set, the composition of the two functions is again in the set.

2. The composition of functions is an associative operation.

3. The identity function is a member of the set.

4. Given any function in the set, its inverse exists and is an element of the set.

To prove that the set S of symmetries of F forms a group, we need to verify that S has all four of these properties. Since the composition of isometries is associative, the second condition is automatically true.

The third condition is true since the identity is clearly a symmetry of any figure.

Let f, g be two symmetries of F. Since $f(F) = F$ and $g(F) = F$, then $g(f(F)) = g(F) = F$ and $g \circ f$ is a symmetry of F, and the first condition holds.

Since a symmetry f is a transformation, then it must have an inverse f^{-1}. Since $f(F) = F$, then $F = f^{-1}(F)$ and the inverse is a symmetry. □

Why is it important that the symmetries of a figure form a group? Groups are a fundamental concept in abstract algebra and would seem to have little relation to geometry. In fact, there is a very deep connection between algebra and geometry. As M. A. Armstrong states in the preface to *Groups and Symmetry*, "groups measure symmetry" [3]. Groups reveal to us the algebraic structure of the symmetries of an object, whether those symmetries are the geometric transformations of a pentagon, or the permutations of the letters in a word, or the configurations of a molecule. By studying this algebraic structure, we can gain deeper insight into the geometry of the figures under consideration.

To completely delve into this beautiful connection between algebra and geometry would take us far afield of this brief survey of geometry. We will, however, try to classify some special types of symmetries that often appear in art and nature, symmetries which are also isometries. In the following sections, we will assume that the symmetries under study are Euclidean isometries—reflections, translations, rotations, and glide reflections.

6.1 FINITE PLANE SYMMETRY GROUPS

We will first look at those symmetry groups of an object that are *finite*, that is, those groups of symmetries that have a finite number of elements. What can be said about symmetries in a finite group? Suppose that f is a symmetry in a finite symmetry group and consider the repeated compositions of f with itself, $f, f^2 = f \circ f, f^3, \ldots$ This set cannot contain all different elements, as then the group would be infinite. Thus, for some $i \neq j$, we have $f^i = f^j$ and $f^i \circ f^{-j} = id$, where $f^{-j} = (f^{-1})^j$ and id is the identity function. We can assume that $i > j$.

> **Definition 6.1.** *We say that a symmetry f has* finite order *in a symmetry group G if for some positive integer n, we have $f^n = id$.*

We then have the following:

> **Lemma 6.2.** *All of the symmetries in a finite symmetry group are of finite order.*

What are the symmetries of finite order? Consider a translation given by $T(x, y) = (x + v_1, y + v_2)$, with $(v_1, v_2) \neq (0, 0)$. Clearly, $T^n(x, y) = (x + nv_1, y + nv_2)$, and thus $T^n(x, y) = (x, y)$ iff $(v_1, v_2) = (0, 0)$. So, the only translations of finite order are the identity translations. Similarly, we can show that a non-trivial glide reflection must have infinite order.

Thus, the symmetries of finite order consist solely of reflections and rotations. But, which reflections and rotations?

Before we answer that question, let's consider the symmetry group of a simple geometric figure, an equilateral triangle. Which rotations and reflections will preserve the triangle?

A rotation R by 120 degrees clockwise about the centroid of the triangle will map the triangle back
to itself, permuting the labels on
the vertices. R^2 will be a rotation of 240 degrees and will also
leave the triangle invariant. The
rotation R^3 will yield the identity isometry. The effects of applying R, R^2, and R^3 are shown at
right, with $R(\triangle ABC)$ the right-
most triangle and $R^2(\triangle ABC)$ the
left-most triangle.

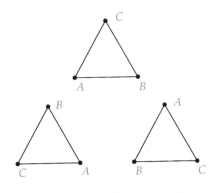

What about reflections? If we
construct the perpendicular bisector for a side of the triangle, then
that bisector will pass through the
opposite vertex. Reflection across
the bisector will just interchange
the other two vertices and thus
will preserve the triangle. There
are three such reflections.

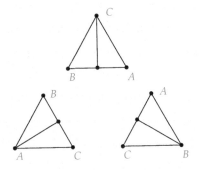

Are there any other possible isometries? To answer this question,
let's consider the effect of an isometry on the labels A, B, C of the original triangle. An isometry must preserve the configuration of the triangle,
so it must move the labels of the vertices around, leaving the shape of
the triangle unchanged. How many labellings of the triangle are there?
Pick one of the vertices to label. We have three possible labels for that
vertex. After labeling this vertex, pick another to label. There are two
possible labels for that vertex, leaving just one choice for the last vertex.
Thus, there is a maximum of six labellings for the triangle, and therefore a maximum of six isometries. Since we have exhibited six different
isometries (three rotations and three reflections), these must form the
complete symmetry group for the triangle.

Let's note several interesting things about this symmetry group.
First, the rotations by themselves form a group, $G = \{id = R^3, R_{120} = R, R_{240} = R^2\}$, and all the rotations in this group have the same center.
Second, the reflections all have lines of symmetry passing through the

center of the rotations, and there are exactly as many reflections as rotations. We will now show that these properties are shared by *all* finite symmetry groups.

> **Theorem 6.3.** *Let R_1, R_2 be two rotations in a finite symmetry group G. Then R_1, R_2 must have the same center of rotation.*

Proof: Let a, b be the centers of rotation for R_1, R_2. Suppose that $a \neq b$. Then we know that $R_1 \circ R_2$ is either a rotation or translation (see Table 5.3 in Chapter 5). Clearly, it cannot be a non-identity translation, as a finite symmetry group has no non-trivial translations. If $R_1 \circ R_2 = id$, then $R_1 = R_2^{-1}$ and both have the same center.

Let α, β be the angles of rotation for R_1, R_2. Then we know that

$$R_1 \circ R_2 = R_{a,\alpha} \circ R_{b,\beta} = R_{c,\alpha+\beta}$$

for some point $c \neq a$ or b. Now $R_2^{-1} = R_{b,-\beta}$ is also in G and so $R_{b,-\beta} \circ R_{c,\alpha+\beta} = R_{d,\alpha}$ is in G for some point d.

We claim that $d \neq a$. For if $d = a$, then

$$R_{b,-\beta} \circ R_{c,\alpha+\beta} = R_{a,\alpha}$$

Or, equivalently,

$$R_{b,-\beta} \circ R_{a,\alpha} \circ R_{b,\beta} = R_{a,\alpha}$$

Then

$$R_{a,\alpha} \circ R_{b,\beta} = R_{b,\beta} \circ R_{a,\alpha}$$

Applying both sides to the point b, we get

$$R_{a,\alpha}(b) = R_{b,\beta} \circ R_{a,\alpha}(b)$$

Thus, $R_{a,\alpha}(b)$ would be a fixed point of $R_{b,\beta}$. But, the only fixed point of a rotation is its center, and so $R_{a,\alpha}(b) = b$. But, this implies that $a = b$, which is a contradiction.

Since $d \neq a$, then $R_{d,\alpha} \circ R_{a,-\alpha}$ is in G, but this composition is a translation.

Thus, it must be that both rotations have the same center. □

> **Theorem 6.4.** *Among all rotations in a finite symmetry group G, let $R_{a,\alpha}$ be the one with the smallest positive angle α. Then α divides 360, and if $R_{a,\beta}$ is another rotation in G, then $\beta = n\alpha$ for some integer n. That is, all rotations are multiples of some minimal angle.*

Proof: Since $R_{a,\alpha}$ is of finite order, then $R_{a,\alpha}^n = R_{a,n\alpha} = id$ for some n. Thus, $n\alpha = 360k$ for some integer k. If $n\alpha \neq 360$, then let i be the largest integer such that $i\alpha < 360$. Then $0 < 360 - i\alpha < \alpha$. But, then $R_{a,360} \circ R_{a,-i\alpha} = R_{a,360-i\alpha}$ must be in G, which would mean that there was a smaller rotation angle than α in G. Thus, $n\alpha = 360$.

Now suppose $R_{a,\beta}$ is another rotation in G, different from $R_{a,\alpha}$. We know that $\alpha < \beta$. Let j be the largest integer such that $j\alpha \leq \beta$. Suppose $j\alpha \neq \beta$. Then, $0 < \beta - j\alpha < \alpha$. Therefore, $\beta - j\alpha$ will be a rotation angle in G smaller than α and again we get a contradiction. □

We conclude that all rotations in a finite symmetry group are generated from a particular rotation having the smallest angle. Groups generated from a single element are called *cyclic groups*.

> **Definition 6.2.** *A group G is called* cyclic *if all of its members can be written as compositions of a single member with itself. That is, $G = \{id, f, f^2, f^3, \ldots\}$.*

> **Corollary 6.5.** *The set of rotations in a finite symmetry group form a cyclic group.*

Now, let's consider how rotations and reflections interact in a finite symmetry group.

> **Theorem 6.6.** *If a finite symmetry group G contains a rotation and a reflection, then the line of symmetry for the reflection must pass through the center of the rotation.*

Proof: Let r_l and $R_{a,\alpha}$ be the reflection and rotation. We know that the composition of these is either a reflection or glide reflection. Since G cannot contain glide reflections, then $r_l \circ R_{a,\alpha}$ is a reflection, say r_m. Then $r_l \circ r_m = R_{a,\alpha}$ and l, m must pass through a. □

Theorem 6.7. *Let G be a finite symmetry group with n rotations (counting the identity as a rotation). If G has at least one reflection, say r_l, then it has exactly n reflections, which can be represented as $r_l \circ R_{a,i\alpha}$, $i = 0, \ldots, n-1$, where $R_{a,\alpha}$ generates the rotations of G.*

Proof: Consider the set $r_l \circ R_{a,i\alpha}$, with $i = 0, \ldots, n-1$. Clearly, all of the elements of this set are reflections. Also, no two elements of this set can be equivalent. Finally, all reflections in G are represented in this set, since if r_m is in G, then $r_l \circ r_m$ is a rotation and so must be some $R_{a,i\alpha}$. It follows that $r_m = r_l \circ R_{a,i\alpha}$. $\square$

Definition 6.3. *A finite symmetry group generated by a rotation and a reflection is called a* dihedral group. *If the group has n distinct rotations it will be denoted by D_n.*

A finite symmetry group is thus either a cyclic group or a dihedral group. This result has been known historically as *Leonardo's Theorem* in honor of Leonardo Da Vinci (1452–1519). According to Hermann Weyl in his book *Symmetry* [41, page 66], Leonardo was perhaps the first person to systematically study the symmetries of a figure in his architectural design of central buildings with symmetric attachments.

Exercise 6.1.1. *Find three examples in nature that have different finite symmetry groups. Sketch these and give the specific elements in their symmetry groups.*

Exercise 6.1.2. *Find the symmetry group for a square.*

Exercise 6.1.3. *Find the symmetry group for a regular pentagon.*

Exercise 6.1.4. *Show that the symmetry group for a regular n-gon must be finite.*

Exercise 6.1.5. *Show that the symmetry group for a regular n-gon must be the dihedral group D_n.*

Exercise 6.1.6. *Show that the dihedral group D_n can be generated by two reflections, that is, any element of the group can be expressed as a product of terms involving only these two reflections.*

Exercise 6.1.7. *Show that the number of symmetries of a regular n-gon is equal to the product of the number of symmetries fixing a side of the n-gon times the number of sides to which that particular side can be switched.*

Exercise 6.1.8. *Find a formula for the number of symmetries of a regular polyhedron by generalizing the result of the last exercise. Use this to find the number of symmetries for a regular tetrahedron (four faces and four vertices) and a cube.*

6.2 FRIEZE GROUPS

Planar symmetry groups that are infinite must necessarily contain translations and/or glide reflections. In this section we will consider symmetry groups having translations in just one direction.

> **Definition 6.4.** *A frieze group G is a planar symmetry group with all translations in the same direction. Also, there exists a translation T_v in G such that v is of minimal (non-zero) length among all translations of G.*

It turns out that all translations in a frieze group G will be generated from a single translation $T \in G$, which we will call the *fundamental translation* for G.

> **Lemma 6.8.** *Let G be a frieze group. Then there exists a translation T in G such that if T' is any (non-identity) translation in G, then $T' = T^n$ for some non-zero integer n.*

Proof: The definition of frieze groups guarantees that there is a translation T in G, with translation vector v of minimal length. We claim that T is the fundamental translation of G. For if T' is any other (non-identity) translation in G, with translation vector v', then $v' \parallel v$ and so $v' = nv$ for some (non-zero) real number n.

If n is not an integer, let k be the nearest integer to n. Then $T^{-k} = (T^{-1})^k$ must be in G and since $T^{-k} \circ T' = T^{-k} \circ T^n = T^{n-k}$, then T^{n-k} is in G. But T^{n-k} has translation vector of length $| (n - k) | \, \|v\|$, which is less than the length of v, as $| (n - k) |< 1$.

Thus, n is an integer and all translations are integer powers of T. $\square$

A frieze *pattern* is a pattern that is invariant under a frieze group. Such a pattern is generally composed of repetitions of a single pattern, or *motif*, in a horizontal direction as in Figure 6.1.

Figure 6.1

Frieze groups have translations generated by a single translation T. By Exercise 5.3.8, frieze groups have invariant lines that are all pair-wise parallel, and also each is parallel to the translation vector of T. What else can be said of the invariant lines of a frieze group?

Theorem 6.9. *Let G be a frieze group with fundamental translation T. If l is invariant under T, then $S(l)$ is parallel (or equal) to l for all symmetries S in G.*

Proof: Consider the conjugation of T by S: $S \circ T \circ S^{-1}$. By Exercise 5.7.4, we know that $S \circ T \circ S^{-1}$ is a translation and so $S \circ T \circ S^{-1} = T^k$ for some positive integer k. In particular, $S \circ T = T^k \circ S$.

Let v be the translation vector for T. Then since $S \circ T = T^k \circ S$, the direction of the line through $S(P)$ and $S(T(P))$ will be the same as that of the line through $S(P)$ and $T^k(S(P))$. This direction is given by the vector kv. Also, the direction for the line through P and $T(P)$ is given by the vector v. Thus, for any point P, we have either $S(\overleftrightarrow{P\,T(P)}) \parallel \overleftrightarrow{P\,T(P)}$, or these two lines are coincident.

For a line l invariant under T, let P be a point on l. Using the results of the last paragraph, we see that $S(l) \parallel l$ or $S(l) = l$. $\square$

The next three results specify the types of non-translational symmetries that are possible in a frieze group.

Corollary 6.10. *Let G be a frieze group with fundamental translation T_v. The only reflections possible in G are reflections along lines that are either parallel to v or perpendicular to v. Additionally, G can have at most one reflection r_n along a line parallel to v. The line of reflection for r_n is thus the unique invariant line of T_v and r_n.*

Proof: We know by Exercise 5.3.8 that the invariant lines for T_v are parallel to the displacement vector v of T_v. Let l be one such invariant line and let r be a reflection in G. Then, by Theorem 6.9, $r(l)$ is parallel to l or $r(l) = l$.

If $r(l) = l$, then l is an invariant line for a reflection and $r = r_l$ or $r = r_u$, where u is perpendicular to l (Exercise 5.2.12).

If $r(l)||l$ then $r = r_n$ for some n parallel to l (thus parallel to v). To see this, let A be on l. Then, the reflection line for r is the perpendicular bisector of $\overline{Ar(A)}$, which must be perpendicular to l.

Suppose there were two reflections r_n and r'_n with both n and n' parallel to v. Then, by Corollary 5.11, $r_n \circ r'_n = T_w$, with $w \perp v$. This is impossible, as all translations in G are a multiple of v. □

> **Corollary 6.11.** *Let G be a frieze group with fundamental translation T_v. The only rotations possible in G are half-turns about points on a single invariant line of T_v. This line is thus the unique invariant line of T_v and all half-turns in G.*

Proof: As with the previous proof, let l be an invariant line of T_v and let R be a rotation in G. By Theorem 6.9, $R(l)$ is parallel to l or $R(l) = l$. In either case, we see from Exercises 5.4.5 and 5.4.6 in section 5.4 that R must be a rotation of 180 degrees.

Suppose there were two half-turns R_A and R_B about points $A \neq B$. By Exercise 5.4.13 we know that the composition $R_a \cdot R_b = T_w$, a translation with vector w from A to B. This translation vector must be in the direction of v. Since this is true for all possible rotations R_B, then the centers of all half-turns lie on a line in the direction of v. Such a line is invariant under T_v. □

> **Corollary 6.12.** *Let G be a frieze group with fundamental translation T_v. The only glide reflections possible in G are glides along a single invariant line of T_v. This line is thus the unique invariant line of T_v and all glides in G.*

Proof Let l be an invariant line of T_v and let g be a glide reflection in G. By Theorem 6.9, $g(l)$ is parallel to l or $g(l) = l$. We know that $g = r \cdot T'$ for some reflection r and translation T'. Since g^2 is a translation, then $g = r \cdot T_{\frac{kv}{2}}$. Now, $g(l) = r \cdot T_{\frac{kv}{2}}(l) = r(l)$. So, $r(l) = l$ or $r(l)||l$. As in the proof of Corollary 6.10 we conclude that $r = r_n$ for $n||l$ or $r = r_u$ for u perpendicular to l.

If $r = r_u$ with u perpendicular to l (and thus to v), then $g = r_u \cdot T_{\frac{kv}{2}}$

is a reflection by Theorem 5.21. This contradicts the fact that g is a glide reflection. So, $g = r_n \cdot T_{\frac{kv}{2}}$ where n is parallel to l.

Suppose there were two glides g_1 and g_2 about two different lines $n_1 \neq n_2$. By the previous paragraph, both lines must be parallel to l. Then $g_1 \cdot g_2 = r_{n_1} \cdot T_{\frac{k_1 v}{2}} \cdot r_{n_2} \cdot T_{\frac{k_2 v}{2}}$. We can switch the order of reflection and translation to get $g_1 \cdot g_2 = T_{\frac{k_1 v}{2}} \cdot r_{n_1} \cdot r_{n_2} \cdot T_{\frac{k_2 v}{2}}$. Since the composition of two reflections along parallel lines is a translation perpendicular to the lines, we have $g_1 \cdot g_2 = T_{\frac{k_1 v}{2}} \cdot T_w \cdot T_{\frac{k_2 v}{2}}$, where w is perpendicular to v. Thus, there is a translation in G with vector $cv + w$. This is impossible, as this vector is not in the direction of v. $\square$

The next result tells us that the "special" invariant lines of the preceding three corollaries are actually a single, *unique* invariant line.

Theorem 6.13. *Let G be a frieze group with fundamental translation T_v. The unique invariant lines described in the preceding corollaries are all coincident, i.e., they are the same line.*

Proof: Suppose that G had a half-turn H_A about point A on invariant line n_H, and suppose G also had a reflection r_n with n parallel to v. Then, by Exercise 5.7.2 we have $r_n \cdot H_A \cdot r_n = H_{r_n(A)}$. Then, $r_n(A)$ must lie on n_H. Now, n and n_H are either parallel or the same. If they are parallel, $r_n(A)$ cannot lie on n_H. Thus, $n = n_H$.

Suppose that G had a half-turn H_A about point A on invariant line n_H, and suppose G also had a glide g along a line n parallel to v. Then, by Exercise 5.7.2 we have $g \cdot H_A \cdot g^{-1} = H_{g(A)}$. By similar reasoning as above, we have that $n = n_H$.

Suppose that G had a reflection r_n with n parallel to v, and suppose G also had a glide $g = r_{n'} \cdot T_{kv}$ along a line n' parallel to v. Then, $r_n \cdot g = r_n \cdot r_{n'} \cdot T_{kv}$. If $n \neq n'$ then, $n \| n'$ and $r_n \cdot r_{n'} = T_w$. Then, $T_w \cdot T_{kv}$ is in G with w perpendicular to v. This is impossible. So, $n = n'$. $\square$

We thus conclude that if a frieze group G has either a half-turn, or a glide, or a reflection along a line parallel to the fundamental translation vector, then the group has a unique line that is fixed by all symmetries of the group.

> **Definition 6.5.** *If the frieze group G has exactly one line invariant under all the elements of G, we call that line the* midline *for G (or any pattern invariant under G).*

We summarize the work above as follows:

> **Theorem 6.14.** *For a frieze group G, the only symmetries possible are those generated by the fundamental translation T (with vector v), a unique reflection about the midline, reflections perpendicular to v, half-turns (180-degree rotations) about points on the midline, and glide reflections along the midline.*

The non-translational isometries in a frieze group can all be generated using the fundamental translation, as follows.

> **Theorem 6.15.** *Let G be a frieze group. Then*
>
> (i) *All half-turns in G are generated by a single half-turn in G and powers (repeated compositions) of T.*
>
> (ii) *All perpendicular reflections are generated from a single perpendicular reflection in G and powers of T.*
>
> (iii) *All glide reflections are generated by a single reflection (or glide reflection) in G that is parallel to the direction of translation, together with powers of T. Also, if v is the translation vector for T, then a glide reflection must have a glide vector equal to kv or half of kv for some positive integer k.*

Proof: For statement (i), let H_A, H_B be two half-turns in G. Then by Exercise 5.4.13, we know that $H_B \circ H_A$ is a translation, thus $H_B \circ H_A = T^k$ for some positive integer k. Since $H_A^{-1} = H_A$, then $H_B = T^k \circ H_A$.

The proofs of statements (ii) and (iii) are left as exercises. □

Thus, all frieze groups (of non-trivial translational symmetry) are generated by a translation $\tau = T_v$ and one or more of the following isometries: r_m (reflection across the midline m), r_u (with u perpendicular to m), H (half-turn rotation about O on m), and γ (glide reflection along

m with glide vector equal to $\frac{v}{2}$). We omit the case of glides with glide vector v as these can be generated by τ and r_m.

In principle, this would give us a total of 2^4 possible frieze groups, each frieze group generated by τ and a subset of the four isometries r_m, r_u, H, and γ. However, many of these combinations will generate the same group. For example, we can choose u such that $H \circ \gamma = r_u$ and $r_u \circ \gamma = H$, and so the group generated by τ, γ, H must be the same as the group generated by τ, γ, r_u.

It turns out that there are only seven different frieze groups. We will list them by their generators. For example, the group listed as $< \tau, r_m >$ is the group generated by all possible compositions of these two isometries (compositions such as τ^3, r_m^5, $r_m^2 \circ \tau \circ r_m^4$, etc.).

1. $< \tau >$

2. $< \tau, r_m >$

3. $< \tau, r_u >$

4. $< \tau, \gamma >=< \gamma >$

5. $< \tau, H >$

6. $< \tau, r_m, H >$

7. $< \tau, \gamma, H >=< \gamma, H >$

(For a complete proof of this result see [31, page 392] or [38, page 190].)

Exercise 6.2.1. *Show that the groups generated by τ, γ, H and γ, H are the same. [Hint: To show two sets equivalent, show that each can be a subset of the other.]*

Exercise 6.2.2. *Show that the groups generated by τ, r_m, H and τ, r_u, H are the same, assuming we choose r_u so that it intersects m at the center of H.*

Exercise 6.2.3. *Prove statement (ii) of Theorem 6.15. [Hint: Consider the composition of reflections.]*

Exercise 6.2.4. *Prove statement (iii) of Theorem 6.15.*

Exercise 6.2.5. *Show that if a frieze group has glide reflections, then the group must have a glide reflection g with glide vector of v or $\frac{v}{2}$, where v is the translation vector for the group.*

Exercise 6.2.6. *Show that if H_A and H_B are two half-turns of a frieze group G, then $AB = kv$ or $AB = kv + \frac{v}{2}$, where v is the translation vector for T and k is a positive integer. [Hint: Consider the action of $H_B \circ H_A$ on A.]*

Exercise 6.2.7. *Show that if r_u and r_v are two reflections of a frieze group G that are perpendicular to the translation vector v of T, then $AB = kv$ or $AB = kv + \frac{v}{2}$, where A, B are the intersection points of u, v, with an invariant line of G and k is a positive integer. [Hint: Consider the action of $r_v \circ r_u$ on A.]*

Definition 6.6. *A subgroup K of a group G is a non-empty subset of elements of G that is itself a group.*

Exercise 6.2.8. *Show that all elements of $< T, r_u >$ are either translations or reflections perpendicular to m. Use this to show that none of $< T, r_m >$, $< T, H >$, or $< T, \gamma >$ are subgroups of $< T, r_u >$. [Hint: Use Table 5.3.]*

Exercise 6.2.9. *Show that all elements of $< T, H >$ are either translations or half-turns. Use this to show that none of $< T, r_m >$, $< T, r_u >$, or $< T, \gamma >$ are subgroups of $< T, H >$.*

Exercise 6.2.10. *Show that all elements of $< T, \gamma >$ are either translations or glide reflections (with a glide vector of $kv + \frac{v}{2}$). Use this to show that none of $< T, r_m >$, $< T, r_u >$, or $< T, H >$ are subgroups of $< T, \gamma >$.*

Exercise 6.2.11. *Show that all elements of $< T, r_m >$ are either translations, or r_m, or glide reflections (with a glide vector of kv). Use this to show that none of $< T, \gamma >$, $< T, r_u >$, or $< T, H >$ are subgroups of $< T, r_m >$.*

Exercise 6.2.12. *Draw a diagram showing which of the seven frieze groups are subgroups of the others. [Hint: Use the preceding exercises.]*

Exercise 6.2.13. *In the figure below, there are seven frieze patterns, one for each of the seven frieze groups. Match each pattern to the frieze group that is its symmetry group.*

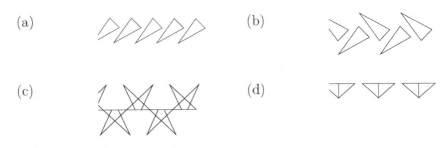

(a) (b)

(c) (d)

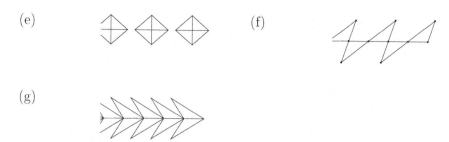

(e)   (f)

(g)

Exercise 6.2.14. *Design two frieze patterns having different frieze groups and list the group for each of your patterns.*

6.3   WALLPAPER GROUPS

A second type of infinite planar symmetry group is a generalization of frieze groups into two dimensions.

Definition 6.7. *A* wallpaper group *is a planar symmetry group containing translations in two different directions, that is, translations with non-parallel translation vectors. Also, there exists a translation T_v in G such that v is of minimal (non-zero) length among all translations of G.*

The condition on minimal length translations will ensure that all rotations in G are of finite order.

Theorem 6.16. *Let G be a group having a translation T_v such that v has least length among all translation vectors in G. Then all rotations of G must have finite order.*

Proof: Suppose G contains a rotation $R_{a,\alpha}$ of infinite order. Then for all integers $k > 0$, we have that $R_{a,\alpha}^k \neq id$, and thus $k\alpha \pmod{360} \neq 0$.

Consider the set of numbers $S = \{k\alpha \pmod{360}\}$. Since S is an infinite set, then α cannot be rational, and if we split the interval from $[0, 360]$ into n equal sub-intervals of length $\delta = \frac{360}{n}$, then one of these sub-intervals must have an infinite number of elements of S. Suppose the interval $[i\delta, (i+1)\delta]$ has an infinite number of elements of S. In particular, the interval has two elements $k_1\alpha \pmod{360}$ and $k_2\alpha \pmod{360}$, with neither value equal to an endpoint of the interval, and $k_1\alpha \pmod{360} < k_2\alpha \pmod{360}$. Then $(k_2 - k_1)\alpha \pmod{360}$ will be an element of S in the interval $[0, \delta]$.

Thus, we can assume that for any δ close to 0, we can find a k such that $k\alpha \pmod{360} < \delta$. So if a symmetry group G contains a rotation $R_{a,\alpha}$ of infinite order, then it must have rotations of arbitrarily small angle $R_{a,\delta}$.

We know that T_v has translation vector v of smallest length among all possible translation vectors in G. Consider $R_{a,\delta} \circ T_v \circ R_{a,-\delta}$. This must be an element of G. It is left as an exercise to show that $R_{a,\delta} \circ T_v \circ R_{a,-\delta}$ is the translation $T_{R_{a,\delta}(v)}$. Then the translation $T_{R_{a,\delta}(v)} \circ T_{-v} = T_{R_{a,\delta}(v)-v}$ must be in G. However, for δ arbitrarily small, the length of the vector $R_{a,\delta}(v) - v$, which is the translation vector for $T_{R_{a,\delta}(v)-v}$, will be arbitrarily small, which contradicts the hypothesis of the theorem. □

Definition 6.8. *A discrete symmetry group is one that has translations of minimal (non-zero) length.*

Thus, frieze groups and wallpaper groups are by definition discrete groups of planar symmetries.

A wallpaper *pattern* is a pattern that is invariant under a wallpaper group. Such a pattern is generally composed of repetitions of a single pattern, or *motif*, in two different directions, as shown at right.

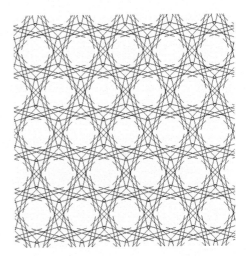

Wallpaper groups are certainly not trivial to classify. However, we can make a few important observations as to their structure.

First of all, every wallpaper group generates a *lattice* in the plane.

Definition 6.9. *A lattice L spanned by two vectors v, w in the plane is the set of all integer combinations of v and w. That is, $L = \{sv + tw \,|\, s, t \in Z\}$, where Z is the set of integers.*

Theorem 6.17. *Let $\mathcal{T}$ be the set of all translations of a wallpaper group G. Then, the set of all translations of the origin by elements of $\mathcal{T}$ forms a lattice that is spanned by v and w, where v is a vector of minimal length in $\mathcal{T}$ and w is a vector of minimal length in a different direction from v.*

Proof: First, we note that all translations in $\mathcal{T}$ must be generated by integer combinations of v and w. For suppose T_z was a translation with z not an integer combination of v and w. The two vectors v, w define a parallelogram in the plane and the set of all integer combinations of v, w will divide up the plane into congruent parallelograms.

Let $P = T_z(O) = T_z$. Then P must lie in one of these parallelograms. Let Q be the corner of the parallelogram containing P that is closest to P. Then the vector $P - Q$ cannot be congruent to any of the edges of the parallelogram, since if it were, then $P - Q = T_z(O) - T_{s_1 v + t_1 w}(0) = T_{s_2 v + t_2 w}(0)$, and z would be an integer combination of v and w.

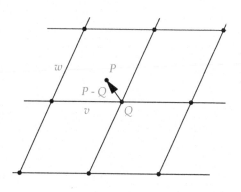

Also, the length of $P - Q$ must be less than the length of w.

To prove this we connect the midpoints of the sides of the parallelogram, yielding four congruent sub-parallelograms. The length of $P - Q$ must be less than the maximum distance between points in one of these sub-parallelograms, which occurs between opposite vertices. (The proof is left as an exercise.)

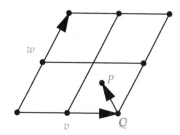

By the triangle inequality, the length of one side of a triangle is less than the sum of the lengths of the other sides. Thus, the distance between opposite vertices of one of the sub-parallelograms must be less than $\frac{\|v\|}{2} + \frac{\|w\|}{2} < 2\frac{\|w\|}{2} = \|w\|$, and so the length of $P - Q$ is also less than $\|w\|$.

Now, if P is inside the parallelogram, or on any side other than that parallel to v, we would get a translation vector that is smaller than w and non-parallel with v, which contradicts the choice of w. Thus, $P - Q$ must lie along a side parallel to v. However, this contradicts the choice of v.

So, all translations have the form $T_z = T_{sv+tw}$ with s, t integers, and the proof is complete. $\square$

What kinds of lattices are possible for a wallpaper group? In the parallelogram spanned by v and w, we have that $v - w$ and $v + w$ are the diagonals. By replacing w by $-w$ we could switch the order of these diagonals, which means we can assume $\|v - w\| \leq \|v + w\|$. Also, neither of these diagonals can be smaller than w because of how w is defined. Thus, we know that

$$\|v\| \leq \|w\|$$

And

$$\|w\| \leq \|v - w\|$$

And

$$\|v - w\| \leq \|v + w\|$$

Thus, $\|v\| \leq \|w\| \leq \|v - w\| \leq \|v + w\|$. Therefore, there are eight possible lattices:

1. **Oblique**
$\|v\| < \|w\| < \|v - w\| < \|v + w\|$. The lattice is made of skew parallelograms.

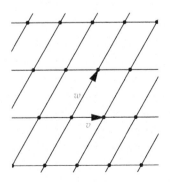

2. Rectangular

$\|v\| < \|w\| < \|v - w\| = \|v + w\|$. If the diagonals of the parallelogram are congruent, then the parallelogram must be a (non-square) rectangle.

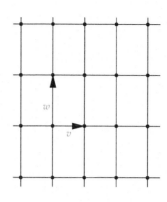

3. Centered Rectangular

$\|v\| < \|w\| = \|v - w\| < \|v + w\|$. If $\|w\| = \|v - w\|$, then these sides form an isosceles triangle and the head of w will be the center of a rectangle built on every other row of the lattice, as shown.

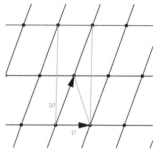

4. Rhombal

$\|v\| = \|w\| < \|v - w\| < \|v + w\|$. Again, we get an isosceles triangle, this time with sides being v and w. This case is essentially the same as the centered rectangle, with the rectangular sides built on the vectors $v - w$ and $v + w$, and the center of the rectangle at a lattice point.

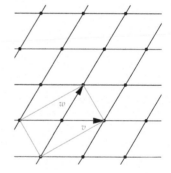

5. $\|v\| < \|w\| = \|v - w\| = \|v + w\|$. This is not a possible configuration for a lattice. Since $\|v - w\|^2 = (v - w) \bullet (v - w) = v \bullet v - 2v \bullet w + w \bullet w$, and $\|v + w\|^2 = v \bullet v + 2v \bullet w + w \bullet w$, then we would have that $v \bullet w = 0$. Then, $\|w\| = \|v - w\|$ would imply that $\|v\| = 0$, which is impossible.

6. **Square**
$\|v\| = \|w\| < \|v - w\| = \|v + w\|$.

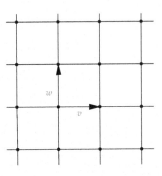

7. **Hexagonal**
$\|v\| = \|w\| = \|v - w\| < \|v + w\|$.

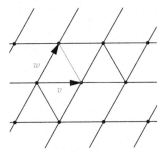

8. $\|v\| = \|w\| = \|v - w\| = \|v+w\|$. This is not a possible configuration for a lattice.

Thus, from eight possible lattice types, we see that there are really only five different lattices possible for a wallpaper group.

> **Theorem 6.18.** *If L is a lattice in the plane, then it is either oblique, rectangular, centered rectangular (which includes rhombal), square, or hexagonal.*

Just as in the case of frieze groups, the possible rotational symmetries of a wallpaper group are restricted to specific values.

> **Theorem 6.19.** *The rotations in a wallpaper group G must map elements of the lattice of the group back to the lattice. Furthermore, angles of rotation in a wallpaper group can only be 60, 90, 120, or 180 degrees.*

Proof: Let $sv + tw$ be a point in the lattice. Then T_{sv+tw} is in G. Let $R_{a,\alpha}$ be a rotation in G. Consider $R_{a,\alpha} \circ T_{sv+tw} \circ R_{a,-\alpha}$. In Exercise 5.7.6, we showed that this composition is the same as $T_{R_{a,\alpha}(sv+tw)}$.

Since the composition of elements in G must yield an element of G, then $T_{R_{a,\alpha}(sv+tw)}$ is in G, and if O is the origin, then $T_{R_{a,\alpha}(sv+tw)}(O) = R_{a,\alpha}(sv + tw)$ is in the lattice. Thus, the rotation maps points of the lattice to other points of the lattice. Also, since there are only a finite number of lattice points to which the rotation can map a given lattice point, then the order of the rotation must be finite.

Let v be the minimal length translation vector among all translations and let P be a point in the lattice. Consider $R_{a,\alpha}(P)$ (labeled "$R(P)$" in Figure 6.2).

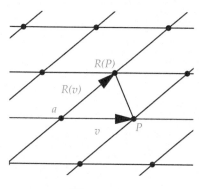

Figure 6.2

The vector $P - R_{a,\alpha}(P)$ will be a translation vector in G (difference of two lattice vectors must be a translation vector). If the rotation angle α is less than 60 degrees, then this vector will have *shorter* length than v. This is impossible due to how v was chosen. The only possible rotation angles are then angles α with $60 \leq \alpha < 360$. (We will ignore the trivial case of rotation by 360.)

In our discussion of finite symmetry groups, we proved that if a rotation has finite order, then the angle of rotation must evenly divide 360. Then the only possible angles for a wallpaper rotation are $60 = \frac{360}{6}$, $72 = \frac{360}{5}$, $90 = \frac{360}{4}$, $120 = \frac{360}{3}$, and $180 = \frac{360}{2}$.

Rotations by 60 and 120 are possible in hexagonal lattices. Rotations by 90 and 180 degrees are possible in square and rectangular lattices. It is left as an exercise to show that rotations of 72 degrees are impossible. □

This theorem has become known as the "Crystallographic Restriction." Many crystals have the property that if you slice them along a plane, the atoms of the crystal along that plane form a wallpaper pattern.

Thus, the possible crystals having wallpaper symmetry is "restricted" to those generated by wallpaper groups.

How many wallpaper groups are there? It turns out that there are exactly 17 wallpaper groups. A careful proof of this result can be carried out by examining the five lattice types we have discussed and finding the symmetry groups that preserve each type. For details see [3, Chapter 26].

The 17 groups have traditionally been listed with a special notation consisting of the symbols p, c, m, and g, and the integers 1, 2, 3, 4, and 6. This is the crystallographic notation adopted by the International Union of Crystallography (IUC) in 1952.

In the IUC system the letter p stands for *primitive*. A lattice is generated from a polygonal cell that is translated to form the complete lattice. In the case of oblique, rectangular, square, and hexagonal lattices, the cell is precisely the original parallelogram formed by the vectors v and w and is, in this sense, a "primitive" cell. In the case of the centered-rectangle lattice, the cell is a rectangle, together with its center point, with the rectangle larger than the original parallelogram and not primitive. Thus, lattice types can be divided into two classes: primitive ones designated by the letter p and non-primitive ones designated by the letter c.

Other symmetries for a general wallpaper group will include reflections, rotations, and glide reflections. The letter m is used to symbolize a reflection and g symbolizes a glide reflection. The numbers 1, 2, 3, 4, and 6 are used to represent the orders of rotation for a group (e.g., 3 would signify a rotation of $\frac{360}{3} = 120$).

In the IUC system, a wallpaper group is designated by a string of letters and numbers. First, there is the letter p (for oblique, rectangular, square, or hexagonal lattices) or c (for centered-rectangular lattices). Then a set of non-translational generators for the group will be listed. These generators include reflections, glide reflections, and/or rotations. The two fundamental translations are not listed, but are understood to also be generators for the group.

For example, "p1" would symbolize the simplest wallpaper group, where the lattice is oblique and there are no non-translational symmetries.

What would "p2mm" symbolize? The lattice is primitive; there is a rotational symmetry of 180 degrees; and there are two reflection symmetries. Since the reflections are listed separately, neither can be generated from the other by combining with the fundamental translations.

Thus, the reflection lines of symmetry cannot be parallel and the two reflections will generate a rotation, which must be of 180 degrees. The two reflection lines will then be in perpendicular directions and the lattice must be rectangular, centered-rectangular, or square. In fact, the lattice must be rectangular. (The proof is left as an exercise.)

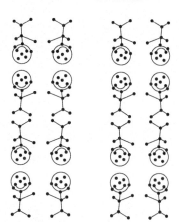

In Appendix E there is a complete listing of all 17 wallpaper groups.

Exercise 6.3.1. *Sketch the lattices spanned by the following pairs of vectors, and specify which of the five types of lattices each pair will generate.*

(a) $\vec{v} = (2, 0)$, $\vec{w} = (2, 4)$

(b) $\vec{v} = (1, \sqrt{3})$, $\vec{w} = (1, -\sqrt{3})$

(c) $\vec{v} = (-1, 1)$, $\vec{w} = (1, 1)$

Exercise 6.3.2. *Show that the transformation $f(x, y) = (-x, y + 1)$ is a glide reflection and that a symmetry group generated by f and the translation $T(x, y) = (x + 1, y + 1)$ must be a wallpaper group. Which lattice type will this group have?*

Exercise 6.3.3. *Show that the symmetry group generated by the glide f, defined in the previous exercise, and the translation $T(x, y) = (x, y + 1)$ will not be a wallpaper group.*

Exercise 6.3.4. *Let G be a wallpaper group, and let H be the subset of all symmetries fixing a point of the lattice for G. If H is a subgroup of G of order 4 and all elements of H are of order 2, show that H must be the group generated by a half-turn and two reflections about perpendicular lines.*

Exercise 6.3.5. *With the same assumptions about G and H as in the preceding exercise, show that the lattice for G must be rectangular, centered-rectangular, or square.*

Exercise 6.3.6. *Show that the lattice for the wallpaper group p2mm must be rectangular. [Hint: Use the previous two exercises.]*

Exercise 6.3.7. *For the wallpaper group p2mm, with mm representing generating reflections along the two translation vectors, prove that there are also reflections along lines through the midpoints of the lattice segments, and find formulas for these reflections in terms of the generators of p2mm.*

Exercise 6.3.8. *Show that the wallpaper group p2mm, with mm representing generating reflections along the two translation vectors, is the same group as pm'm', where m'm' represents two reflections through midpoints of the translation vectors. [Hint: Use the preceding exercise.]*

Exercise 6.3.9. *Show that a wallpaper pattern with translation symmetry given by two perpendicular vectors v, w, and wallpaper group p2mm, can be generated by applying the symmetries of the group to the shaded area of the rectangle spanned by v and w shown. Thus, this area can be thought of as a generating area for the pattern. [Hint: Use the previous exercise.]*

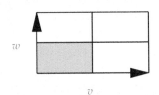

Exercise 6.3.10. *(This problem assumes familiarity with the process of finding the matrix representation of a transformation with respect to a basis.) Let L be a lattice spanned by vectors $\vec{v}$ and $\vec{w}$. Let $R_{A,\alpha}$ be an element of a wallpaper group G for L. By Theorem 6.16, $R_{A,\alpha}$ has finite order, say n. Using the idea of conjugation, show that $R_{O,\alpha}$ is also in G. The matrix for $R_{O,\alpha}$, in the standard basis, is then*

$$\begin{bmatrix} \cos(\frac{360}{n}) & -\sin(\frac{360}{n}) \\ \sin(\frac{360}{n}) & \cos(\frac{360}{n}) \end{bmatrix}$$

Show that the matrix for $R_{O,\alpha}$ with respect to the basis $\{\vec{v}, \vec{w}\}$ will have integer entries. Then, use the fact that the trace of a matrix is invariant under change of bases to give an alternate proof of the second part of Theorem 6.19.

Exercise 6.3.11. *Show that in a parallelogram, the maximum distance between any two points on or inside the parallelogram occurs between opposite vertices of the parallelogram. [Hint: Let $A = lv + mw$ and $B = sv + tw$ be two points in the parallelogram. Show that these are farthest apart when s, t, l, and m are all 0 or 1.]*

Exercise 6.3.12. *Show that a wallpaper group cannot have rotations of 72 degrees about a point. [Hint: Use an argument similar to that used in Theorem 6.19 and consider the angle between the vector $-v$ and a double rotation of v by $R_{a,72}$.]*

Exercise 6.3.13. *Sketch a planar pattern that has a symmetry group that is not discrete. (Blank or completely filled sheets of paper are not allowed.)*

6.4 TILING THE PLANE

A long time ago, I chanced upon this domain [of regular division of the plane] in one of my wanderings; I saw a high wall and as I had a premonition of an enigma, something that might be hidden behind the wall, I climbed over with some difficulty. However, on the other side I landed in a wilderness and had to cut my way through with great effort until—by a circuitous route—I came to the open gate, the open gate of mathematics.

—Maurits Cornelis (M. C.) Escher (1898–1972)

6.4.1 Escher

Much of the renewed interest in geometric design and analysis in the modern era can be traced to the artistic creations of M. C. Escher. While he did not prove new theorems in geometry, he did use geometric insights to create fascinating periodic designs like the design in Figure 6.3.

A beautiful book that describes Escher's artwork and the mathematics that underlies it is *M. C. Escher: Visions of Symmetry* [37], by Doris Schattschneider.

One of Escher's favorite themes was that of a tessellation of the plane by geometric shapes. A *tessellation* or tiling is a covering of the plane by repeated copies of a shape such that there are no gaps left uncovered and the copied shapes never overlap. In Figure 6.3 we see the beginnings of a tessellation of the plane by a three-sided shape.

A tessellation is produced by repeating a basic figure, or set of figures, throughout the plane. Repetitions of the basic tile(s) are carried out by isometries of the plane. Thus, every tiling has associated with it a group of symmetries that map the tiling back to itself. For example, in Figure 6.3 we see that this tiling is invariant under a rotation of 60 degrees and, thus, it must have a symmetry group equal to one of the two wallpaper groups p6 or p6m (see Appendix E). Since the tiling is not invariant under a reflection, then it must have a symmetry group of p6.

Escher received inspiration for his tiling designs from two sources:

the intricate designs of Arab artists and the work of mathematicians on classifying planar symmetry groups.

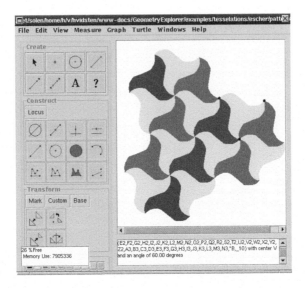

Figure 6.3 A tiling in the spirit of Escher

At the age of 24, Escher made his first trip to the Alhambra, a fourteenth century palace in Grenada, Spain. He found almost every wall, floor, and ceiling surface covered with abstract geometric tilings. In Figure 6.4 we see reproductions of the tilings that Escher saw during his visit to the Alhambra.

Figure 6.4 Alhambra tilings, created by Lawrence D'Oliveiro <ldo@geek-central.gen.nz> (Creative Commons License)

Note that the pattern in the upper-left corner of Figure 6.4 is essentially the same as the pattern in Figure 6.3.

The wide variety of tiling patterns and tile shapes exhibited at the Alhambra inspired Escher to investigate the different ways that one could tile the plane in a systematic fashion. Escher found the answer to this question through the work of the mathematician George Pólya. Pólya, in a 1924 article in the journal *Zeitschrift für Kristallographie*, gave a complete classification of the discrete symmetry patterns of the plane, with patterns repeated in two directions, namely, the wallpaper patterns. All 17 patterns were exhibited in the Alhambra tilings. This mathematical revelation was the "open gate" through which Escher was able to carefully design and produce his famous and imaginative tilings.

6.4.2 Regular Tessellations of the Plane

The simplest tessellations of the plane are made with a single tile that has the shape of a regular polygon. A regular polygon has the property that all the sides have the same length and all interior angles created by adjacent sides are congruent. We will call a regular polygon with n sides a *regular n-gon*.

A regular 3-gon is an equilateral triangle, a regular 4-gon is a square, and so on. Which regular polygons tile the plane?

Let's consider the simplest n-gon tile, the equilateral triangle.

In the figure at the right, we have a tiling by equilateral triangles in which all triangles meet at common vertices.

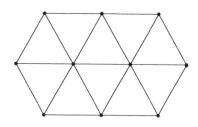

In the new tiling at the right, we have shifted the top row of triangles a bit. This configuration will still lead to a tiling of the plane, although triangles no longer share common vertices.

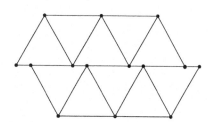

> **Definition 6.10.** *We will call a tessellation* regular *if it is made from copies of a single regular n-gon with all n-gons meeting at common vertices.*

The second triangular tiling above is not a regular tiling. How many regular tilings are there? Clearly, the example above shows that there is a regular tiling with triangles. We can easily create a regular tiling with squares. The triangle tiling also shows that regular hexagonal tilings are possible. In fact, these three are the *only* regular tilings.

It is not hard to see why this is the case. At a common vertex of a regular tiling, suppose that there are k regular n-gons meeting at the vertex. Then, an angle of 360 degrees will be split into k parts by the edges coming out of this vertex. Thus, the interior angle of the n-gon must be $\frac{360}{k}$. On the other hand, suppose we take a regular n-gon, find its central point and draw edges from this point to the vertices of the n-gon. In the example at the right, we have done this for the regular hexagon. The point labeled "A" is the central point of the hexagon.

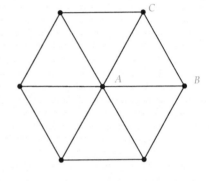

For a regular n-gon, this triangulation will yield n isosceles triangles, each with angle sum of 180 degrees. If we add up the sum of the angles in all of the triangles in the figure, we get a total angle sum of $180n$.

On the other hand, if we add up only those triangle angles defined at the central point, then the sum of these will have to be 360. For each of our isosceles triangles, the other two angles at the base of the triangle will be congruent (the angles at B and C in the hexagon example). Let's call these angles α. Then, equating the total triangle sum of $180n$ with the sum of the angles at the center and the sum of the base angles of

each isosceles triangle, we get

$$180n = 360 + 2n\alpha$$

and thus,

$$2\alpha = 180 - \frac{360}{n}$$

Now, 2α is also the interior angle of each n-gon meeting at a vertex of a regular tessellation. We know that this interior angle must be $\frac{360}{k}$, for k n-gons meeting at a vertex of the tessellation. Thus, we have that

$$\frac{360}{k} = 180 - \frac{360}{n}$$

If we divide both sides by 180 and multiply by nk, we get

$$nk - 2k - 2n = 0$$

If we add 4 to both sides, we can factor this as

$$(n-2)(k-2) = 4$$

There are only three integer possibilities for n and k, namely 6, 4, and 3. These three possibilities directly correspond to the three regular tessellations with equilateral triangles, squares, and regular hexagons.

6.5 PROJECT 9 - CONSTRUCTING TESSELLATIONS

Escher modeled many of his tilings after those he saw at the Alhambra. In this project we will look at how to create one of these Moorish tilings, find its symmetry group, and construct other tilings with a specified symmetry group.

The tiling that we will construct is based on the dart-like shape found in the tiling at the lower left in Figure 6.4. (Notes on how to use particular dynamic geometry software packages for this project can be found at http://www.gac.edu/~hvidsten/geom-text.)

Start up your geometry software and create a segment $\overline{AB}$ in the Canvas. Attach a point C to the segment.

To construct a square on $\overline{CB}$, first set C as a center of rotation with a rotation angle of 90 degrees. Then rotate B to get point D.

Set point D as a new center of rotation and rotate point C 90 degrees to get point E. Connect segments as shown.

To make the point of our dart, create a ray from point C vertically through point D and attach a point F to this ray.

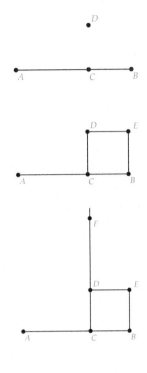

Next, hide the ray, hide $\overline{CD}$, and connect segments as shown.

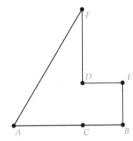

This completes the construction of half of the dart. Select points A, B, E, D, and F and create a filled polygon to color our half-dart. Define a mirror reflection across $\overline{AF}$. Select the half-dart by clicking inside the filled area and then reflect the half-dart to get the entire dart.

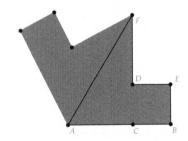

Now, set E as a center of rotation of 90 degrees. Select the entire dart and rotate it three times. In the figure at the right, we have changed the color of each component dart so that we can see the pieces better. Also, we have rescaled the Canvas, since the image can grow quite large.

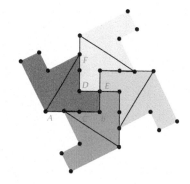

To make this a valid tiling (no gaps), we move point F (the "point" of the dart) to a position where it directly matches the base of the dart to its right, as shown in the figure.

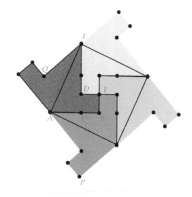

We have now constructed a basic "tile" that can be translated to completely cover the plane.

To translate this basic tile we define a translation from P to Q. Then we translate the whole 4-dart region by this translation. It is clear that the 4-dart region will tile the plane if we continue to translate it vertically and horizontally.

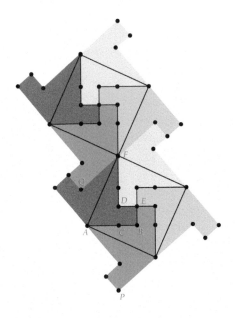

Figure 6.5

Exercise 6.5.1. *Determine the symmetry group for this figure. [Hint: It must be one of the 17 wallpaper groups. Why? Use Figure 6.4 for a more complete tiling picture.]*

Exercise 6.5.2. *Choose one of the wallpaper groups and construct a tiling (other than a regular tiling) with this symmetry group. You can do this on the computer or by using paper cutouts if you wish.*

For the project report give a thorough answer to the first question, illustrating all symmetries on the dart tiling. For the second exercise, describe your chosen symmetry group, illustrate the basic lattice cell for this group, and describe the steps you took to construct your tiling.

CHAPTER 7

Hyperbolic Geometry

I have discovered such wonderful things that I was amazed
... out of nothing I have created a strange new universe.

– János Bolyai (1802–1860), from a letter to his father, 1823

7.1 BACKGROUND AND HISTORY

Euclid's development of planar geometry was based on five postulates
(or axioms):

1. Between any two distinct points, a segment can be constructed.

2. Segments can be extended indefinitely.

3. Given a point and a distance, a circle can be constructed with the
   point as center and the distance as radius.

4. All right angles are congruent.

5. Given two lines in the plane, if a third line l crosses the given lines
   such that the two interior angles on one side of l are less than two
   right angles, then the two lines if continued indefinitely will meet
   on that side of l where the angles are less than two right angles.

Postulates 1–4 seem very intuitive and self-evident. To construct geometric figures, one needs to construct segments, extend them, and construct circles. Also, geometry should be uniform so that angles do not
change as objects are moved about.

The fifth postulate, the so-called *parallel postulate*, seems overly complex for an axiom. It is not at all self-evident or obvious and reads more
like a theorem.

In fact, many mathematicians tried to find simpler postulates, ones that were more intuitively believable, to replace Euclid's fifth, with the hope that the fifth postulate could then be proved from the first four postulates and the new postulate.

We have already considered one of these substitutes, *Playfair's Postulate*:

> Given a line and a point not on the line, it is possible to construct exactly one line through the given point parallel to the line.

This postulate is certainly simpler to state and easier to understand when compared to Euclid's fifth postulate. However, in Chapter 2 we saw that Playfair's Postulate is logically equivalent to Euclid's fifth postulate, and so replacing Euclid's fifth postulate with Playfair's Postulate does not really simplify Euclid's axiomatic system.

Other mathematicians attempted to prove Euclid's fifth postulate as a *theorem* solely on the basis of the first four postulates. One popular method of proof was to assume the logical opposite of Euclid's fifth postulate (or Playfair's Postulate). If one could show the logical opposite to be false, or if one could obtain a contradiction to a known result by using the opposite, then Euclid's fifth postulate would be true as a *theorem* based on the first four postulates.

Giovanni Girolamo Saccheri (1667–1773) and Johann Lambert (1728–1777) both used this method of attack to prove Euclid's fifth postulate. Saccheri's work focused on quadrilaterals whose base angles are right angles and whose base-adjacent sides are congruent. In Euclidean geometry, such quadrilaterals must be rectangles, that is, the top (or summit) angles must be right angles. The proof of this result depends on Euclid's fifth postulate. Saccheri supposed that the top angles were either greater than or less than a right angle. He was able to show that the hypothesis of the obtuse angle resulted in a contradiction to theorems based on the first four Euclidean postulates. However, he was unable to derive any contradictions using the hypothesis of the acute angle. Lambert, likewise, studied the hypothesis of the acute angle. In fact, Lambert spent a good fraction of his life working on the problem of the parallel postulate. Even though both men discovered many important results of what has become known as *non-Euclidean geometry*, geometry based on a negation of Euclid's fifth postulate, neither could accept the possibility of non-Euclidean geometry. Saccheri eventually resorted to simply asserting that such a geometry was impossible:

The hypothesis of the acute angle is absolutely false, because [it is] repugnant to the nature of the straight line! [16, page 125]

In the 1800s several mathematicians experimented with negating Playfair's Postulate, assuming that a non-Euclidean fifth postulate could be consistent with the other four. This was a revolutionary idea in the history of mathematics. János Bolyai (1802–1860), Carl Friedrich Gauss (1777–1855), and Nikolai Lobachevsky (1792–1856) explored an axiomatic geometry based on the first four Euclidean postulates plus the following postulate:

Given a line and a point not on the line, it is possible to construct *more than one* line through the given point parallel to the line.

This postulate has become known as the *Bolyai–Lobachevskian Postulate*. Felix Klein, who was instrumental in classifying non-Euclidean geometries based on a surface with a particular conic section, called this the *hyperbolic postulate*. A geometry constructed from the first four Euclidean postulates, plus the Bolyai–Lobachevskian Postulate, is known as *Bolyai–Lobachevskian geometry*, or *Hyperbolic geometry*.

Gauss, one of the greatest mathematicians of all time, was perhaps the first to believe that Hyperbolic geometry could be consistent. Harold Wolfe in *Non-Euclidean Geometry* describes how Gauss wrote a letter to a friend about his work.

The theorems of this geometry appear to be paradoxical and, to the uninitiated, absurd; but calm, steady reflection reveals that they contain nothing at all impossible. For example, the three angles of a triangle become as small as one wishes, if only the sides are taken large enough; yet the area of the triangle can never exceed a definite limit, regardless of how great the sides are taken, nor indeed can it ever reach it. All my efforts to discover a contradiction, an inconsistency, in this Non-Euclidean geometry have been without success. [43, page 47]

In fact Gauss, Bolyai, and Lobachevsky developed the basic results of Hyperbolic geometry at approximately the same time, though Gauss developed his results before the other two, but refused to publish them.

What these three mathematicians constructed was a set of theorems based on the first four Euclidean postulates plus the hyperbolic postulate. This did not mean, however, that they showed Hyperbolic geometry to be a *consistent* system. There was still the possibility that one of the three just missed finding a theorem in Hyperbolic geometry that would lead to a contradiction of a result based on the first four Euclidean postulates.

The consistency of Hyperbolic geometry was demonstrated by Eugenio Beltrami (1835–1900), Felix Klein (1849–1925), and Henri Poincaré (1854–1912) in the late 1800s to early 1900s. They created *models* of Hyperbolic geometry inside Euclidean geometry, with perhaps strange definitions of points, lines, circles, and angles, but models nonetheless. In each of their models, they showed that Euclid's first four postulates were true and that the hyperbolic postulate was true as well. Since each model was created within Euclidean geometry, if Hyperbolic geometry had an internal contradictory statement, then that statement, when translated into its Euclidean environment, would be an internal contradiction in Euclidean geometry! Thus, if one believed Euclidean geometry was consistent, then Hyperbolic geometry was equally as consistent.

In the next section we will introduce two models of Hyperbolic geometry, the Poincaré model and the Klein model. Then, in the following sections of the chapter, we will look at results that are true in *any* model satisfying the five axioms of Hyperbolic geometry.

7.2   MODELS OF HYPERBOLIC GEOMETRY

7.2.1   Poincaré Model

In the Poincaré model for 2-dimensional Hyperbolic geometry, a point is defined to be any point interior to the unit disk. That is, any point $P = (x, y)$, with $x^2 + y^2 < 1$. The collection of all such points will be called the *Poincaré disk*.

Definition 7.1. *A* hyperbolic line *(or* Poincaré line*) is a Euclidean arc, or Euclidean line segment, within the Poincaré disk that meets the boundary circle at right angles.*

At right are two "lines" in the
Poincaré model.

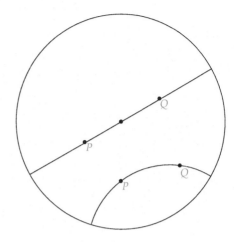

To verify that this actually is a model for Hyperbolic geometry, we
need to show that Euclid's first four postulates, plus the hyperbolic
postulate, are satisfied in this model.

We will start with the first two postulates of Euclid—that unique
segments can always be constructed through two points and that segments can always be extended. By segments we will mean subsets of the
Poincaré lines as defined thus far.

Given two points P and Q, suppose they lie on a diameter of the
boundary circle in the Poincaré disk. Then, as shown in the figure above,
we can construct the Euclidean segment $\overline{PQ}$ along the diameter. Since
this diameter meets the boundary circle at right angles, then $\overline{PQ}$ will lie
on a Poincaré (hyperbolic) line and so will be a hyperbolic segment.

Now suppose that P and Q do not lie on a diameter. Then by the
work we did on orthogonal circles at the end of Chapter 2, there is a
unique circle through P and Q that meets the boundary circle at right
angles. We find this circle by constructing the inverse point P' to P
with respect to the boundary circle. The circle through P, P', and Q
will then meet the boundary circle at right angles. The portion of this
circle within the unit disk will then be a hyperbolic segment through P
and Q.

To verify Euclid's second postulate, that lines (hyperbolic) can always be extended, we first note that the points of our geometry are not
allowed to be on the boundary circle, by the definition of the Poincaré
model. This allows us to extend any hyperbolic segment.

For example, let X be an intersection point (Euclidean point) of the Poincaré line through two hyperbolic points P and Q with the boundary circle. Then, since P cannot be on the boundary, the distance along the circle arc from P to X will always be positive, and thus we can find another point Y between these two points with $\overline{YQ}$ extending $\overline{PQ}$.

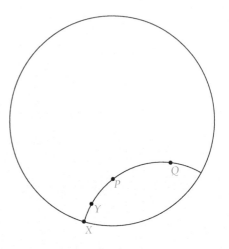

Hyperbolic Distance

To define circles for the third postulate, we need a notion of distance. Since the boundary of the Poincaré disk is not reachable in Hyperbolic geometry, we want a definition of distance such that the distance goes to infinity as we approach the boundary of the Poincaré disk.

In the figure at the right, we have two points P and Q in the Poincaré disk. There is a unique hyperbolic line (Euclidean arc $RPQS$) on which P and Q lie that meets the boundary of the disk at points R and S.

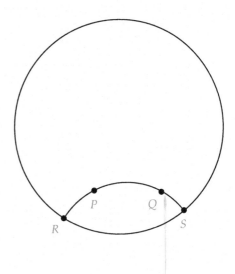

Definition 7.2. *The* hyperbolic distance *from P to Q is*

$$d_P(P,Q) = \left| \ln\left(\frac{(PS)}{(PR)}\frac{(QR)}{(QS)}\right) \right| \qquad (7.1)$$

where R and S are the points where the hyperbolic line through P and Q meets the boundary circle, PS is the Euclidean distance between P and S, and likewise for PR, QR, and QS.

This function satisfies the critical defining properties of a distance function. It is non-negative and equal to zero only when $P = Q$. It is additive along lines, and it satisfies the triangle inequality (for triangles constructed of hyperbolic segments). These properties will be proved in detail in the last section of this chapter.

One thing that is clear from looking at the form of the distance function is that as P or Q approach the points on the boundary (R or S), the fraction inside the log function goes to ∞ or 0, and thus the distance function itself goes to infinity.

We can now define hyperbolic circles.

Definition 7.3. *A hyperbolic circle c of radius r centered at a point O in the Poincaré disk is the set of points in the Poincaré disk whose hyperbolic distance to O is r.*

Here are some hyperbolic circles with their associated hyperbolic centers.

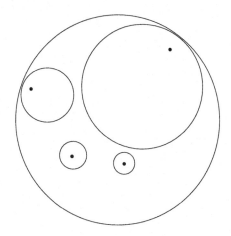

We must now verify that circles always exist. To construct the circle of radius r at O, we note that through any line passing through O, we can find points that are r units away (measured in the hyperbolic distance function). This is because no matter how close O might be to the boundary points R or S, we can always find points between O and those boundary points whose distance to O will grow without bound.

For the fourth postulate, we will define angles just as they are defined in Euclidean geometry. We use the Euclidean tangent lines to Poincaré lines (i.e., Euclidean arcs) in the Poincaré model to determine angles. That is, the angle determined by two hyperbolic lines will be the angle made by their Euclidean tangents. Since angles inherit their Euclidean meaning, the fourth postulate is automatically true.

For the fifth postulate (the hyperbolic postulate), consider a line l and a point P not on l as shown at right.

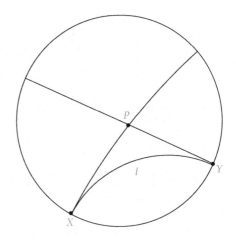

Let X and Y be the intersection points of l with the boundary circle. Then by Theorem 2.43 we know that there are two circular arcs, one through P and X and one through P and Y, that are orthogonal to the circle boundary. Also, neither of these can intersect l at a point inside the boundary circle. For, suppose that the arc through P and X intersected l at a point Q inside the circle. Then Q and X would be on l and also on the arc through P and X. By the uniqueness part of Theorem 2.43, these two arcs must be coincident. But, this is impossible as P was assumed not to lie on l. Thus, the two arcs through P will be two hyperbolic lines that do not intersect l inside the boundary circle and by definition are parallel lines to l.

We see that all of the first four Euclidean postulates hold in this geometry and the hyperbolic parallel postulate holds as well. We conclude that this strange geometry in the Poincaré disk is just as logically consistent as Euclidean geometry based on Euclid's postulates. If there were contradictory results about lines, circles, and points in this new geometry, they would have to be equally contradictory in Euclid's planar geometry, in which this geometry is embedded.

Now, for Hyperbolic geometry to be truly consistent relative to Euclidean Geometry, we must be a bit careful. It is not enough to show that the Poincaré model satisfies the first four Euclidean postulates. We have already seen in earlier chapters that Euclid's axiomatic system had some problems. What we really need to verify is that the Poincaré model satisfies the axioms of Hilbert, which are covered in the on-line chapter on Hilbert's geometry. For the sake of exposition, we will not go into the details of that verification here, but refer the reader instead to the discussion on Hyperbolic and Elliptic geometry at the end of the on-line chapter on Hilbert's axioms.

We note that there is really nothing special about using the unit disk in the Poincaré model. We could just as well have used any circle in the plane and defined lines as diameters or circular arcs that meet the boundary circle at right angles.

7.2.2 Mini-Project - The Klein Model

The Poincaré model preserves the Euclidean notion of angle, but at the expense of defining lines in a fairly strange manner. Is there a model of Hyperbolic geometry, built within Euclidean geometry, that preserves *both* the Euclidean definition of lines *and* the Euclidean notion of angle? Unfortunately, this is impossible. If we had such a model, and $\triangle ABC$ was any triangle, then the angle sum of the triangle would be 180 degrees, which is a property that is equivalent to the parallel postulate of Euclidean geometry [Exercise 2.1.8 in Chapter 2].

A natural question to ask is whether it is possible to find a model of Hyperbolic geometry, built within Euclidean geometry, that preserves *just* the Euclidean notion of lines.

In this project we will investigate a model first put forward by Felix Klein, where hyperbolic lines are segments of Euclidean lines. Klein's model starts out with the same set of points we used for the Poincaré model, the set of points inside the unit disk.

However, lines will be defined differently. A hyperbolic line (or Klein line) in this model will be any chord of the boundary circle (minus its points on the boundary circle).

Here is a collection of Klein lines.

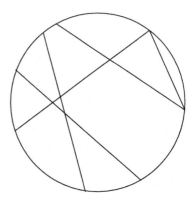

Exercise 7.2.1. *Show that Euclid's first two postulates are satisfied in this model.*

In order to verify Euclid's third postulate, we will need to define a distance function.

Definition 7.4. *The hyperbolic distance from P to Q in the Klein model is*

$$d_K(P,Q) \;=\; \frac{1}{2}\left|\ln(\frac{(PS)\,(QR)}{(PR)\,(QS)})\right| \qquad (7.2)$$

where R and S are the points where the hyperbolic line (chord of the circle) through P and Q meets the boundary circle.

Note the similarity of this definition to the definition of distance in the Poincaré model. We will show at the end of this chapter that the Klein and Poincaré models are *isomorphic*. That is, there is a one-to-one map between the models that preserves lines and angles and also preserves the distance functions.

Just as we did in the Poincaré model, we now define a circle as the set of points a given (hyperbolic) distance from a center point.

Exercise 7.2.2. *Show that Euclid's third postulate is satisifed with this definition of circles. [Hint: Use the continuity of the logarithm function, as well as the fact that the logarithm is an unbounded function.]*

Euclid's fourth postulate deals with right angles. Let's skip this postulate for now and consider the hyperbolic postulate. It is clear that given a line and a point not on the line, there are many parallels (non-intersecting lines) to the given line through the point. Draw some pictures on a piece of paper to convince yourself of this fact.

Now, let's return to the question of angles and, in particular, right angles. What we need is a notion of *perpendicularity* of lines meeting at a point. Let's start with the simplest case, where one of the lines, say l, is a diameter of the Klein disk. Suppose we define another line m to be (hyperbolically) perpendicular to l at a point P if it is perpendicular to l in the Euclidean sense.

Shown here are several Klein lines perpendicular to the Klein line l, which is a diameter of the boundary circle.

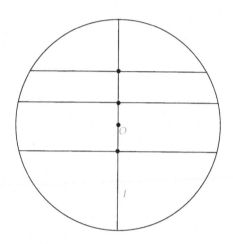

It is clear that we cannot extend this definition to non-diameter chords directly. If we did so, then right angles would have the same meaning in the Klein model as they do in the Euclidean plane, which would mean that parallels would have to satisfy the Euclidean parallel postulate.

The best we can hope for is an extension of *some* property of perpendiculars to a diameter. If we consider the extended plane, with the point at infinity attached, then all the perpendicular lines to the diameter l in the figure above meet at the point at infinity. The point at infinity is the *inverse* point to the origin O, with respect to the unit circle (as was discussed at the end of Chapter 2). Also, O has a unique position on l—it is the Euclidean midpoint of the chord defining l.

If we move l to a new position, say to line l', so that l' is no longer a diameter, then it makes sense that the perpendiculars to l would also move to new perpendiculars to l', but in such a way that they still intersected at the inverse point of the midpoint of the chord for l'. This inverse point is called the *pole* of the chord.

> **Definition 7.5.** *The* pole *of chord $\overline{AB}$ in a circle c is the inverse point of the midpoint of $\overline{AB}$ with respect to the circle.*

From our work in Chapter 2, we know that the pole of chord AB is also the intersection of the tangents at A and B to the circle.

Here we see a Klein line with Euclidean midpoint M and tangents at A and B (which are not actually points in the Klein model) meeting at pole P. Notice that the three chords ($m1$, $m2$, and $m3$) inside the circle have the property that, when extended, they pass through the pole. It makes sense to extend our definition of perpendicularity to state that these three chords will be perpendicular (in the hyperbolic sense) to the given line ($\overline{AB}$) at the points of intersection.

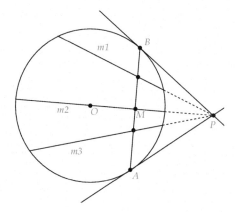

> **Definition 7.6.** *A line m is perpendicular to a line l (in the Klein model) if the Euclidean line for m passes through the pole P of l.*

Exercise 7.2.3. *Use this definition of perpendicularity to sketch some perpendicular lines in the Klein model. Then, use this definition to show that the common perpendicular to two parallel Klein lines exists in most cases. That is, show that there is a Klein line that meets two given parallel Klein lines at right angles, except in one special case of parallels. Describe this special case.*

Given a Klein line l (defined by chord $\overline{AB}$) and a point P not on l, there are two chords $\overline{BC}$ and $\overline{AD}$, both passing through P and parallel to l (only intersections are on the boundary). These two parallels possess the interesting property of dividing the set of all lines through P into two subsets: those that intersect l and those that are parallel to l. These special parallels ($\overline{AD}$ and $\overline{BC}$) will be called *limiting parallels* to l at P. (A precise definition of this property will come later.)

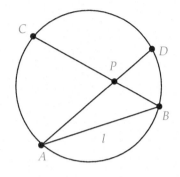

From P drop a perpendicular to l at Q as shown. Consider the hyperbolic angle $\angle QPT$, where T is a point on the hyperbolic ray from P to B. This angle will be called the *angle of parallelism* for l at P.

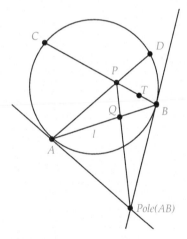

Exercise 7.2.4. *Considering the figure above, explain why the angle made by QPT (the angle of parallelism) cannot be a right angle. Then use this result, and the fact that $\overline{AD}$ and $\overline{BC}$ are limiting parallels, to show that the angle of parallelism cannot be greater than a right angle.*

7.3 BASIC RESULTS IN HYPERBOLIC GEOMETRY

We will now look at some basic results concerning lines, triangles, circles, and the like, that hold in all models of Hyperbolic geometry.

Just as we use diagrams to aid in understanding the proofs of results in Euclidean geometry, we will use the Poincaré model (or the Klein model) to draw diagrams to aid in understanding Hyperbolic geometry. However, we must be careful. Too much reliance on figures and diagrams can lead to hidden assumptions. We must be careful to argue solely from the postulates or from theorems based on the postulates.

One aid in our study of Hyperbolic geometry will be the fact that all results in Euclidean geometry that do not depend on the parallel postulate (those of *neutral geometry*) can be assumed in Hyperbolic geometry immediately. For example, we can assume that results about congruence of triangles, such as SAS, will hold in Hyperbolic geometry. In fact, we can assume the first 28 propositions in Book I of Euclid (see Appendix B).

Also, we can assume results on isometries, including reflections and rotations, found in sections 5.1, 5.2, and 5.4, as they do not depend on the parallel postulate. These results will hold in Hyperbolic geometry, assuming that we have a distance function that is well defined. We also assume basic properties of betweenness and continuity of distance and angle. We saw earlier that these assumptions must be added to Euclid's axiomatic system to ensure completeness of that system, so it is reasonable to assume these properties in Hyperbolic geometry as well.

We will start our study with the main area in which Hyperbolic geometry distinguishes itself from Euclidean geometry—the area of parallels.

7.3.1 Parallels in Hyperbolic Geometry

As we saw in the Klein model in the last project, if we have a hyperbolic line l and a point P not on l, there are always two parallel lines m, n through P with special properties. This is true in any model of Hyperbolic geometry.

Theorem 7.1. *(Fundamental Theorem of Parallels in Hyperbolic Geometry) Given a hyperbolic line l and a point P not on l, there are exactly two parallel lines m, n through P that have the following properties:*

1. *Every line through P lying within the angle made by one of the parallels m, n and the perpendicular from P to l must intersect l while all other lines through P are parallel to l.*

2. *m, n make equal acute angles with the perpendicular from P to l.*

Proof: Drop a perpendicular to l through P, intersecting l at Q.

Consider all angles with side $\overline{PQ}$. The set of these angles will be divided into those angles $\angle QPA$ where $\overrightarrow{PA}$ intersects l and those where it does not. By continuity of angle measure, there must be an angle that separates the angles where $\overrightarrow{PA}$ intersects l from those where it does not. Let $\angle QPC$ be this angle.

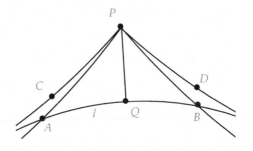

Let $\overrightarrow{PD}$ be the reflection of $\overrightarrow{PC}$ across $\overleftrightarrow{PQ}$. Since reflections preserve parallelism, we have that $\overrightarrow{PD}$ must separate intersecting lines from parallels. Also, reflections preserve angle, so $\angle QPD$ must be congruent to $\angle QPC$. This finishes the proof of part (1) of the theorem.

We now show that both angles are less than 90 degrees. It suffices to show that neither can be a right angle. Suppose that $\angle QPC$ is a right angle. Then by the preceding paragraph, we know that $\angle QPD$ must be a right angle. Points C, P, D are then collinear and make up one parallel to l through P by Euclid's Proposition 27, which does not depend on Euclid's fifth postulate. By the hyperbolic parallel postulate, there must be another line m' through P parallel to l. But, m' has to lie within one of the two right angles $\angle QPC$ or $\angle QPD$, which would be a contradiction to what was just proved about angles $\angle QPC$ and $\angle QPD$ separating intersecting and non-intersecting lines. □

Definition 7.7. *The two special parallels defined in the previous theorem are called* limiting parallels *(also called* asymptotic parallels *or* sensed parallels*) to l through P. These will be lines through P that separate intersecting and non-intersecting lines to l. There will be a* right *and a* left *limiting parallel to l through P. Other lines through P that do not intersect l will be called* ultraparallels *(or* divergent parallels*) to l. The angle made by a limiting parallel with the perpendicular from P to l is called the* angle of parallelism at P.

The properties of limiting parallels have no counterpart in Euclidean geometry and thus it is hard to develop an intuition for them. In the project on the Klein model, we saw how to construct limiting parallels to a given Klein line. It may be helpful to review that construction to have a concrete mental picture of how limiting parallels work in hyperbolic geometry.

In the next few theorems, we will review some of the basic results on limiting parallels. We will prove these results in a *synthetic* fashion, independent of any model. The proofs will focus on the case of right-limiting parallels. The proofs for left-limiting parallels follow by symmetry of the arguments used.

Theorem 7.2. *Let $\overleftrightarrow{PP'}$ be the right-limiting parallel to a line l through P. Then $\overleftrightarrow{PP'}$ is also the right-limiting parallel to l through P'.*

Proof: There are two cases to consider.

In the first case point P' is to the right of P on $\overleftrightarrow{PP'}$ (Figure 7.1). Drop perpendiculars from P and P' to l at Q and Q'. Then by Euclid's Prop. 27, $\overleftrightarrow{P'Q'} \parallel \overleftrightarrow{PQ}$, and we know that all points on $\overleftrightarrow{P'Q'}$ will be on the same side of $\overleftrightarrow{PQ}$.

Let R be a point to the right of P' on $\overleftrightarrow{PP'}$. We need to show that every ray $\overrightarrow{P'S}$ lying within $\angle Q'P'R$ will intersect l.

Since S is interior to $\angle QPR$, we know that ray $\overrightarrow{PS}$ will intersect l at some point T. $\overrightarrow{PS}$ will also intersect $\overline{P'Q}$ at a point U, since this ray cannot intersect either of the other two sides of triangle $\triangle PQP'$.

Now, consider $\triangle QUT$. $\overrightarrow{P'S}$ intersects side $\overline{UT}$ and does not intersect side $\overline{QU}$ (S and U must be on opposite sides of $\overleftrightarrow{P'Q'}$). Thus, $\overrightarrow{P'S}$ must intersect side $\overline{QT}$ and thus intersects l.

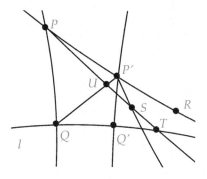

Figure 7.1

In the second case, point P' is to the left of P on $\overleftrightarrow{PP'}$. A similar proof can be given in this case. □

Note that in this proof we are using the fact that a line intersecting a triangle at a side must intersect one of the other sides. This fact is known as *Pasch's axiom* and is independent of Euclid's original set of five postulates. Since we normally assume this axiom in Euclidean geometry, and it does not depend on any parallel properties, we will likewise assume it in Hyperbolic geometry.

The next theorem tells us that the property of being a limiting parallel is a *symmetric* property. That is, we can talk of a pair of lines being limiting parallels to each other without any ambiguity.

Theorem 7.3. *If line m is a right-limiting parallel to l, then l is conversely a right-limiting parallel to m.*

Proof: Let $m = \overleftrightarrow{PD}$ be the right-limiting parallel to l through P. Drop a perpendicular from P to l at Q (Figure 7.2). Also, drop a perpendicular from Q to m at R. We know that R must be to the right of P, since if it were to the left, then we would get a contradiction to the exterior angle theorem for $\triangle QRP$. Let B be a point on l that is on the same side of $\overline{PQ}$ as D. To show that l is limiting parallel to m, we need to show that any ray interior to $\angle BQR$ must intersect m.

Let $\overrightarrow{QE}$ be interior to $\angle BQR$. We will show that $\overrightarrow{QE}$ must intersect m. Drop a perpendicular from P to $\overleftrightarrow{QE}$ at F. By an exterior angle theorem argument, we know that F will lie on the same side of Q on $\overleftrightarrow{QE}$ as E.

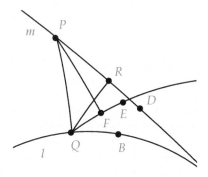

Figure 7.2

Also, since in $\triangle PQF$ the greater side lies opposite the greater angle, we know that $PQ > PF$.

Now, we can use the rotation results from Chapter 5 to rotate $\overline{PF}$, $\overleftrightarrow{PD}$, and $\overleftrightarrow{FE}$ about P by the angle $\theta = \angle FPQ$, as shown in Figure 7.3. Since $PQ > PF$, F will rotate to a point F' on $\overline{PQ}$ and $\overleftrightarrow{FE}$ will rotate to a line $\overleftrightarrow{F'E'}$ that is ultraparallel to l, because of the right angles at Q and F'. Also, $\overleftrightarrow{PD}$ will rotate to a line $\overleftrightarrow{PD'}$ that is interior to $\angle QPR$ and so will intersect l at some point G.

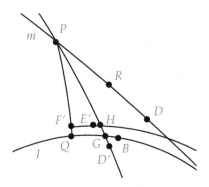

Figure 7.3

Since $\overleftrightarrow{F'E'}$ intersects $\triangle PQG$ and does not intersect $\overline{QG}$, it must intersect $\overline{PG}$ at some point H. Rotating $\overleftrightarrow{F'E'}$ and $\overrightarrow{PD'}$ about P by $-\theta$ shows that $\overrightarrow{QE}$ will intersect m at the rotated value of H. $\square$

The next theorem tells us that the property of being a limiting parallel is a *transitive* property. That is, if l is limiting parallel to m and m is limiting parallel to n, then l is limiting parallel to n.

> **Theorem 7.4.** *If two lines are limiting parallel (in the same direction) to a third line, then they must be limiting parallel to each other.*

Proof: There are two cases to consider.

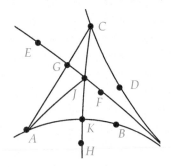

In the first case suppose $\overleftrightarrow{AB}$ and $\overleftrightarrow{CD}$ are both right-limiting parallel to $\overleftrightarrow{EF}$ with points A and C on opposite sides of $\overleftrightarrow{EF}$. Then, $\overline{AC}$ will intersect $\overleftrightarrow{EF}$ at some point G. Let $\overrightarrow{CH}$ be any ray interior to $\angle ACD$. Then, since $\overrightarrow{CD}$ is right-limiting parallel to $\overleftrightarrow{EF}$, we have that $\overrightarrow{CH}$ will intersect $\overleftrightarrow{EF}$ at some point J.

Connect A and J. Since limiting parallelism is symmetric and $\overleftrightarrow{AB}$ is right-limiting parallel to $\overleftrightarrow{EF}$, then $\overleftrightarrow{EF}$ is right-limiting parallel to $\overleftrightarrow{AB}$ and is limiting parallel at all points, including J. Thus, $\overrightarrow{CJ}$ will intersect $\overleftrightarrow{AB}$ at some point K.

Since $\overleftrightarrow{AB}$ and $\overleftrightarrow{CD}$ do not intersect (they are on opposite sides of $\overleftrightarrow{EF}$), and for all rays $\overrightarrow{CH}$ interior to $\angle ACD$ we have that $\overrightarrow{CH}$ intersects $\overleftrightarrow{AB}$, then $\overleftrightarrow{CD}$ is right-limiting parallel to $\overleftrightarrow{AB}$.

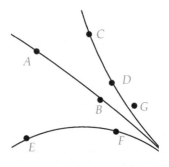

For the second case we assume that $\overleftrightarrow{AB}$ and $\overleftrightarrow{CD}$ are both on the same side of $\overleftrightarrow{EF}$. Suppose that $\overleftrightarrow{AB}$ and $\overleftrightarrow{CD}$ are not right-limiting parallel to each other. Without loss of generality, we may assume that C and E are on opposite sides of $\overleftrightarrow{AB}$. Let $\overleftrightarrow{CG}$ be the right-limiting parallel to $\overleftrightarrow{AB}$.

By the first part of this proof, $\overleftrightarrow{CG}$ and $\overleftrightarrow{EF}$ must be right-limiting parallel. But, $\overleftrightarrow{CD}$ is already the right-limiting parallel to $\overleftrightarrow{EF}$ at C, and thus G must be on $\overleftrightarrow{CD}$ and $\overleftrightarrow{CD}$ is right-limiting parallel to $\overleftrightarrow{AB}$. □

7.3.2 Omega Points and Triangles

In the previous section we saw that given a line l and a point P not on l, there were two special lines called the right- and left-limiting parallels to l through P. These separated the set of lines through P that intersect l from those that did not intersect l.

In the Poincaré and Klein models, these limiting parallels actually meet at points on the boundary circle. While these boundary points are not actually valid points in the geometry, it is still useful to consider the boundary points as representing the special relationship that limiting parallels have with the given line.

We will call these special points *omega points*, or *ideal points*. Thus, a given line will have in addition to its set of "ordinary" points a special pair of omega points. All limiting parallels to the given line will pass through these omega points.

Definition 7.8. *Given a line l, the* right [left] omega point *to l represents the set of all right- [left-] limiting parallels to l. We say two lines intersect at an omega point if one is right- [left-] limiting parallel to the other. If Ω is an omega point of l and P is an ordinary point, then by the* line through P and Ω *we will mean the line through P that is right [left-] limiting parallel to l.*

Definition 7.9. *Given two omega lines $\overleftrightarrow{P\Omega}$ and $\overleftrightarrow{Q\Omega}$, the omega triangle defined by P, Q, and Ω is the set of points on $\overline{PQ}$ and the two rays defined by the limiting parallels $\overleftrightarrow{P\Omega}$ and $\overleftrightarrow{Q\Omega}$.*

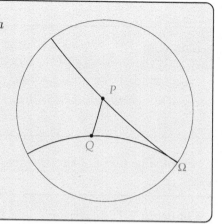

It is interesting to note that these omega triangles, while not true triangles, still share many important properties of ordinary triangles.

Theorem 7.5. *(Pasch's Axiom for Omega Triangles, Part I) If a line passes through a vertex P or Q of an omega triangle $PQ\Omega$, and passes through an interior point of the triangle, then it must intersect the opposite side. If a line passes through Ω, and an interior point, it must intersect side $\overline{PQ}$.*

Proof: The proof of the first part of the theorem is a direct consequence of the fact that $\overleftrightarrow{P\Omega}$ and $\overleftrightarrow{Q\Omega}$ are limiting parallels and will be left as an exercise.

For the proof of the second part, suppose that a line passes through Ω and through a point X within the omega triangle.

By *passing through* Ω we mean that the line through X is a limiting parallel to $\overleftrightarrow{P\Omega}$ (or $\overleftrightarrow{Q\Omega}$). By the first part of this proof, we know that $\overleftrightarrow{PX}$ will intersect $\overleftrightarrow{Q\Omega}$ at some point Y. By Pasch's axiom, we know that the line $\overleftrightarrow{X\Omega}$ will intersect side $\overline{PQ}$ of $\triangle PQY$.

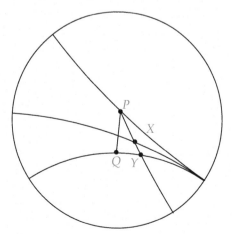

☐

> **Theorem 7.6.** *(Pasch's Axiom for Omega Triangles, Part II) If a line intersects one of the sides of an omega triangle $PQ\Omega$ and passes through an interior point of the triangle, but does not pass through a vertex, then it will intersect exactly one of the other two sides.*

Proof: The proof of this theorem is left as a series of exercises below.

☐

> **Theorem 7.7.** *(Exterior Angle Theorem for Omega Triangles) The exterior angles of an omega triangle $PQ\Omega$ made by extending $\overline{PQ}$ are greater than their respective opposite interior angles.*

Proof: Extend $\overline{PQ}$ to R. It suffices to show that $\angle RQ\Omega$ is greater than $\angle QP\Omega$.

We can find a point X on the right side of $\overline{PQ}$ such that $\angle RQX \cong \angle QP\Omega$. Suppose $\overrightarrow{QX}$ intersected $\overleftrightarrow{P\Omega}$ at S. Then $\angle RQX$ would be an exterior angle to $\triangle PQS$ and would equal the opposite interior angle in this triangle, which contradicts the exterior angle theorem for ordinary triangles.

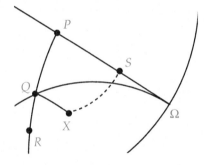

Could $\overline{QX}$ lie on $\overleftrightarrow{Q\Omega}$? Suppose it does.

Let M be the midpoint of $\overline{PQ}$ and drop a perpendicular from M to $\overleftrightarrow{Q\Omega}$ at N. Now, on the line $\overleftrightarrow{P\Omega}$, we can find a point L, on the other side of $\overline{QP}$ from N, with $\overline{QN} \cong \overline{PL}$. If $\overline{QX}$ lies on $\overleftrightarrow{Q\Omega}$, then $\angle NQM \cong \angle LPM$ (both are supplementary to congruent angles). Thus, by SAS, $\triangle NQM$ and $\triangle LPM$ are congruent and $\angle MLP$ is a right angle. Since the angles at M are congruent, L, M, and N lie on a line. But, this would imply that $\overline{LN}$ is perpendicular to both $\overleftrightarrow{P\Omega}$ and $\overleftrightarrow{Q\Omega}$ and that the angle of parallelism is 90 degrees, which is impossible.

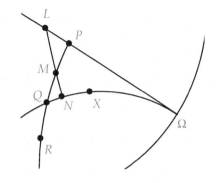

Thus, $\angle RQX$ is less than angle $\angle RQ\Omega$ and since $\angle RQX \cong \angle QP\Omega$, we are finished with the proof. □

Theorem 7.8. *(Omega Triangle Congruence) If $\overline{PQ}$ and $\overline{P'Q'}$ are congruent and $\angle PQ\Omega$ is congruent to $\angle P'Q'\Omega'$, then $\angle QP\Omega$ is congruent to $\angle Q'P'\Omega'$ (Figure 7.4).*

Proof: Suppose one of the angles is greater, say $\angle QP\Omega$. We can find R interior to $\angle QP\Omega$ such that $\angle QPR \cong \angle Q'P'\Omega'$. Then $\overrightarrow{PR}$ will intersect $\overleftrightarrow{Q\Omega}$ at some point S. On $\overrightarrow{Q'\Omega'}$ we can find S' with $\overline{QS} \cong \overline{Q'S'}$. Triangles $\triangle PQS$ and $\triangle P'Q'S'$ are then congruent by SAS and $\angle Q'P'S' \cong \angle QPS$. Also, $\angle QPS \cong \angle Q'P'\Omega'$. Thus, $\angle Q'P'S' \cong \angle Q'P'\Omega'$, which is impossible.

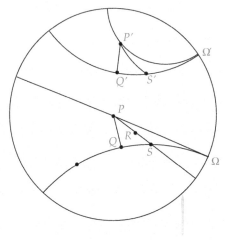

Figure 7.4

□

Exercise 7.3.1. *Illustrate Theorem 7.3 in the Klein model and explain why it must be true using the properties of that model.*

Exercise 7.3.2. *Illustrate Theorem 7.4 in the Klein model and explain why it must be true using the properties of that model.*

Exercise 7.3.3. *Let l be a hyperbolic line and let r_l be reflection across that line. Use the parallelism properties of reflections to show that r_l maps limiting parallels of l to other limiting parallels of l. Use this to show that r_l fixes omega points of l.*

Exercise 7.3.4. *Show that the omega points of a hyperbolic line l are the only omega points fixed by reflection across l. [Hint: If a reflection r_l fixes an omega point of a line $l' \neq l$, show that r_l must fix l' and get a contradiction.]*

Exercise 7.3.5. *Show that if a rotation R fixes an omega point, then it must be the identity rotation. [Hint: Use the preceding exercise.]*

Exercise 7.3.6. *Show that for an omega triangle $PQ\Omega$, the sum of the angles $\angle PQ\Omega$ and $\angle QP\Omega$ is always less than 180 degrees.*

Exercise 7.3.7. *Show that if a line intersects an omega triangle $PQ\Omega$ at one of the vertices P or Q, then it must intersect the opposite side.*

Exercise 7.3.8. *(Partial Proof of Theorem 7.6)*

Suppose a line intersects side $\overleftrightarrow{P\Omega}$ of an omega triangle $PQ\Omega$ but does not pass through a vertex. Let R be the point of intersection and connect R to Q. Use Pasch's axiom and Theorem 7.5 to show that a line through R that does not pass through P or Q must intersect $\overline{PQ}$ or $\overleftrightarrow{Q\Omega}$.

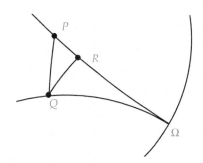

Exercise 7.3.9. *(Partial Proof of Theorem 7.6)*

Suppose a line intersects side $\overline{PQ}$ of an omega triangle $PQ\Omega$ but does not pass through a vertex. Let R be the point of intersection. We can find the limiting parallel $\overrightarrow{R\Omega}$. Show that a line through R must intersect either $\overleftrightarrow{P\Omega}$ or $\overleftrightarrow{Q\Omega}$. [Hint: Use Theorem 7.5.]

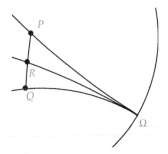

Exercise 7.3.10. *In this exercise we will prove Angle-Angle Congruence for omega triangles. Suppose $PQ\Omega$ and $P'Q'\Omega'$ are two omega triangles with $\angle PQ\Omega \cong \angle P'Q'\Omega'$ and $\angle QP\Omega \cong \angle Q'P'\Omega'$. Show that $\overline{PQ} \cong \overline{P'Q'}$. [Hint: Suppose the segments are not congruent. See if you can derive a contradiction to Exercise 7.3.6.]*

Exercise 7.3.11. *Let $\overline{PQ}$ be a segment of length h. Let l be a perpendicular to $\overline{PQ}$ at Q and $\overrightarrow{PR}$ the limiting parallel to l at P. Define the angle of parallelism to be $a(h) = \angle QPR$, where the angle is measured in degrees. Use a theorem on omega triangles to show this definition is well defined, that it depends only on h.*

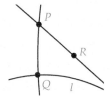

Exercise 7.3.12. *Show that if $h < h'$, then $a(h) > a(h')$. That is, the function $a(h)$ is order-reversing.*

Exercise 7.3.13. *Show that $a(h)$ is a one-to-one function on the set of positive real numbers h.*

Exercise 7.3.14. *In Euclidean and Hyperbolic geometry, angles are said to be* absolute. *One can* construct *a particular right angle and from there all other angles. In Euclidean geometry, however, length is not absolute. One must choose a segment to be of unit length, and then other lengths can be measured. Measures of length are thus* arbitrary *in Euclidean geometry. Assuming that it is always possible to carry out the construction described in the definition of $a(h)$, use the preceding exercise to show that length is* absolute *in hyperbolic geometry.*

7.4 PROJECT 10 - THE SACCHERI QUADRILATERAL

Girolamo Saccheri was a Jesuit priest who, like Gauss and others mentioned at the beginning of this chapter, tried to negate Playfair's Postulate and find a contradiction to known results based on the first four Euclidean postulates. His goal in this work was not to study Hyperbolic geometry, but rather to prove Euclid's fifth postulate as a *theorem*. Just before he died in 1733, he published *Euclides ab Omni Naevo Vindicatus* ("Euclid Freed of Every Flaw"), in which he summarized his work on negating the parallel postulate by the "hypothesis of the acute angle."

Saccheri's idea was to study quadrilaterals whose base angles are right angles and whose base-adjacent sides are congruent. Of course, in Euclidean geometry, such quadrilaterals must be rectangles; that is, the top (or summit) angles must be right angles. Saccheri negated the parallel postulate by assuming the summit angles were less than 90 degrees. This was the "hypothesis of the acute angle." Saccheri's attempt to prove the parallel postulate ultimately failed because he could not find a contradiction to the acute angle hypothesis.

To see why Saccheri was unable to find a contradiction, let's consider his quadrilaterals in Hyperbolic geometry.

We will be using the *Poincaré Disk* model of hyperbolic geometry for our exploration of the Saccheri quadrilateral. Before you start this project, go to http://www.gac.edu/~hvidsten/geom-text for instructions on how to download the hyperbolic geometry system for your particular software environment.

Start up your geometry software and open a *Poincaré Disk* window. Create a segment $\overline{AB}$ on the screen. This will be the base of our quadrilateral.

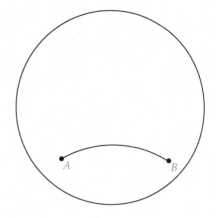

Construct the perpendicular to $\overline{AB}$ through point A. Likewise, construct the perpendicular to $\overline{AB}$ through B.

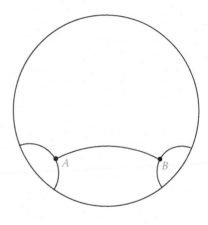

Now, attach a point C along the perpendicular at B as shown.

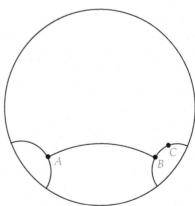

Now, hide the perpendicular line at B and create segment $\overline{BC}$.

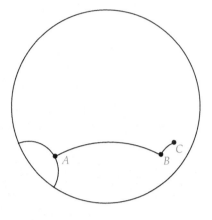

Construct a circle with center point A and radius given by segment $\overline{BC}$. This will create a circle centered at A of *hyperbolic* radius the *hyperbolic* length of $\overline{BC}$. Construct the intersection of the circle and the perpendicular at A.

The upper intersection point, D, is all we need, so hide the circle, the perpendicular, and the lower intersection point. Then, connect A to D and D to C to finish the construction of a Saccheri Quadrilateral.

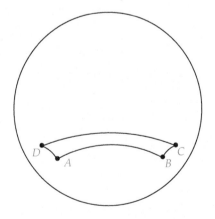

At this point we can use our Saccheri Quadrilateral to study many fascinating properties in Hyperbolic geometry. Let's look at a couple.

Measure the two angles at D and C in the quadrilateral. For this configuration they are equal and less than 90 degrees, which is what they would be in Euclidean geometry. It appears that Saccheri's acute angle hypothesis holds in Hyperbolic geometry, at least for this one case. Move points A and B and check the summit angles.

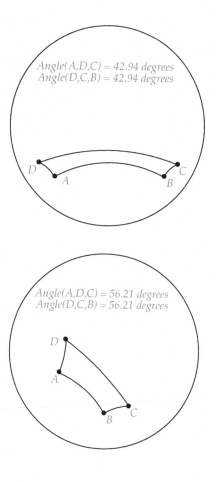

It appears that this result holds for all configurations of our Saccheri Quadrilateral (except perhaps when orientations switch, because our intersection point switches from above the rectangle to below it and the angles become greater than 180 degrees). In fact, this is a theorem in hyperbolic geometry. The summit angles of a Saccheri Quadrilateral are always equal and less than 90 degrees (i.e., are acute).

Saccheri could find no contradiction in assuming that the summit angles of this quadrilateral were acute, because in Hyperbolic geometry (which can be modeled within Euclidean geometry), the summit angles are always acute. If he had been able to find a contradiction, then that would also be a contradiction in Euclidean geometry. However, Saccheri could not believe what his own work was telling him. As mentioned at the beginning of this chapter, he found the hypothesis of the acute angle "repugnant to the nature of the straight line."

Exercise 7.4.1. *Prove that the summit angles of a Saccheri Quadrilateral are always congruent. [Hint: Start by showing the diagonals are congruent.]*

Exercise 7.4.2. *Fill in the "Why?" parts of the following proof that the summit angles of a Saccheri Quad are always acute. Refer to the figure at right.*

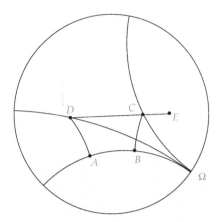

At C we can construct the right-limiting parallel $\overleftrightarrow{C\Omega}$ to $\overline{AB}$. At D we can also construct the right-limiting parallel $\overleftrightarrow{D\Omega}$ to $\overline{AB}$. These two limiting parallels will also be limiting parallel to each other. (Why?)

$\overleftrightarrow{DC}$ must be an ultraparallel (non-intersecting line) to $\overleftrightarrow{AB}$. (Why?) [Hint: Construct midpoints to $\overline{AB}$ and $\overline{CD}$ and use triangles to show that the two lines share a common perpendicular.]

Thus, the right-limiting parallels through D and C, respectively, must lie within the angles $\angle ADC$ and $\angle BCE$, where E is a point on $\overrightarrow{DC}$ to the right of C. Now $\angle AD\Omega$ and $\angle BC\Omega$ are equal in measure. (Why?)

Furthermore, in omega triangle $CD\Omega$, $\angle EC\Omega$ must be greater than interior angle $\angle CD\Omega$. (Why?)

Finally, this proves that $\angle ADC$ and $\angle BCD$ are both less than 90 degrees. (Why?)

For your report give a careful and complete summary of your work done on this project.

7.5 LAMBERT QUADRILATERALS AND TRIANGLES

7.5.1 Lambert Quadrilaterals

In the last section we defined a Saccheri Quadrilateral as a quadrilateral $ABCD$, where the base angles $\angle BAD$ and $\angle CBA$ are right angles and the side lengths AD and BC are congruent (Figure 7.5).

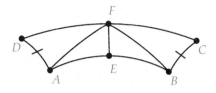

Figure 7.5

Let E and F be the midpoints of the base and summit of a Saccheri quadrilateral. By the results of the last section, we know that $\triangle ADF$ and $\triangle BCF$ are congruent by SAS and thus $\overline{AF} \cong \overline{BF}$. Thus, $\triangle AEF$ and $\triangle BEF$ are congruent by SSS and the angle at E must be a right angle. A similar argument will show that $\angle DFE$ and $\angle CFE$ are also right angles. Thus,

Theorem 7.9. *The segment joining the midpoints of the base and summit of a Saccheri quadrilateral makes right angles with the base and summit.*

If we look at the two quadrilaterals $AEFD$ and $BEFC$, they both share the property of having three right angles.

Definition 7.10. *A* Lambert Quadrilateral *is a quadrilateral having three right angles.*

The midpoint construction just described gives a natural way to associate a Lambert quadrilateral with a given Saccheri quadrilateral. We can also create a Saccheri quadrilateral from a given Lambert quadrilateral.

Theorem 7.10. *Let $ABDC$ be a Lambert quadrilateral with right angles at A, B, and C. If we extend $\overline{AB}$ and $\overline{CD}$, we can find points E and F such that $\overline{AB} \cong \overline{AE}$ and $\overline{CD} \cong \overline{CF}$ (Figure 7.6). Then $EBDF$ is a Saccheri quadrilateral.*

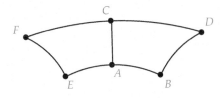

Figure 7.6

Proof: The proof is left as an exercise. □

Corollary 7.11. *In a Lambert Quadrilateral the fourth angle (the one not specified to be a right angle) must be acute.*

Proof: By the preceding theorem, we can embed a given Lambert quadrilateral in a Saccheri quadrilateral and we know the summit angles of a Saccheri quadrilateral are acute. □

Corollary 7.12. *Rectangles do not exist in Hyperbolic geometry.*

Proof: This is an immediate consequence of the preceding corollary. □

Here is another interesting fact concerning Lambert Quads.

Theorem 7.13. *In a Lambert quadrilateral the sides adjoining the acute angle are greater than the opposite sides.*

Proof: Given a Lambert quadrilateral $ABDC$ with right angles at A, B, and C, suppose that $DB < AC$ (Figure 7.7). Then there is a point E on the line through B, D with D between B and E such that $\overline{BE} \cong \overline{AC}$. It follows that $ABEC$ is a Saccheri quadrilateral, and $\angle ACE \cong \angle BEC$, and both angles are acute.

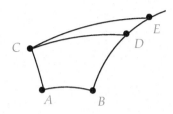

Figure 7.7

However, since A and E are on opposite sides of $\overleftrightarrow{CD}$ (A and B are on the same side and B and E are on opposite sides), then $\overline{CD}$ lies within $\angle ACE$ and thus $\angle ACE$ contains $\angle ACD$. So, $\angle ACE$ must be greater than a right angle. This contradicts the fact that $\angle ACE$ must be acute.

Thus, $DB \geq AC$. If $DB = AC$, then we would have a Saccheri quadrilateral $ABDC$ and the summit angles would be congruent, which is impossible. We must have, then, that $DB > AC$. A similar argument shows $CD > AB$. □

7.5.2   Triangles in Hyperbolic Geometry

Here are two basic (but still amazing) facts about triangles in hyperbolic geometry.

> **Theorem 7.14.** *The angle sum for any hyperbolic right triangle is always less than 180 degrees.*

Proof: Let $\triangle ABC$ be a right triangle with a right angle at A. Let D be the midpoint of $\overline{BC}$ (Figure 7.8).

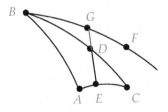

Figure 7.8

Drop a perpendicular from D to $\overleftrightarrow{AC}$ intersecting at E. Then E must lie between A and C. For by Pasch's Axiom we know that the perpendicular must pass through one of the two sides $\overline{AB}$ or $\overline{AC}$. Clearly, it cannot pass through A, B, or C. Suppose it passed through $\overline{AB}$ at X. Then $\triangle XEA$ would have two right angles, contradicting the absolute geometry result that the sum of two angles in a triangle must be less than two right angles (Euclid, Book I, Prop. 17).

Now, we can find a point F such that $\angle DCE$ is congruent to $\angle DBF$. On $\overrightarrow{BF}$ we can find a point G such that $\overline{BG} \cong \overline{EC}$. By SAS we have that $\triangle ECD \cong \triangle GBD$. Thus, E, D, and G are collinear and $\angle DGB$ is a right angle.

Quadrilateral $ABGE$ is thus a Lambert Quadrilateral and $\angle ABG$ must be acute. Since $\angle DCE$ is congruent to $\angle DBF$, then $m\angle DCE + m\angle ABD = m\angle DBF + m\angle ABD < 90$. Thus, the sum of the angles in $\triangle ABC$ is less than two right angles. □

Theorem 7.15. *The angle sum for any hyperbolic triangle is less than 180 degrees.*

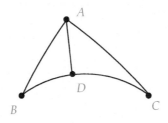

Figure 7.9

Proof: Let ABC be a triangle (Figure 7.9). We have proved the result if one of the angles is right. Thus, suppose none of the three angles are right angles. Then two must be acute, as otherwise we would have the sum of two angles being greater than or equal to 180 degrees.

Let the angles at B and C be acute. Drop a perpendicular from A down to $\overleftrightarrow{BC}$, intersecting at D. Then D is between B and C. (If it intersected elsewhere on the line, we could construct a triangle with two angles more than 180 degrees.)

Now, $\triangle ABC$ can be looked at as two right triangles $\triangle ABD$ and $\triangle ADC$. Using the previous result, we know that the sum of the angles in $\triangle ABD$ is less than 180, as is the sum of the angles in $\triangle ADC$. So the sum of the angles in the two triangles together is less than 360. But clearly, this sum is the same as the sum of the angles in $\triangle ABC$ plus the two right angles at D. Thus, the sum of the angles in $\triangle ABC$ is less than $360 - 180 = 180$. □

Definition 7.11. *Given $\triangle ABC$ we call the difference between 180 degrees and the angle sum of $\triangle ABC$ the* defect *of $\triangle ABC$.*

Corollary 7.16. *The sum of the angles of any quadrilateral is less than 360 degrees.*

Proof: The proof is left as an exercise. □

Definition 7.12. *Given quadrilateral $ABCD$ the* defect *is equal to 360 degrees minus the sum of the angles in the quadrilateral.*

Theorem 7.17. *Given $\triangle ABC$ let line l intersect sides $\overline{AB}$ and $\overline{AC}$ at points D and E, respectively. Then the defect of $\triangle ABC$ is equal to the sum of the defects of $\triangle AED$ and quadrilateral $EDBC$.*

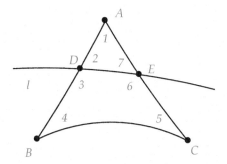

Figure 7.10

Proof: If l intersects at any of A, B, or C, the result is clear. Suppose that D and E are interior to $\overline{AB}$ and $\overline{AC}$. Label all interior angles as shown in Figure 7.10. Then

$$defect(\triangle ABC) = 180 - (\angle 1 + \angle 4 + \angle 5)$$

And,

$$defect(\triangle ADE) + defect(DECB) = 180 - (\angle 1 + \angle 2 + \angle 7)$$
$$+ 360 - (\angle 3 + \angle 4 + \angle 5 + \angle 6)$$

Since $\angle 2 + \angle 3 = 180$ and $\angle 6 + \angle 7 = 180$, the result follows. □

We can use this result to prove one of the most amazing facts about triangles in Hyperbolic geometry—similar triangles are congruent!

Theorem 7.18. *(AAA Congruence) If two triangles have corresponding angles congruent, then the triangles are congruent (Figure 7.11).*

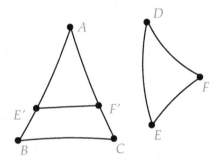

Figure 7.11

Proof: Let $\triangle ABC$ and $\triangle DEF$ be given with $\angle ABC \cong \angle DEF$, $\angle BCA \cong \angle EFD$, and $\angle CAB \cong \angle FDE$.

If any pair of sides between the two is congruent, then by ASA the triangles would be congruent.

So, either a pair of sides in $\triangle ABC$ is larger than the corresponding pair in $\triangle DEF$ or is smaller than the corresponding pair. Without loss of generality we can assume $\overline{AB}$ and $\overline{AC}$ are larger than $\overline{DE}$ and $\overline{DF}$.

Then we can find points E' and F' on $\overline{AB}$ and $\overline{AC}$ so that $\overline{AE'} \cong \overline{DE}$ and $\overline{AF'} \cong \overline{DF}$. By SAS, $\triangle AE'F' \cong \triangle DEF$, and the two triangles have the same defect.

But, since ΔDEF and ΔABC have the same defect, we have that $\Delta AE'F'$ and ΔABC have the same defect. From the previous theorem, the defect of quadrilateral $E'F'CB$ is then zero, which is impossible.

Thus, all pairs of sides are congruent. □

Exercise 7.5.1. *Prove Theorem 7.10.*

Exercise 7.5.2. *Prove Corollary 7.16.*

Exercise 7.5.3. *Prove that the summit is always larger than the base in a Saccheri Quadrilateral.*

Exercise 7.5.4. *Show that two Saccheri Quadrilaterals with congruent summits and congruent summit angles must be congruent quadrilaterals; that is, the bases must be congruent and the sides must be congruent. [Hint: Suppose they were not congruent. Show that you can then construct a rectangle using the quadrilateral with the longer sides.]*

Exercise 7.5.5. *Let l and m intersect at O at an acute angle. Let $A, B \neq O$ be points on l and drop perpendiculars to m from A and B, intersecting m at A', B'. If $OA < OB$, show that $AA' < BB'$.*

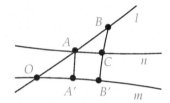

[Hint: Argue that the perpendicular to $\overline{AA'}$ at A must intersect $\overline{BB'}$ and use Lambert Quadrilateral results.]

Exercise 7.5.6. *Show that parallel lines cannot be everywhere equidistant in Hyperbolic geometry. [Hint: Suppose l, m are two parallel lines and that at points A, B, C on l, the distance to m (measured along a perpendicular to m) is the same. Derive a contradiction using Saccheri Quadrilaterals.]*

Exercise 7.5.7. *Let l be a line and m a limiting parallel to l through a point P. Show that the perpendicular distance from P to l decreases as you move P along m in the direction of the omega point of l. [Hint: Use the angle of parallelism results from Exercises 7.3.11 and 7.3.12.]*

Exercise 7.5.8. *Show that if l and m are limiting parallel, then they cannot have a common perpendicular.*

Exercise 7.5.9. *Show that two hyperbolic lines cannot have more than one common perpendicular.*

Exercise 7.5.10. *In the Poincaré model, show that two parallel lines that are not limiting parallel must have a common perpendicular. [Hint: Argue that you can assume one line is the x-axis, and then show that you can find the point on the other line (using Euclidean circle geometry) to form a common perpendicular.] Note: This result is true for any model of Hyperbolic geometry. For the proof, see [16, page 158].*

Exercise 7.5.11. *Prove that two Saccheri Quadrilaterals with congruent bases and congruent summit angles must be congruent. [Hint: Suppose they were not congruent. Show that you can then construct a quadrilateral having angle sum equal to 360.]*

Exercise 7.5.12. *Given* $\triangle ABC$ *let* l *be a cevian line (a line through a vertex and an opposite side). The cevian line divides the triangle into two sub-triangles. Show that the defect of* $\triangle ABC$ *is equal to the sum of the defects of the component sub-triangles. (This result suggests that the defect works much like the concept of area for a hyperbolic triangle.)*

Exercise 7.5.13. *Is it possible to construct scale models of a figure in Hyperbolic geometry? Briefly explain your answer.*

7.6   AREA IN HYPERBOLIC GEOMETRY

In Chapter 2 we saw that areas in Euclidean geometry were defined in terms of figures that were *equivalent*. Two figures are equivalent if the figures can be split up (or subdivided) into a finite number of pieces so that pairs of corresponding pieces are congruent. By using the notion of equivalence, we were able to base all Euclidean area calculations on the simple figure of a rectangle.

In Hyperbolic geometry this is unfortunately not possible, as rectangles do not exist! Thus, we need to be a bit more careful in building up a notion of area. We will start with some defining axioms of how area works.

> **Area Axiom I** If A, B, C are distinct and not collinear, then the area of triangle ABC is positive.
>
> **Area Axiom II** The area of equivalent sets must be the same.
>
> **Area Axiom III** The area of the union of disjoint sets is the sum of the separate areas.

These are reasonable axioms for area in Hyperbolic geometry. Note that Axiom II automatically implies that the area of congruent sets is the same.

Since area is axiomatically based on triangles, we will need the following result.

Theorem 7.19. *Two triangles ABC and A'B'C' that have two sides congruent, and the same defect, are equivalent and thus have the same area.*

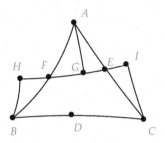

Figure 7.12

Proof: Suppose that $\overline{BC} \cong \overline{B'C'}$. Let D, E, and F be the midpoints of sides $\overline{BC}$, $\overline{AC}$, and $\overline{AB}$, as shown in Figure 7.12. Construct the line through F and E and drop perpendiculars to this line from A, B, and C meeting the line at G, H, and I, respectively.

Now, right triangles $\triangle BHF$ and $\triangle AGF$ will be congruent by AAS. Similarly, $\triangle AGE$ and $\triangle CIE$ will be congruent. Thus, $\overline{BH} \cong \overline{AG} \cong \overline{CI}$ and $BHIC$ is a Saccheri Quadrilateral. Also, it is clear that this quadrilateral is equivalent to the original triangle $\triangle ABC$ (move appropriate pieces around).

Now, it might be the case that the positions of G, H, and I are switched, as shown in Figure 7.13. However, one can easily check that in all configurations, $BHIC$ is a Saccheri Quadrilateral (with summit $\overline{BC}$) that is equivalent to $\triangle ABC$.

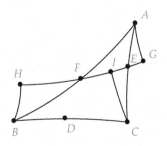

Figure 7.13

Consider the summit angles of $BHIC$. The sum of these two summit angles must equal the angle sum of $\triangle ABC$, and thus the summit angles are each half of the angle sum of $\triangle ABC$.

Similarly, we can construct a Saccheri Quadrilateral for $\triangle A'B'C'$ that is equivalent to $\triangle A'B'C'$ and with summit $\overline{B'C'}$.

Since $\overline{BC} \cong \overline{B'C'}$ and the summit angles for both quadrilaterals are congruent (equal defects implies equal angle sums), then by Exercise 7.5.4 we know that the two quadrilaterals are congruent. Thus, the triangles are equivalent. □

We can prove an even more general result.

Theorem 7.20. *Any two triangles with the same defect are equivalent and thus have the same area.*

Proof: If one side of $\triangle ABC$ is congruent to a side of $\triangle A'B'C'$, then we can use the previous theorem to show the result. So, suppose no side of one matches a side of the other. We can assume side $\overline{A'C'}$ is greater than $\overline{AC}$ (Figure 7.14). As in the last theorem, let E, F be midpoints of $\overline{AC}$ and $\overline{AB}$. Construct the perpendiculars from B and C to the line through E and F, and construct the Saccheri Quadrilateral $BHIC$.

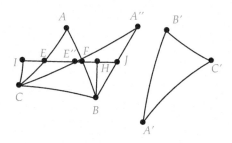

Figure 7.14

Since $A'C' > AC$, we can find a point E'' on $\overleftrightarrow{EF}$ such that the length of $\overline{CE''}$ is half that of $\overline{A'C'}$ and $E'' \neq E$. Extend $\overline{CE''}$ to a point A'' such that $\overline{A''C} \cong \overline{A'C'}$.

Clearly, $\overleftrightarrow{EF}$ cuts $\overline{A''C}$ at its midpoint E''. It is left as an exercise to show that $\overleftrightarrow{EF}$ cuts $\overline{A''B}$ at its midpoint, point J. Then, as in the previous theorem, $\triangle A''BC$ will be equivalent to the Saccheri Quadrilateral $BHIC$ and thus also equivalent to $\triangle ABC$. Since triangles $\triangle A''BC$ and

$\triangle A'B'C'$ share a congruent side, they must be equivalent, and hence the original two triangles are equivalent. □

We have shown that any two triangles having the same defect can be split into pieces that can be made congruent. Is the converse also true? Suppose we have two equivalent triangles. That is, the two triangles can be subdivided into sub-triangles with corresponding pairs congruent. Then for each pair, the defect will be the same. Also, if a set of sub-triangles forms a polygonal shape, it is easy to see that the defect of the polygon is equal to the sum of the defects of the triangles of which it is comprised. Thus, the defects of the original triangles will be the sum of the defects of all sub-triangles, which match pair-wise. We summarize this as follows.

Theorem 7.21. *Any two triangles that are equivalent (and thus have the same area) must have the same defect.*

Our conclusion from this exploration of hyperbolic area is that the defect and area share exactly the same properties. This tells us that the area must be a function of the defect. Since the area and defect are both linear functions in terms of triangle subdivisions, the area should be a linear function of the defect, and it must be positive. Therefore, $area = k^2 \, defect + c$, for some constants k and c. But, the area of the empty set must be zero, and thus $area = k^2 \, defect$. We can fix k by choosing some triangle to have unit area. We summarize this discussion in the following theorem.

Theorem 7.22. *If we have defined an area function for Hyperbolic geometry satisfying Axioms I–III, then there is a positive constant k such that for any triangle $\triangle ABC$, we have*

$$area(\triangle ABC) = k^2 \, defect(\triangle ABC)$$

Note that our discussion does not give a complete proof of this result, only an argument for the reasonableness of the theorem. For a complete, rigorous proof see the on-line chapter on hyperbolic calculations found at http://www.gac.edu/~hvidsten/geom-text or look in [32, pages 351–352].

Since the defect measures how much the angle sum of a triangle is *below* 180 and the lowest the angle sum can be is zero, we have the following corollary.

Corollary 7.23. *In Hyperbolic geometry the area of a triangle is at most* $180k^2$ *(or* πk^2*, if we use radian measure for angles).*

Exercise 7.6.1. *Show that in Theorem 7.20* $\overleftrightarrow{EF}$ *cuts* $\overline{A''B}$ *at its midpoint J. [Hint: Suppose that the intersection of* $\overleftrightarrow{EF}$ *with* $\overline{A''B}$ *is not the midpoint. Connect* E'' *to the midpoint of* $\overline{A''B}$*; construct a second Saccheri quadrilateral; and then show that you can use the perpendicular bisector of* $\overline{BC}$ *to construct a triangle with more than* 180 *degrees.]*

Exercise 7.6.2. *Show that there is no finite triangle in Hyperbolic geometry that achieves the maximum area bound.*

Exercise 7.6.3. *It is possible that our universe is hyperbolic in its geometry. Could you use the measurements of triangle areas on earth to determine if the universe were hyperbolic? Why or why not?*

7.7 PROJECT 11 - TILING THE HYPERBOLIC PLANE

In Euclidean geometry there are just three different regular tessellations of the Euclidean plane—the ones generated by equilateral triangles, by squares, and by regular hexagons. How many regular tilings are there in Hyperbolic geometry?

We can argue in a similar fashion as we did in the Euclidean tilings of Chapter 6. If we have k hyperbolic regular n-gons meeting at a common vertex of a tiling, then the interior angles of the n-gons would be $\frac{360}{k}$. Also, we can find a central point and triangulate each n-gon. The triangles in the triangulation will consist of isosceles triangles with base angles of $\alpha = \frac{360}{2k}$. The total angle sum of all the triangles in the triangulation will be *less* than $180n$. At the same time, this angle sum can be split into the angles around the center point (which add to 360) plus the base angles of the isosceles triangles (which add to $2n\alpha$). Thus, we get that

$$360 + 2n\alpha < 180n$$

Or,

$$2\alpha < 180 - \frac{360}{n}$$

Since the angle formed by adjacent edges of the n-gon is $2\alpha = \frac{360}{k}$, we get

$$\frac{360}{k} < 180 - \frac{360}{n}$$

Dividing both sides by 360 and then rearranging, we get

$$\frac{1}{n} + \frac{1}{k} < \frac{1}{2}$$

If there is a regular tessellation by n-gons meeting k at a vertex in Hyperbolic geometry, then $\frac{1}{n} + \frac{1}{k} < \frac{1}{2}$. On the other hand, if this inequality is true, then a tiling with n-gons meeting k at a vertex must exist. Thus, this inequality completely characterizes regular hyperbolic tilings. We will call a regular hyperbolic tessellation of n-gons meeting k at a vertex a *(n,k)* tiling.

As an example, let's see how to generate a $(5, 4)$ tiling. In a $(5,4)$ tiling, we have regular pentagons meeting four at a vertex. How do we construct regular pentagons of this kind? First of all, it is clear that the interior angle of the pentagon must be 90 degrees $\left(\frac{360}{4}\right)$. If we take such a pentagon and triangulate it via triangles constructed to a central point, the angles about the central point will be 72 degrees and the base angles of the isosceles triangles will be 45 degrees (half the interior angle). Thus, to build the pentagon we need to construct a hyperbolic triangle with angles of 72, 45, and 45.

In Euclidean geometry, there are an infinite number of triangles that have a specified set of three angles, and these triangles are all similar to each other. In Hyperbolic geometry, two triangles with congruent pairs of angles must be congruent themselves!

Thus, we know that a hyperbolic triangle with angles of 72, 45, and 45 degrees must be unique. To start this project, you will need to download a file which contains the construction of this triangle. Download this file, built for your particular dynamic geometry software environment, from http://www.gac.edu/~hvidsten/geom-text.

Start up your geometry software and open the downloaded file. You should see a set of three points A, B, and C that are the vertices of our desired triangle. To make sure that this triangle is correct, let's measure the angles.

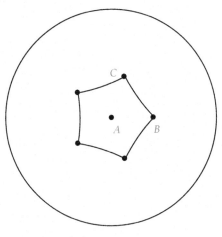

Looks good. Hide the angle measurements and connect C and B with a segment. We will construct a pentagon using this segment as a side. Set A as a center of rotation of 72 degrees and rotate segment $\overline{BC}$ four times to get a regular pentagon.

Finally, set point C as a center of rotation of 90 degrees. Rotate the pentagon three times, yielding four pentagons meeting at right angles!

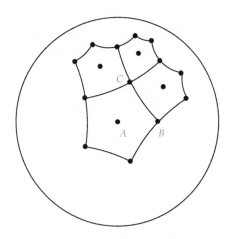

If we continue to rotate the pentagon about exterior points in this figure, we see that a tiling of the hyperbolic plane is indeed possible with regular pentagons meeting at right angles.

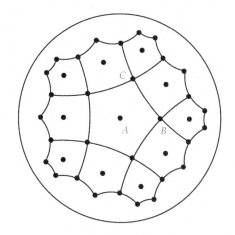

However, if we move the point at the origin, we see that the tiling breaks up in a quite nasty way. Why is this the case? The problem here is that by translating the origin point, we have created compound translations and rotations for other parts of the figure. In Hyperbolic geometry compositions of translations are not necessarily translations again as they are in Euclidean geometry.

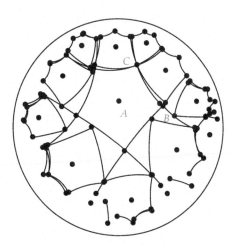

Exercise 7.7.1. *In Elliptic geometry the sum of the angles in a triangle is more than 180 degrees. Show that it is possible to have a (3, 3) tiling in Elliptic geometry.*

Exercise 7.7.2. *There are only three regular tilings (up to scaling) in Euclidean planar geometry. How many regular tilings are there in Hyperbolic (two-dimensional) geometry?*

Exercise 7.7.3. *Describe and illustrate the steps you would take to construct a (6,5) tiling in the Poincaré Plane.*

For your report give a careful and complete summary of your work done in this project.

CHAPTER 8

Elliptic Geometry

To see a World in a Grain of Sand
And a Heaven in a Wild Flower,
Hold Infinity in the palm of your hand
And Eternity in an hour.

– William Blake (1757–1827)

8.1 BACKGROUND AND HISTORY

Our discussion of non-Euclidean geometry in Chapter 7 started with a review of Euclid's axioms for planar geometry:

We noted that the fifth postulate, the *parallel postulate*, seemed more like a theorem —it did not seem self-evident or obvious. We can replace the fifth postulate with the logically equivalent statement called *Playfair's Postulate*:

> Given a line and a point not on the line, it is possible to construct exactly one line through the given point parallel to the line.

Historically, mathematicians have attempted to prove Euclid's fifth postulate (or, equivalently, Playfair's Axiom) as a *theorem* solely on the basis of the first four postulates. One mathematician, Girolamo Saccheri, didn't try to prove the fifth postulate directly, but instead tried to prove it by the method of contradiction. He looked at special figures in the plane which we now call Saccheri Quadrilaterals. We studied these in detail in Project 7.3.2 in Chapter 7. Saccheri quadrilaterals are quadrilaterals whose base angles are right angles and whose base-adjacent sides are congruent. In Exercise 2.2.14 of Chapter 2 we proved that the statement "The summit angles of a Saccheri quadrilateral are right angles" is logically equivalent to Playfair's axiom, assuming the first four Euclidean axioms hold.

Saccheri tried to prove Euclid's fifth postulate by working with this new equivalent statement about the summit angles of a Saccheri quadrilateral being right angles. His proof by contradiction assumed an axiomatic basis of the first four Euclidean axioms together with the logical negation of the summit angles statement. There are two different logical negations to the assumption that the summit angles are right angles:

1. The summit angles of a Saccheri quadrilateral are *acute* angles.

2. The summit angles of a Saccheri quadrilateral are *obtuse* angles.

Saccheri was never able to reach a contradiction when assuming the hypothesis of the acute angle. The hypothesis of the acute angle can be developed into a logically consistent geometry —Hyperbolic geometry —which was the focus of Chapter 7.

In the case of the obtuse angle hypothesis, Saccheri was able to achieve a contradiction of a known result in Neutral geometry (the geometry which only depends on the first four Euclidean axioms). If the summit angles are obtuse, then the angle sum in a Saccheri quadrilateral is more than 360 degrees. If one then divides the quadrilateral into two triangles (by using one of the diagonals) one of the triangles must have an angle sum of more than 180 degrees. This contradicts the neutral geometry result that the sum of the angles in a triangle can be no more than 180 degrees. It may appear that there is no hope for a consistent geometry based on the obtuse angle hypothesis. But, let's take a closer look at the contradiction.

The proof of the neutral geometry result that the angle sum of a triangle is no more than 180 degrees ultimately relies on Euclid's Proposition **I-17**, that the sum of any *two* angles in a triangle is less than 180 degrees. In turn, the proof of Proposition **I-17** is a direct consequence of Euclid's Proposition **I-16**, the Exterior Angle Theorem. None of the propositions preceding Euclid **I-16** deal specifically with comparisons of angle measure in triangles. If there is any chance of creating a geometry with the obtuse hypothesis, then, it must be built using only those results that depend on the first 15 Euclidean Propositions. Such a geometry will be called "Elliptic" geometry.

In the rest of this chapter we will investigate Elliptic geometry as our second example of a Non-Euclidean geometry. We will develop this geometry by using results from Neutral geometry that can be proven without using the Exterior Angle Theorem, or propositions whose proofs depend on this theorem. One property we will have to be careful about is *betweenness*, as it is the betweenness properties of Neutral geometry that one uses to prove the Exterior Angle Theorem. In particular, the proof relies on the betweenness properties that ensure that lines can be indefinitely (infinitely) extended.

We will also make use of reflections and rotations, found in sections 5.1, 5.2, and 5.4. The construction of reflections and rotations does not depend on the parallel postulate, nor on results based on the Exterior Angle Theorem. However, many results about these isometries are proven using standard betweenness properties of Neutral geometry. For example, the proof of Theorem 5.1 relies on such betweenness properties. As was mentioned at the end of section 5.1, this betweenness issue can be remedied by providing new axioms for betweenness in Elliptic geometry (see the on-line chapter on Hilbert's axioms). Refer to the discussions at the end of sections 5.1, 5.2, and 5.4 for specific information on the results concerning rotations and reflections that can be used in Elliptic geometry.

We will assume Pasch's Axiom and that there is a distance function defined for pairs of points. We will also assume there is an angle measure function, and that distance and angle measure are continuous functions. We saw earlier that these assumptions must be added to Euclid's axiomatic system (and to Hyperbolic geometry) to ensure completeness of those systems, so it is reasonable to assume these properties in Elliptic geometry as well.

Finally, we will assume the SAS, SSS, and ASA triangle congruence theorems. As shown in the on-line chapter on Hilbert's axioms, these can be proven without using the Exterior Angle Theorem.

8.2 PERPENDICULARS AND POLES IN ELLIPTIC GEOMETRY

As mentioned in the introduction to this chapter, the parallel postulate has been the focus of mathematicians since the time of Euclid. A simple, but logically equivalent, form of this postulate is Playfair's Postulate:

Given a line and a point not on the line, it is possible to construct exactly one line through the given point parallel to the line.

In Chapter 7, we developed a consistent geometry (Hyperbolic geometry) based on the assumption that there are *multiple* lines through a point parallel to a given line. In this chapter we will see that there is a logically consistent third geometry where there are no parallels. In this geometry, we replace the parallel postulate with the *Elliptic Parallel Axiom*.

Given a line and a point not on the line, it is not possible to construct any line through the given point parallel to the line.

In developing Elliptic geometry, we will use as our axiomatic basis the first four Euclidean axioms plus the Elliptic Parallel axiom. We will also assume the Euclidean Propositions before the Exterior Angle Theorem and we will assume that segment length and angle measure are continuous. Neutral results about reflections and rotations will be used, as well as the SAS, SSS, and ASA triangle congruence properties. The reasonableness of all of these assumptions was discussed in the introduction to this chapter.

We start our development of Elliptic geometry with a basic result about perpendiculars to a line.

Theorem 8.1. *Given a line l, if two perpendiculars to l (at points P and Q) intersect at a point O, then all perpendiculars to l intersect at O. Also, for every point P on l, the distance from P to O is constant.*

Proof: Let lines m and n be perpendicular to l at P and Q.

Using a ASA argument, we know that triangles $\triangle PQO$ and $\triangle QPO$ are congruent. Thus, $\overline{OP} \cong \overline{OQ}$. Let R be the midpoint of $\overline{PQ}$. By SSS, $\triangle PRO \cong \triangle QRO$. Thus, the angle at R is a right angle and $\overline{RO} \cong \overline{PO}$.

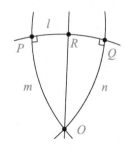

Now, let R_2 be the midpoint of $\overline{PR}$ and R_3 the midpoint of $\overline{QP}$. Then, by an argument analogous to that above, we can show that $\overline{OR_2}$ and $\overline{OR_3}$ meet l at right angles and both segments have length equal to the length of $\overline{PO}$.

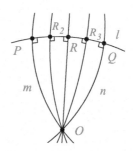

It is clear that we can continue bisecting successive segments, generating a sequence of points $R_2, R_3, R_4, \ldots$ such that at each point R_k we have that $\overline{OR_k}$ meets m at right angles. After we bisect segments successively n times, the points $R_2, R_3, R_4, \ldots$ will generate a sequence of segments whose length is $\frac{1}{2^n}$ of the original length of $\overline{PQ}$. These *dyadic segments* are used in the on-line chapter on Hilbert's axioms to demonstrate the continuity of segment length. A careful reading of this development of segment length reveals that it uses *betweenness* properties. However, it does not use the "infinite extent" property referred to above when we discussed the problems of using the Exterior Angle Theorem in Elliptic geometry. We can thus safely assume the continuity of segment length (and angle measure). Using a limiting process of dyadic segments to approach closer and closer to points on $\overline{PQ}$, we can assume that for *any* point R on $\overline{PQ}$, if we construct the segment $\overline{OR}$ then this segment meets line l at right angles. Also, for all R_k we have that $\overline{R_kO} \cong \overline{PO}$.

What about other points on l? Let Q' be a point on l on the other side of Q from P such that $\overline{PQ} \cong \overline{QQ'}$. Then, by SAS we have $\triangle OPQ \cong \triangle OQ'Q$, and thus $\angle OQ'Q$ is a right angle. By the argument above, for S any point on $\overline{QQ'}$, we have that $\overline{OS}$ meets l at right angles and $\overline{SO} \cong \overline{PO}$.

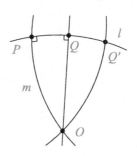

For any other point T on l, we know that we can reach T by successfully constructing new segments on l congruent to $\overline{PQ}$ until we reach a segment that contains T. The argument above then shows that $\overline{OS}$ meets l at right angles and $\overline{TO} \cong \overline{PO}$. $\square$

The astute reader may have noticed something odd about this proof. In the figures accompanying the proof we illustrated lines by curved arcs. This was done for two reasons —to keep the reader from intuitively reasoning about Elliptic geometry in Euclidean terms, and to foreshadow the work we will do in finding a model for this geometry (in that model, Elliptic lines *are* parts of Euclidean circles).

The point O in the preceding theorem is called a "pole" of the line l.

> **Definition 8.1.** *If the perpendiculars to a given line are concurrent at a point O, then O is called the* pole *of the line.*

The following result shows that the relationship of lines to poles is homogeneous throughout Elliptic geometry.

> **Theorem 8.2.** *The distance (measured along a perpendicular) from a line to its pole is the same for all lines.*

Proof: Let $\overleftrightarrow{AB}$ and $\overleftrightarrow{CD}$ be two lines.

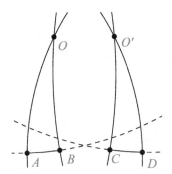

We can assume that $\overline{AB}$ and $\overline{CD}$ have the same length. At A and B construct the perpendiculars to $\overline{AB}$ and at C and D construct the perpendiculars to $\overline{CD}$. Then, the perpendiculars at A and B will intersect at some point O and the perpendiculars at C and D will intersect at O'. By ASA congruence, $\triangle ABO \cong \triangle CDO'$. The result follows.

□

This theorem allows us to define a *universal* polar distance.

> **Definition 8.2.** *The* polar distance *in Elliptic geometry is the distance (measured along a perpendicular) from any line to its pole.*

The next result explains why the Euclidean (and Neutral) assumption of the infinite extent of lines is not possible in Elliptic geometry.

Theorem 8.3. *All lines in Elliptic geometry have a finite length equal to twice the polar distance.*

Proof: Let l be a line and let A be a point on l. At A construct the perpendicular m to l. Let $B \neq A$ be a point on m. At B construct the perpendicular n to m. Then l and n meet at the polar point O of m.

However, extending rays $\overrightarrow{OA}$ and $\overrightarrow{OB}$ across m we also get the polar point O' at the same distance from m. Since two lines can only intersect in a single point, it must be that $O = O'$ and the line $l = \overleftrightarrow{OA}$ consists of the points from O to $O' = O$. That is, the line returns back on itself. Clearly the length of l is twice the polar distance of m.

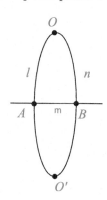

□

We started this chapter with a discussion of the historical importance of Saccheri quadrilaterals in the development of non-Euclidean geometry. In order to analyze Saccheri quadrilaterals in Elliptic geometry, we will need the next result on right triangles.

Theorem 8.4. *Let $\triangle ABC$ be a right triangle with right angle at A. Let q be the polar distance. Then*

1. *Angle B (C) is less than 90 degrees if and only if the length of $\overline{AC}$ (respectively $\overline{AB}$) is less than q.*

2. *Angle B (C) is a right angle if and only if the length of $\overline{AC}$ (respectively $\overline{AB}$) is equal to q.*

3. *Angle B (C) is more than 90 degrees if and only if the length of $\overline{AC}$ (respectively $\overline{AB}$) is more than q.*

Proof: We start with the first part of the theorem.

Let O be the polar point of $\overleftrightarrow{AB}$. Construct $\overline{OA}$ and $\overline{OB}$. Then, $\angle OBA$ is a right angle. If $\angle CBA$ is less than a right angle, then ray $\overrightarrow{BC}$ must intersect side $\overline{OA}$ in triangle $\triangle OAB$. So, $\overline{AC} < \overline{AO} = q$.

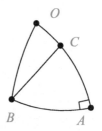

Conversely, if $\overline{AC} < q = \overline{AO}$, then, ray $\overrightarrow{BC}$ is interior to $\angle OBA$ and thus less than a right angle.

For the second part of the theorem, suppose that the angle at B is a right angle. Then, $O = C$ and the length of $\overline{AC}$ is q. Conversely, if the length of $\overline{AC}$ is q, then $O = C$ and the angle at B must be the same as $\angle OBA$ which is a right angle.

The third claim of the theorem is immediate once the first two claims are proven. □

We can now analyze Saccheri quadrilateral summit angles.

Theorem 8.5. *The summit angles in a Saccheri quadrilateral are obtuse.*

Proof: Let $ABCD$ be a Saccheri quadrilateral.

Assume the quadrilateral has base right angles at A and B. Let M and N be the midpoints of $\overline{AB}$ and $\overline{CD}$.

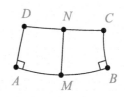

Then, $\overleftrightarrow{MN}$ meets the base and summit lines at right angles. (Exercise) Now, construct lines $\overleftrightarrow{NC}$ and $\overleftrightarrow{MB}$. These will intersect at the polar point O.

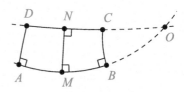

Also, segments $\overline{MO}$ and $\overline{NO}$ both have length q. So, $\overline{BO}$ has length less than q and by the previous theorem, $\angle BCO$ is less than 90, so $\angle NCB$, which is a summit angle, must be more than 90. □

Exercise 8.2.1. *Show that the line joining the midpoints of the base and summit of a Saccheri Quadrilateral is perpendicular to the summit and base. [Hint: use triangles]*

Exercise 8.2.2. *A* Lambert Quadrilateral *is a quadrilateral having three right angles. Let ABDC be a Lambert quadrilateral with right angles at A, B, and C. If we extend $\overline{AB}$ and $\overline{CD}$, we can find points E and F such that $\overline{AB} \cong \overline{AE}$ and $\overline{CD} \cong \overline{CF}$ Show that EBDF is a Saccheri quadrilateral.*

Exercise 8.2.3. *Show that in a Lambert quadrilateral the fourth angle (the one that is not a right angle) is always obtuse.*

Exercise 8.2.4. *Show that in a Lambert quadrilateral ABCD each side adjacent to the fourth angle has length smaller than the opposite side. [Hint: Look at the proof of Theorem 7.13]*

Exercise 8.2.5. *Let l and m be two distinct lines. Suppose that line n is a common perpendicular to l and m, i.e., n meets l and m at right angles. Let O be the intersection point of l and m. Show that O is the polar point of n.*

8.3 PROJECT 12 - MODELS OF ELLIPTIC GEOMETRY

We have started to build up some basic results in Elliptic geometry. Before we go further, we should stop and ask the question of *consistency*. As was the case in Hyperbolic geometry, what we will be looking for is a model for Elliptic geometry that is built within Euclidean geometry and that satisfies the axioms of Elliptic geometry. We are looking for a model that satisfies the first four Euclidean axioms (but without the infinite extent property of lines) and the Elliptic Parallel Postulate —that there are no parallels.

Since we know that all lines are bounded in Elliptic geometry, it is not feasible to create a model that is built from the entire Euclidean plane (unless we *drastically* change what distance means). So, we are looking for another Euclidean space (or geometric object) that is *bounded* and yet has no *boundary*. Why no boundary? We know from the previous section that Elliptic lines, though finite, do not have boundaries —they can be extended indefinitely (think of a circle, for example).

8.3.1 Double Elliptic Model

What clues do we have to construct a model of Elliptic geometry? In the proof of Theorem 8.3 we saw that if we reflected the polar construction of a line across the line itself, we would get a second polar point. We ruled this out at the time, based on the (axiomatic) assumption that two lines intersect in a single point. But, as a matter of speculation, what would a dual intersection property imply? If a model satisfies the dual intersection property, it would have to be a two-dimensional shape that has the property that there are two special points (poles) for each pair of lines, depending on how we define lines. We call this *Double Elliptic geometry*.

To model Double Elliptic geometry, we want a shape that is bounded, has no boundary, and has double poles. A quick survey of common shapes reveals a likely candidate —the sphere. We know that elliptic lines must always intersect at poles that are a maximum distance away from each other along the lines. Thus, the two poles, called *antipodes* must lie along a diameter of the sphere. The definition of an elliptic line that fits this requirement is that an elliptic line is a great circle on the sphere. A great circle is a circle on the sphere that results from slicing the sphere with a plane through the origin. Clearly, any two such great circles (elliptic lines) will always intersect at antipodal points (Figure 8.1).

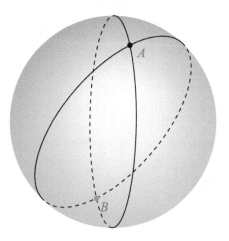

Figure 8.1 Spherical great circles

We note here that this is the geometry we explored in Exercise 1.6.7 at the end of Chapter 1. In that exercise we called this geometry "Spherical geometry," as it is based on the sphere. We can now see that this geometry is closely related to Elliptic geometry. However, in the Elliptic geometry model we are striving to create, two distinct lines can only intersect at a single point. So, Spherical geometry cannot exactly model Elliptic geometry.

Before we investigate how to modify the sphere model to make a valid model for Elliptic geometry, let's take a short excursion into Spherical geometry, just for fun!

8.3.2 Spherical Lunes

In Euclidean and Hyperbolic geometry, one needs a minimum of three non-collinear line segments to define a polygonal area. In Spherical geometry, two spherical lines can define an area called a *lune* (also called a *biangle*).

To investigate this interesting shape we will make use of the Spherical geometry capabilities of the software program *Geometry Explorer*. If you are not already using this program, go to http://www.gac.edu/~hvidsten/geom-text for instructions on how to download and use this software.

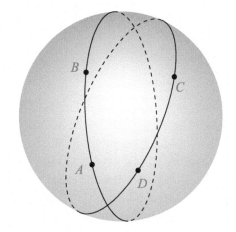

Start *Geometry Explorer* and open a Spherical geometry window. (Choose **New** (**File** menu) and then choose "Spherical" in the dialog box that pops up). Create lines $\overleftrightarrow{AB}$ and $\overleftrightarrow{CD}$ on the screen. Note that the two lines meet at antipodal points and that the two lines divide the sphere into four areas. Each of these is called a *lune*. Select the two lines and click the Filled Polygon button in the Construct panel (third button in third row).

One of the four lunes has been created. The term lune comes from the Latin word for moon – *luna.* The crescent shape of the moon at night is formed from the shadow of the earth intersecting with the boundary great circle of the moon.

Click on point B and move it around. As you move B the lune will change in size, but will always be defined by the two spherical "lines."

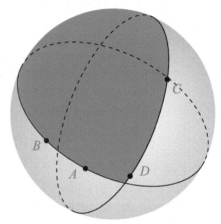

If you move B too far, however, the lune will switch to another of the four possible shapes defined by the two spherical lines. This is due to the internal algorithm that *Geometry Explorer* uses to determine the orientation and choice of lune.

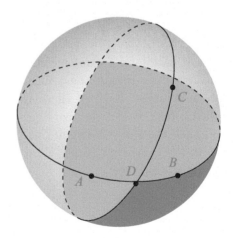

Lunes define areas in Spherical geometry. Let's push a bit further and calculate these lunar areas.

Undo the moves from the previous example so that the lune looks like it did originally. Then, click on the inside of the lune to select it and choose **Area** (**Measure** menu). The area measurement should appear in the window. We will be making another measurement, so to give ourselves some room, click the period key "." to zoom out a bit.

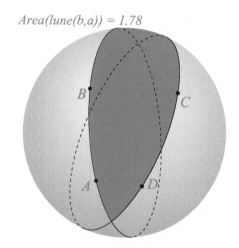

Area(lune(b,a)) = 1.78

The lunar area as a single number is not that impressive. We know the lunar area is positive and that it must be less than the total area of the sphere (4π, assuming a unit radius). But, is there any simple way to connect the lunar area to other geometric quantities of the lune? One natural geometric value defined by a lune is the angle made by the intersecting lines.

To calculate the angle made by the intersection of our lines, select the two lines and click the Intersect button in the Construct Panel. Intersection points E and F will be created. Select points D, F, and A (in that order) and choose **Angle** (**Measure** menu). The new measurement will appear. Note that angles are measured in degrees.

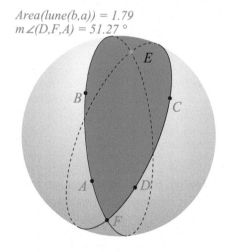

Area(lune(b,a)) = 1.79
m∠(D,F,A) = 51.27 °

In Spherical geometry a much more insightful unit for angle measurement is radians. To change the angle unit to radians, right-click on the angle text and choose **Properties...** from the pop-up Menu. Once the Properties Dialog box comes up, select the Style tab and then select "Radians" from the Measure Units selection box. Click "Okay" to change the display of the angle's measure to radians.

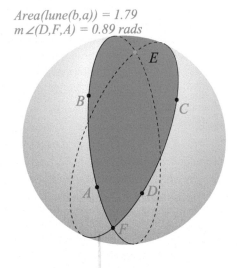

Area(lune(b,a)) = 1.79
m∠(D,F,A) = 0.89 rads

If we move point B around and look at the relationship between the area and the angle defined by the lune, a conjecture might arise. Take a minute and play with this configuration and see if you can guess the relationship between the angle and the area.

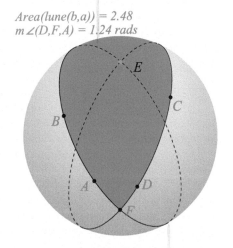

Area(lune(b,a)) = 2.48
m∠(D,F,A) = 1.24 rads

Exercise 8.3.1. *Prove the result suggested by the activity above. That is, show that if the sphere has radius of 1, then the area of a lune is always twice the radian measure of the angle determined by the intersection of the defining spherical lines. [Hint: Start by considering the case where the angle is $\frac{\pi}{2}, \frac{\pi}{4}, \frac{3\pi}{4}$, etc.]*

Exercise 8.3.2. *Write down the corresponding formula for the area of a lune if the radius of the sphere is R.*

While Spherical geometry is a fascinating subject, we must return to the topic of this chapter —Elliptic geometry. The sphere model cannot work as a model for Elliptic geometry due to the fact that lines meet at two points. But, can we modify the model so that we can have lines meeting at only one point? As mentioned in the proof of Theorem 8.3, if we require that lines meet at a single point, then it must be the case that the two poles from the spherical geometry model must actually be the same point. That is, we have to identify these two seemingly different points as representing the same Elliptic geometry point. The geometry based on this model is called *Single Elliptic geometry*

8.3.3 Single Elliptic Geometry

Our first true model of Elliptic geometry will be based on the sphere, but with the modifications discussed above. We have the following definitions for point and line.

> **Definition 8.3.** *An* elliptic point *will represent an antipodal pair of points on the sphere. An* elliptic line *will be a great circle on the sphere (with antipodal points identified). We say that point P lies on line l if P (or its antipodal counterpart) lies on l.*

The Elliptic geometry based on this model is called *Single Elliptic geometry*, as it uses a single copy of each antipodal pair of points on the sphere. To help visualize this model, we will use our Spherical geometry computer environment.

Clear your previous work and construct a line $\overleftrightarrow{AB}$ in the window. The line will be constructed with two points A and B. To see the antipodal points to these two points, select A and B and choose **Antipodal Transformation** (**Misc** menu). The antipodal points C and D will be created.

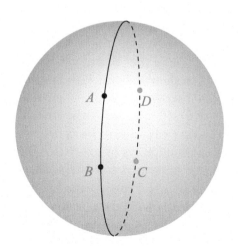

In Elliptic geometry the two spherical points A and C represent the same elliptic point. This may seem a bit strange, but we have seen earlier in this chapter that this identification is forced on us, if we want to preserve the property that two elliptic lines meet in a single point. Thus, in creating our model of Elliptic geometry, we only need to plot one of each pair of antipodal points, and we only need to plot the front half of an elliptic line.

The software program *Geometry Explorer* also has a built-in Elliptic geometry environment. For the rest of this project we will use that environment.

Open an Elliptic geometry window by choosing **New** (**File** menu) and choose Elliptic in the dialog box that pops up. *Geometry Explorer* actually has two models of Elliptic geometry, both based on the sphere. We will first explore the model that is directly based on the antipodal identification of sphere points discussed above. To set this model as our working model, choose **Sphere X-Y Projection** (**Model** menu). Draw a line somewhere on the screen.

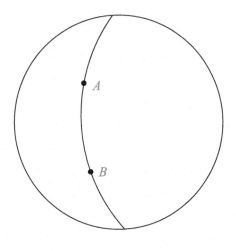

Note the similarity in the two figures above. In the spherical case, we see the entire spherical line and all antipodal points. In the elliptic model we see only one of each pair of antipodal points.

To verify that this model is actually a model for Elliptic geometry, we need to show that Euclid's first four postulates, plus the elliptic parallel axiom, are satisfied in this model.

We will start with the first two postulates of Euclid—that unique segments can always be constructed through two points and that segments can always be extended. By segments we will mean subsets of the elliptic lines as defined thus far. Since the only problem with defining a line between two points on the sphere is if the points are antipodal, then clearly the first Euclidean axiom holds. If we interpret "extension" of a segment to mean continuing to move along the line beyond the segment, then clearly this can be done in our model, although one starts to re-trace the same set of points after awhile.

For the construction of a circle at a point of a certain radius, we would use distance as measured *along great circles*. For example, if we wanted to construct the circle at a point whose radius was π (half the circumference of a unit circle), then that circle would fill the entire Elliptic model, as the circle on the sphere would cover a hemisphere. Thus, not every circle of a given radius can be constructed, but if one restricts the set of possible radii to be between 0 and the longest length of a line (the polar distance), then all circles of those radii can be constructed.

Angles are defined by the Euclidean angle on the sphere, so Euclid's fourth axiom holds. Finally, this model was built to satisfy the Elliptic parallel axiom.

Thus, the single Elliptic model we have constructed is a valid model for Elliptic geometry and this non-Euclidean geometry is equally as consistent as Euclidean geometry.

Let's look at one of Euclid's "Neutral" results (independent of the parallel postulate) —the construction of an equilateral triangle.

Clear the Elliptic geometry window and draw a segment somewhere on the screen. Then, draw circles centered at each of A and B with radius to the other point as shown. Finally select the two circles and construct the intersection points C and D.

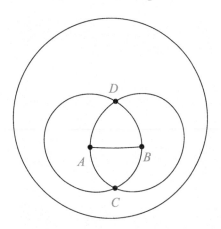

Hide the circles and point C and construct the other two sides of triangle ABD. Then, select these three sides and measure their lengths. Drag the vertices of the triangle around the screen to verify the equality of the three lengths.

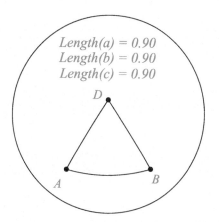

Exercise 8.3.3. *Measure the angles of the triangle. Is it possible to make a triangle with all right angles? With all angles larger than 90 degrees? Is there an upper bound on the angles? If so, what is it?*

Exercise 8.3.4. *Move the points of the triangle so that the three angles become right angles. Look at the lengths of the sides. What conjecture would you make as to the side length (in relation to the polar length) when the angles are right angles? Give a brief argument for your conjecture, based on spherical geometry.*

For your report give a careful and complete summary of your work on this project.

8.4   BASIC RESULTS IN ELLIPTIC GEOMETRY

We will now look at some basic results concerning lines, triangles, circles, and the like, that hold in all models of Elliptic geometry. In studying any new geometry, it is helpful to be able to draw diagrams. Such diagrams are based on specific models. In the previous project we studied one model of Elliptic geometry that was based on the sphere. That model is appealing because it ties directly to the sphere —an object with which we are familiar. However, the model has several drawbacks.

In Figure 8.2 we have several constructions in the sphere model of Elliptic geometry.

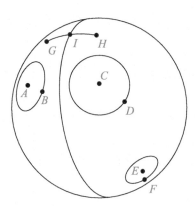

Figure 8.2

The circle defined by C and D looks fairly normal, but the circles centered at A and E are near the edge of the sphere, so when they are projected onto the X-Y plane, they get scrunched into narrow ellipses. Also, the line shown through I on $\overline{GH}$ is actually perpendicular to $\overline{GH}$ at I. This would be apparent on the sphere, but in the model there is a distortion of the angles when projecting onto the X-Y plane.

It would be nice to have a model where circles look like "normal" circles and angles behave like Euclidean angles. Luckily, there is such a model based on *stereographic projection* of the sphere onto the plane.

8.4.1 Stereographic Projection Model

Let S be the unit sphere, as shown in Figure 8.3.

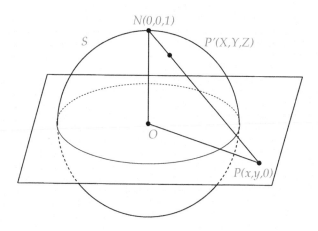

Figure 8.3

Let N be the north pole of the sphere, the point at $(0,0,1)$. The sphere is cut into two equal hemispheres by the X-Y plane. Given a point $P = (x, y)$ in the plane, we map P onto the sphere by joining N to P by a line and finding the intersection point P' of this line with the sphere. Points (x, y) inside the unit disk in the X-Y plane will map to the lower hemisphere, and points outside the unit disk will map to the upper hemisphere.

The map π given by $\pi(P') = P$ identifies points on the sphere with points in the X-Y plane. This map is called the *stereographic projection* of S onto the complex plane.

It can be shown that stereographic projection maps circles on the sphere to circles in the plane and it maps spherical lines (great circles) to Euclidean lines in the plane. (Consult the on-line chapter on advanced analytic geometry found at http://www.gac.edu/~hvidsten/geom-text for details.) Stereographic projection also has the property that it preserves *angles*. This is true of any map of the extended complex plane that takes circles and lines to circles and lines (see [9, page 90] or [23, pages 248–254]).

Stereographic projection of the spherical configuration shown in Figure 8.2 will result in the figure below.

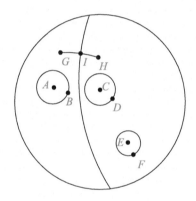

Figure 8.4

Elliptic circles look like Euclidean circles and the right angle at I is visually apparent. Because of the more intuitive nature of the stereographic model, we will use this model as the basis for our diagrams of Elliptic geometry throughout the rest of this chapter. However, we must be careful. As was pointed out when we discussed Hyperbolic geometry in Chapter 7, too much reliance on diagrams can lead to hidden assumptions. We will be careful to argue solely from the postulates of Elliptic geometry and the theorems based on those postulates.

In the rest of this section, we will discuss some of the basic results concerning Elliptic geometry. We start with a review of already proven results concerning the properties of poles and polar length:

1. The perpendiculars to a given line are all concurrent at a point O called the *pole* of the line. (Theorem 8.1)

2. The distance (measured along a perpendicular) from a line to its pole is the same for all lines. (Theorem 8.2)

3. The *polar distance* is the distance (measured along a perpendicular) from any line to its pole. (Definition 8.2)

4. All lines in Elliptic geometry have a finite length equal to twice the polar distance. (Theorem 8.3)

Here is a useful fact about lines and polar points:

Theorem 8.6. *Let q be the polar distance and let l be a line with polar point O. Let P be a point on l. Then P is the unique point on $\overleftrightarrow{OP}$ such that the length of $\overline{OP}$ is equal to q.*

Proof: Suppose there were another point Q on $\overleftrightarrow{OP}$ such that length of $\overline{OQ}$ is equal to q. Let R be a point not on l and not on $\overleftrightarrow{OP}$. Consider triangles $\triangle POR$ and $\triangle QOR$. (Figure 8.5) These are congruent by SAS. Thus, angle $\angle ORP$ is congruent to angle $\angle ORQ$. This means Q must be on $\overleftrightarrow{PR}$. But this line intersects $\overleftrightarrow{OP}$ at a unique point, so $P = Q$.

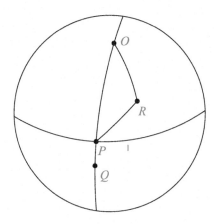

Figure 8.5

□

One of the most important figures in Elliptic geometry is the Saccheri quadrilateral. Let's review some of the results about this figure (and a related figure —the Lambert quadrilateral):

1. The line joining the midpoints of the base and summit of a Saccheri Quadrilateral is perpendicular to the summit and base. (Exercise 8.2.1)

2. The summit angles in a Saccheri quadrilateral are obtuse. (Theorem 8.5)

3. In a Lambert quadrilateral the fourth angle (the one that is not a right angle) is always obtuse. (Exercise 8.2.3)

4. In a Lambert quadrilateral $ABCD$ each side adjacent to the fourth angle (the one that is not a right angle) has length smaller than the opposite side. (Exercise 8.2.4)

Our first result will extend what we know about Saccheri quadrilaterals.

> **Theorem 8.7.** *The length of the summit segment in a Saccheri quadrilateral is always less than the length of the base segment. Also, the length of the segment joining the midpoints of the summit and base is always larger than either of the two sides of the quadrilateral.*

Proof: Let $ABCD$ be a Saccheri quadrilateral with base $\overline{AB}$ and summit $\overline{CD}$. Let E and F be the midpoints of the base and summit, respectively. (Figure 8.6)

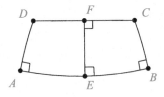

Figure 8.6

Then, $AEFD$ and $EBCF$ are both Lambert quadrilaterals (they both have three right angles). Thus the length of $\overline{CF}$ is less than the length of $\overline{BE}$ and the length of $\overline{DF}$ is less than the length of $\overline{AE}$. Adding lengths shows that the length of the summit is less than the length of the base. Also, we know that the lengths of $\overline{BC}$ and $\overline{AD}$ are less than the length of $\overline{EF}$. This finishes the proof of the theorem. □

8.4.2 Segments and Triangle Congruence in Elliptic Geometry

In Theorem 8.4 we proved that if $\triangle ABC$ is a right triangle with right angle at A, and q is the polar distance, then

1. Angle B (C) is less than 90 degrees if and only if the length of $\overline{AC}$ (respectively $\overline{AB}$) is less than q.

2. Angle B (C) is a right angle if and only if the length of $\overline{AC}$ (respectively $\overline{AB}$) is equal to q.

3. Angle B (C) is more than 90 degrees if and only if the length of $\overline{AC}$ (respectively $\overline{AB}$) is more than q.

When we proved this theorem, we assumed that we had a solid definition of a triangle as consisting of three segments defined by three non-collinear points. We also assumed we had a well-defined notion of the length of a line segment. In both cases, the definitions in question ultimately depended on a well-defined notion of what a *segment* is in Elliptic geometry.

Here is a line defined by points A and B. We have drawn in a solid color the set of points we would normally think of as the *segment* $\overline{AB}$. However, the set of points on the dashed curve could also be reasonably considered the segment from A to B. The problem here is the definition of the *unique* segment defined by two points. Since lines are finite in length and return on themselves, then there are really two different possible choices for the segment defined by A and B.

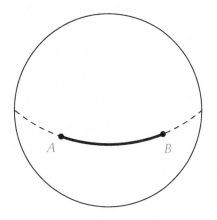

We will resolve this ambiguity by defining a segment to be the smaller of the two choices. The two possible segments must be defined by a modified version of betweenness in Elliptic geometry. This is covered in the on-line chapter on Hilbert's axioms. We assume that there are two subsets of the line through A and B, and that both have endpoints A and B.

Definition 8.4. *The segment defined by $A \neq B$ in the Elliptic plane is defined by the choice of the line subset defined by A and B having the smaller length. If both subsets have the same length, the choice is arbitrary.*

Note that this definition allows us to unambiguously define the length of a segment. It also makes clear what the segments for a triangle are, except in the ambiguous case mentioned in the definition.

In Figure 8.7 we have triangle $\triangle ABC$. We have colored the *interior* of the triangle gray, or at least this is what we expect the interior to look like. However, the three sides of the triangle also define another three-sided region —the region colored blue in the figure. The triangle interior could reasonably be defined as either the shape in gray or the shape in blue; there is an ambiguity to the notion of the interior of a triangle.

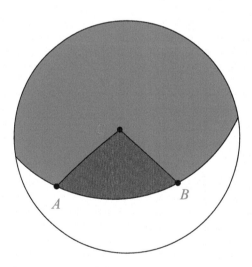

Figure 8.7

We will fix this problem by restricting the angles in a triangle to be less than 180 degrees.

Now that we have a clear idea of what a segment is and what defines a triangle, we can review triangle congruence. SAS congruence is assumed as an axiom (as it was in Euclidean and Hyperbolic geometry). SSS congruence and ASA congruence can be proven without using the Exterior Angle Theorem, so those results are valid in Elliptic geometry, and we have used these triangle congruences several times already.

It is interesting to note that AAS is not a triangle congruence in Elliptic geometry. A counter-example is quite easy to construct. Construct a line through two points A and B (Figure 8.8). At A and B construct the perpendiculars to the line. These will meet at the polar point O. On $\overleftrightarrow{AB}$ construct a point C not on $\overline{AB}$ such that the length of $\overline{AB}$ does not equal the length of $\overline{BC}$. Then triangles $\triangle ABO$ and $\triangle BCO$ satisfy AAS (right angles at A, B, C and equal sides to the polar point), but these two triangles are clearly not congruent.

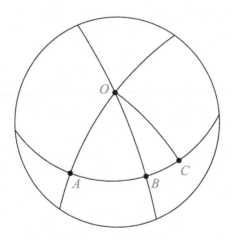

Figure 8.8

Exercise 8.4.1. *Let q be the polar distance. We have shown that for all points on a line l with polar point O the distance (measured along a perpendicular) from a point P on l to O is always q. Show, conversely, that if a point R is a distance q from O then R must be on l.*

Exercise 8.4.2. *Let O be a point. By the previous exercise, the set of points which are a distance q away from O must lie on the line l for which O is the polar point. The line l is called the* polar *of O. Prove that every line passing through O must be a perpendicular to the polar of O and vice-versa.*

Exercise 8.4.3. *Let l and m be two distinct lines. Show that their polar points must be distinct and that the line through the two polar points must be a common perpendicular to both l and m.*

Exercise 8.4.4. *A circle c of radius r and center C is defined as the set of points in the Elliptic plane that are a distance r from the center point C. Show that if the radius of a circle is q, the polar distance, then the circle is also an elliptic line.*

Exercise 8.4.5. *The* interior *of a circle with radius less than the polar distance is defined as the set of points whose distance to the center of the circle is less than the radius. What would be a logical definition for the* exterior *of a circle? Is the interior of a circle well-defined if the radius is equal to the polar distance?*

Exercise 8.4.6. *In Euclidean geometry, given three collinear points there is a unique way to label them as A, B, and C so that the equation $AB+BC = AC$ is true. Give an example to show that this is not always possible in Elliptic geometry.*

Exercise 8.4.7. *Let R be a rotation about the polar point O to a line l. Show that R maps l back to itself (but doesn't necessarily fix each point on l). (Hint: use the fact that a rotation is built from two reflections through the center of rotation.)*

Exercise 8.4.8. *Recall that a* half-turn *is a rotation about a point of 180 degrees. Let l be an elliptic line and let r be reflection across this line. Recall that a reflection can be identified as a non-identity isometry that fixes all points on a line. Let H be the half-turn defined about the polar point to l. Prove that r = H. This amazing result says that in Elliptic geometry all reflections are actually rotations, which means there are no orientation-preserving isometries in planar Elliptic geometry. (Hint: Show that H fixes all points on l.)*

Exercise 8.4.9. *Show that composition of two rotations in the Elliptic plane must be another rotation. (Hint: write the rotations in terms of reflections)*

Exercise 8.4.10. *Recall that every isometry can be built from three reflections. Show that the only possible isometries in the elliptic plane are rotations.*

8.5 TRIANGLES AND AREA IN ELLIPTIC GEOMETRY

We start this section with two basic results about triangles in Elliptic geometry.

Theorem 8.8. *The angle sum for any elliptic right triangle is always greater than 180 degrees.*

Proof:

Let $\triangle ABC$ be a right triangle with a right angle at A. Let D be the midpoint of $\overline{BC}$

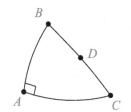

Drop a perpendicular from D to $\overleftrightarrow{AC}$. By Pasch's Axiom we know that the perpendicular must pass through one of the two sides $\overline{AB}$ or $\overline{AC}$. Clearly, it cannot pass through A, B, or C.

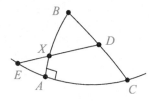

Suppose it passed through $\overline{AB}$ at X. Then we would have two perpendiculars from points on $\overleftrightarrow{AC}$ intersecting at X. This would mean that X is the pole of $\overleftrightarrow{AC}$, which is clearly impossible, as triangle sides have maximal length equal to the polar distance. So, the perpendicular from D to $\overleftrightarrow{AC}$ intersects $\overline{AC}$ at E.

We can find a point F such that $\angle DCE$ is congruent to $\angle DBF$. On $\overrightarrow{BF}$ we can find a point G such that $\overline{BG} \cong \overline{EC}$. By SAS we have that $\triangle ECD \cong \triangle GBD$. Thus, E, D, and G are collinear and $\angle DGB$ is a right angle.

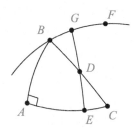

Quadrilateral $ABGE$ is thus a Lambert Quadrilateral and $\angle ABG$ must be obtuse. Since $\angle DCE$ is congruent to $\angle DBF$, then $m\angle DCE + m\angle ABD = m\angle DBF + m\angle ABD = m\angle ABG > 90$. Thus, the sum of the angles in $\triangle ABC$ is greater than two right angles. □

Theorem 8.9. *The angle sum for any elliptic triangle is greater than 180 degrees.*

Proof: Let ABC be a triangle. We have proved the result if one of the angles is right. Thus, suppose none of the three angles are right angles. If two are obtuse, then the result is clear. If two are not obtuse, then two must be acute.

Let the angles at B and C be acute. Construct a perpendicular to $\overleftrightarrow{BC}$ at B and another perpendicular at C. (Figure 8.9). These will meet at the polar point O. Since the angles at B and C in the triangle are acute, then these angles lie within the right angles at B and C made by the two perpendiculars. Since A is the intersection of the defining rays of these angles, then A lies within the triangle $\triangle BCO$. Construct the line $\overleftrightarrow{OA}$. Since this is a line through the polar, it will intersect $\overleftrightarrow{BC}$ at right angles, and by Pasch's axiom, it must intersect side $\overline{BC}$ at some point D.

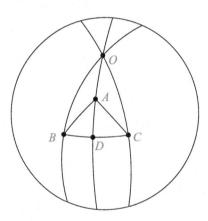

Figure 8.9

Now, $\triangle ABC$ can be looked at as two right triangles $\triangle ABD$ and $\triangle ADC$. Using the previous result, we know that the sum of the angles in $\triangle ABD$ is greater than 180, as is the sum of the angles in $\triangle ADC$. So the sum of the angles in the two triangles together is greater than 360. But clearly, this sum is the same as the sum of the angles in $\triangle ABC$ plus the two right angles at D. Thus, the sum of the angles in $\triangle ABC$ is greater than $360 - 180 = 180$. □

8.5.1 Triangle Excess and AAA

> **Definition 8.5.** *Given $\triangle ABC$ we call the difference between the angle sum of $\triangle ABC$ and 180 degrees the* excess *of $\triangle ABC$.*

> **Corollary 8.10.** *The sum of the angles of any quadrilateral is greater than 360 degrees.*

Proof: The proof is left as an exercise. □

> **Definition 8.6.** *Given quadrilateral ABCD the* excess *is equal to the sum of the angles in the quadrilateral minus 360 degrees.*

> **Theorem 8.11.** *Given $\triangle ABC$ let line l intersect sides $\overline{AB}$ and $\overline{AC}$ at points D and E, respectively. Then the excess of $\triangle ABC$ is equal to the sum of the excesses of $\triangle AED$ and quadrilateral $EDBC$.*

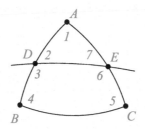

Figure 8.10

Proof: Suppose that D and E are interior to $\overline{AB}$ and $\overline{AC}$. Label all interior angles as shown in Figure 8.10. Then

$$excess(\triangle ABC) = (\angle 1 + \angle 4 + \angle 5) - 180$$

And,

$$excess(\triangle ADE) + excess(DECB) = (\angle 1 + \angle 2 + \angle 7) - 180$$
$$+ (\angle 3 + \angle 4 + \angle 5 + \angle 6) - 360$$

Since $\angle 2 + \angle 3 = 180$ and $\angle 6 + \angle 7 = 180$, the result follows.

If l intersects at any of A, B, or C, a similar argument can be used on the triangles defined. □

We can use this result to prove an amazing facts about triangles in Elliptic geometry—similar triangles are congruent!

> **Theorem 8.12.** *(AAA Congruence) If two triangles have corresponding angles congruent, then the triangles are congruent.*

Proof: Let $\triangle ABC$ and $\triangle DEF$ be given with $\angle ABC \cong \angle DEF$, $\angle BCA \cong \angle EFD$, and $\angle CAB \cong \angle FDE$.

If any pair of sides between the two is congruent, then by ASA the triangles are congruent.

So, either a pair of sides in $\triangle ABC$ is larger than the corresponding pair in $\triangle DEF$ or is smaller than the corresponding pair. Without loss of generality we can assume $\overline{AB}$ and $\overline{AC}$ are larger than $\overline{DE}$ and $\overline{DF}$.

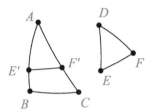

Figure 8.11

Then we can find points E' and F' on $\overline{AB}$ and $\overline{AC}$ so that $\overline{AE'} \cong \overline{DE}$ and $\overline{AF'} \cong \overline{DF}$(Figure 8.11). By SAS, $\triangle AE'F' \cong \triangle DEF$, and the two triangles have the same excess.

But, since $\triangle DEF$ and $\triangle ABC$ have the same excess, we have that $\triangle AE'F'$ and $\triangle ABC$ have the same excess. From the previous theorem, the excess of quadrilateral $E'F'CB$ is then zero, which is impossible.

Thus, all pairs of sides are congruent. □

8.5.2 Area in Elliptic Geometry

Areas in Euclidean geometry were defined in terms of figures that are *equivalent*. Two figures are equivalent if the figures can be split up (or subdivided) into a finite number of pieces so that pairs of corresponding pieces are congruent.

We need to be careful in building up a notion of area in Elliptic geometry in that we cannot assume a basis of rectangles for defining area. We start with some defining axioms of how area works. These axioms are identical to the ones we used in defining area for Hyperbolic geometry.

Area Axiom I If A, B, C are distinct and not collinear, then the area of triangle ABC is positive.

Area Axiom II The area of equivalent sets must be the same.

Area Axiom III The area of the union of disjoint sets is the sum of the separate areas.

Note that Axiom II automatically implies that the area of congruent sets is the same.

Since area is axiomatically based on triangles, we will need the following result.

Theorem 8.13. *Two triangles ABC and $A'B'C'$ that have two sides congruent, and the same excesses, are equivalent and thus have the same area.*

Proof: Suppose that $\overline{BC} \cong \overline{B'C'}$. Let D, E, and F be the midpoints of sides $\overline{BC}$, $\overline{AC}$, and $\overline{AB}$, as shown in Figure 8.12. Construct the line through F and E and drop perpendiculars to this line from A, B, and C meeting the line at G, H, and I, respectively.

We will now show that $\overline{FH} \cong \overline{FG}$. Let R be a half-turn rotation about point F. Then, R will map B to A and H to a point G' on the line through F and G. If G and G' are different points, then this implies that there are two perpendiculars from A to $\overleftrightarrow{FG}$ (at points G and G'), and so A must be the pole of $\overleftrightarrow{FG}$.

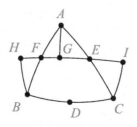

Figure 8.12

But, this implies that side $\overline{AB}$ of triangle $\triangle ABC$ has length larger than the polar distance q. This contradicts the definition of segment in Elliptic geometry. Thus, $G' = G$ and $\overline{FH} \cong \overline{FG}$. A similar argument shows that $\overline{GE} \cong \overline{EI}$.

Thus, right triangles $\triangle BHF$ and $\triangle AGF$ are congruent by SAS and $\triangle AGE$ and $\triangle CIE$ and congruent. This implies that $\overline{BH} \cong \overline{AG} \cong \overline{CI}$, so $BHIC$ is a Saccheri Quadrilateral with base $\overline{HI}$. Also, it is clear that this quadrilateral is equivalent to the original triangle $\triangle ABC$ (move appropriate pieces around).

Now, it might be the case that the positions of G, H, and I are switched, as shown in Figure 8.13. However, one can easily check that in all configurations, $BHIC$ is a Saccheri Quadrilateral (with summit $\overline{BC}$) that is equivalent to $\triangle ABC$.

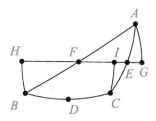

Figure 8.13

Consider the summit angles of $BHIC$. The sum of these two summit angles must equal the angle sum of $\triangle ABC$, and thus the summit angles are each half of the angle sum of $\triangle ABC$.

Similarly, we can construct a Saccheri Quadrilateral for $\triangle A'B'C'$ that is equivalent to $\triangle A'B'C'$ and with summit $\overline{B'C'}$.

Since $\overline{BC} \cong \overline{B'C'}$ and the summit angles for both quadrilaterals are congruent (equal excess values implies equal angle sums), then by Exercise 8.5.2 we know that the two quadrilaterals are congruent. Thus, the triangles are equivalent. □

We can prove an even more general result.

Theorem 8.14. *Any two triangles with the same excesses are equivalent and thus have the same area.*

Proof: If one side of $\triangle ABC$ is congruent to a side of $\triangle A'B'C'$, then we can use the previous theorem to show the result. So, suppose no side of one matches a side of the other. We can assume side $\overline{A'C'}$ is greater than $\overline{AC}$

As in the last theorem, let E, F be midpoints of $\overline{AC}$ and $\overline{AB}$. Construct the perpendiculars from B and C to the line through E and F, and construct the Saccheri Quadrilateral $BIHC$.

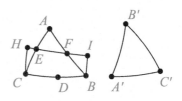

Since $A'C' > AC$, we can find a point A'' along $\overleftrightarrow{CA}$ from A such that the length of $\overline{A''C}$ is equal to the length of $\overline{A'C'}$ and $A'' \neq A$. Let E'' be the midpoint of $\overline{A''C}$.

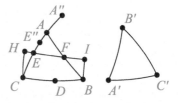

We note here that we are not assuming that lines can be extended indefinitely to find the point A''. Rather, we are using the fact that we can extend a segment until we hit its maximum length, which is the polar distance. Since the length of $\overleftrightarrow{A'C'}$ is less than or equal to this polar distance, we can always extend a smaller segment to match the length of $\overleftrightarrow{A'C'}$.

Now, construct a circle centered at C of radius CE''. Since point E on $\overleftrightarrow{HI}$ is inside this circle, then $\overleftrightarrow{HI}$ must intersect this circle at a point G. (We are assuming continuity of intersections of lines and circles, just as we did in Euclidean and Hyperbolic geometry).

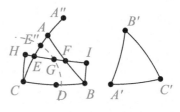

Construct $\overline{CG}$ and extend this segment to a point $A\hat{}$ such that G is the midpoint of $\overline{A\hat{}C}$. Then, $\overline{A\hat{}C} \cong \overline{A'C'}$. Clearly, $\overleftrightarrow{EF}$ cuts $\overline{A\hat{}C}$ at its midpoint G. It is left as an exercise to show that $\overleftrightarrow{EF}$ cuts $\overline{A\hat{}B}$ at its midpoint, point J.

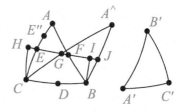

Then, as in the previous theorem, $\Delta A^{\wedge} BC$ will be equivalent to the Saccheri Quadrilateral $BHIC$, which is in turn equivalent to ΔABC. Since triangles $\Delta A^{\wedge} BC$ and $\Delta A'B'C'$ share a congruent side, they must be equivalent, and hence the original two triangles are equivalent. □

We have shown that any two triangles having the same excess values can be split into pieces that can be made congruent. Is the converse also true? Suppose we have two equivalent triangles. That is, the two triangles can be subdivided into sub-triangles with corresponding pairs congruent. Then for each pair, the excess will be the same. Also, if a set of sub-triangles forms a polygonal shape, it is easy to see that the excess of the polygon is equal to the sum of the excesses of the triangles of which it is comprised. Thus, the excesses of the original triangles will be the sum of the excesses of all sub-triangles, which match pair-wise. We summarize this as follows.

Theorem 8.15. *Any two triangles that are equivalent (and thus have the same area) must have the same excesses.*

Our conclusion from this exploration of elliptic area is that the excess and area share exactly the same properties. This tells us that the area must be a function of the excess. Since the area and excess are both linear functions in terms of triangle subdivisions, the area should be a linear function of the excess, and it must be positive. Therefore, $area = k^2 excess + c$, for some constants k and c. But, the area of the empty set must be zero, and thus $area = k^2 excess$. We can fix k by choosing some triangle to have unit area. We summarize this discussion in the following theorem.

Theorem 8.16. *If we have defined an area function for Elliptic geometry satisfying Axioms I–III, then there is a positive constant k such that for any triangle ΔABC, we have*

$$area(\Delta ABC) = k^2 \ excess(\Delta ABC)$$

Note that our discussion does not give a complete proof of this result, only an argument for the reasonableness of the theorem.

Exercise 8.5.1. *Prove Corollary 8.10.*

Exercise 8.5.2. *Show that two Saccheri Quadrilaterals with congruent summits and congruent summit angles must be congruent quadrilaterals; that is,* *the bases must be congruent and the sides must be congruent. [Hint: Suppose they were not congruent. Show that you can then construct a rectangle using the quadrilateral with the longer sides.]*

Exercise 8.5.3. *Show that two distinct elliptic lines cannot have more than one common perpendicular.*

Exercise 8.5.4. *Prove that two Saccheri Quadrilaterals with congruent bases and congruent summit angles must be congruent.*

Exercise 8.5.5. *Show that in Theorem 8.14 $\overleftrightarrow{EF}$ cuts $\overline{A\hat{}B}$ at its midpoint J. [Hint: Suppose that the intersection of $\overleftrightarrow{EF}$ with $\overline{A\hat{}B}$ is not the midpoint. Connect G to the midpoint of $\overline{A\hat{}B}$ and construct a second Saccheri quadrilateral. Let M be the midpoint of $\overline{BC}$. We know that the perpendicular at M will be perpendicular to both Saccheri quadrilateral bases. (Exercise 8.2.1) Use a polar point argument to reach a contradiction.]*

Exercise 8.5.6. *Is it possible to construct scale models of a figure in Elliptic geometry? Briefly explain your answer.*

8.6   PROJECT 13 - ELLIPTIC TILING

In this project we will investigate regular tilings in Elliptic geometry. We saw in section 6.4 that there are just three types of regular tilings of the Euclidean plane—the ones generated by equilateral triangles, by squares, and by regular hexagons. In Project 7.6 we saw that in hyperbolic geometry there was a regular tiling of n-gons meeting k at a vertex if and only if $\frac{1}{n} + \frac{1}{k} < \frac{1}{2}$.

We can investigate Elliptic tilings in a similar manner to how we studied hyperbolic tilings in Project 7.6. Suppose we have k elliptic regular n-gons meeting at a common vertex of a tiling. Then the interior angles of the n-gons would be $\frac{360}{k}$. Also, we can find a central point and triangulate each n-gon. The triangles in the triangulation will consist of isosceles triangles with base angles of $\alpha = \frac{360}{2k}$. The total angle sum of all the triangles in the triangulation will be *greater* than $180n$. At the same time, this angle sum can be split into the angles around the center point (which add to 360) plus the base angles of the isosceles triangles (which add to $2n\alpha$). Thus, we get that

$$360 + 2n\alpha > 180n$$

Or,

$$2\alpha > 180 - \frac{360}{n}$$

Since the angle formed by adjacent edges of the n-gon is $2\alpha = \frac{360}{k}$, we get

$$\frac{360}{k} > 180 - \frac{360}{n}$$

Dividing both sides by 360 and then rearranging, we get

$$\frac{1}{n} + \frac{1}{k} > \frac{1}{2}$$

Thus, if there is a regular tiling by n-gons meeting k at a vertex in Elliptic geometry, then $\frac{1}{n} + \frac{1}{k} > \frac{1}{2}$. On the other hand, if this inequality is true, then a tiling with n-gons meeting k at a vertex must exist. Thus, this inequality completely characterizes regular elliptic tilings. We will call a regular elliptic tiling of n-gons meeting k at a vertex a *(n,k)* tiling. Which such tilings are possible?

Clearly, $n, k > 2$. If $n = 3$, then the only possible values for k are $k = 3, 4, 5$. If $k = 3$, the only possible values for n are $n = 3, 4, 5$. If $k = 4$ or $k = 5$, the only possible values for n are $n = 3$. If $n = 4$ or $n = 5$, the only possible values for k are $k = 3$. If $n > 5$ or $k > 5$, there are no tilings possible. Thus, there are exactly five possible regular elliptic tilings: $(3, 3), (3, 4), (3, 5), (4, 3), (5, 3)$.

As an example, let's see how to generate a $(4, 3)$ tiling. In a $(4, 3)$ tiling, we have squares meeting three at a vertex. How do we construct squares of this kind? First of all, it is clear that the interior angle of the square must be 120 degrees $\left(\frac{360}{3}\right)$. If we take a square and triangulate it via triangles constructed to a central point, the angles about the central point will be 90 degrees and the base angles of the isosceles triangles will be 60 degrees (half the interior angle). Thus, to build the pentagon we need to construct an elliptic triangle with angles of 90, 60, and 60.

In Euclidean geometry, there are an infinite number of triangles that have a specified set of three angles, and these triangles are all similar to each other. In Elliptic geometry, two triangles with congruent pairs of angles must be congruent themselves!

Thus, we know that an elliptic triangle with angles of 90, 60, and 60 degrees must be unique.

In this project we will be using the *Geometry Explorer* software to create tilings of the Elliptic plane. If you are not already using this program, go to http://www.gac.edu/~hvidsten/geom-text for instructions on how to download and use this software.

To begin this project, start up *Geometry Explorer* and open an Elliptic geometry window. (Choose **New** (**File** menu) and then choose "Elliptic" in the dialog box that pops up).

In order to create a $(4, 3)$ tiling, we need to create a 90,60,60 triangle. We begin the construction of this triangle by constructing a point at the origin of our Elliptic model.

Choose **Point At Origin** (**Misc** menu) to create a point at the origin in the Elliptic plane.

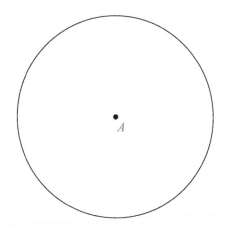

To create our triangle, we will first create the base segment of a triangle with angles of 90, 60, and 60.

Select the point at the origin and choose **Base Pt of Triangle with Angles...**(**Misc** menu). A dialog box will pop up as shown. Note how the angles α, β, and γ are designated. In our example we want $\alpha = 90$ and the other two angles to be 60. Type these values in and hit "Okay."

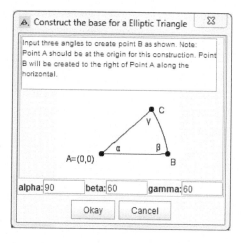

In the Canvas, a new point will be created that corresponds to point B in the dialog box just discussed. To find the third point of our triangle, we need to rotate point B about point A by an angle of 90 degrees. Carry out this rotation to get point C.

Let's measure the three angles just to verify that we have the triangle we want.

Looks good. Next, hide the angle measurements and select A as a center of rotation and define a custom rotation of 90 degrees. Rotate C two times to get points D and E. Then, connect the outer points to form a square.

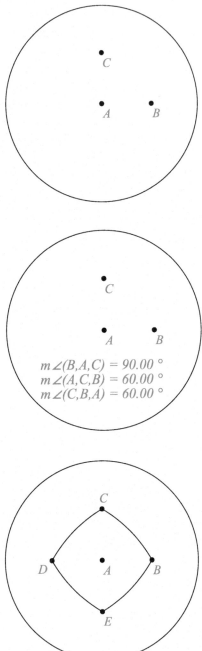

Finally, select point C as a center of rotation and define a new custom rotation of 120 degrees. Then, rotate the square (and point A) two times, yielding three squares meeting at C at angles of 120 degrees! You may want to hide some of the intermediate points that get rotated. Use the **Manage Properties...** (**Edit** menu) to help with this.

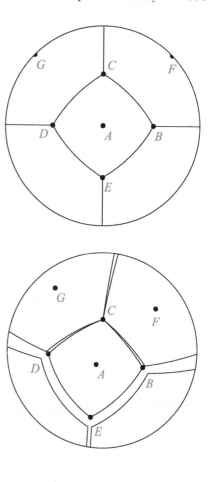

We have now created a tiling of the elliptic plane by squares. Try moving the point at the origin, point A. We see that the tiling breaks up in a quite nasty way. Why is this the case? The problem here is that by translating the origin point, we have created compound translations and rotations for other parts of the figure. In Elliptic geometry compositions of translations are not necessarily translations again as they are in Euclidean geometry.

Looking at our tiling construction, it is not immediately clear how this construction creates a tiling.

In Elliptic geometry once we cross the "boundary" circle at a point, we immediately appear at the antipodal point on the opposite side. Thus, segment $\overline{BC}$ gets rotated to become segments $\overline{CE}$ and $\overline{BD}$ and point A rotates to points F and G. Thus, this tiling of the Elliptic plane consists of just three squares —$BCDE$ (inner point A), $BCDE$ (inner point F), and $BCDE$ (inner point G).

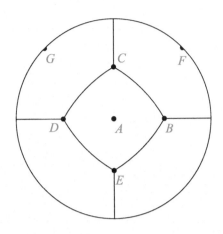

Let's recall that the elliptic plane is built from the sphere by identifying antipodal points. How does our tiling relate to the spherical model?

The standard elliptic model is built by *stereographic projection* from the sphere. We can get a sense of how our tiling looks on the sphere by switching our model to the direct projection of the sphere into the $x - y$ plane. To switch to this model, choose **Sphere X-Y Projection** (**Model** menu). Note how points F and G seem to have moved, but they really haven't —they have just switched to their antipodal counterpart.

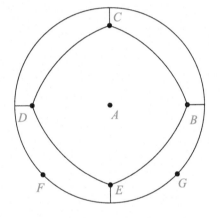

Now, we can imagine how the elliptic tiling will extend to the sphere. Each of the three squares will have to be doubled to undo the antipodal identification from the sphere to the elliptic plane. Thus, we would have six congruent squares on the sphere that divide the sphere into congruent parts. What other common 3-D shape consists of six congruent squares? The cube!

In Figure 8.14 we see how the cube can be "blown up" to form a tiling of the sphere.

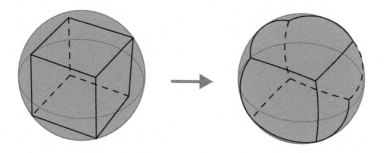

Figure 8.14 Projection of the cube onto the sphere

Exercise 8.6.1. *In Elliptic geometry there are exactly five possible regular elliptic tilings:* $(3,3), (3,4), (3,5), (4,3), (5,3)$. *There are also five Platonic solids in Euclidean geometry. These are the tetrahedron, the cube, the octahedron, the dodecahedron, and the icosahedron. Each of these will generate a spherical tiling by projection onto a sphere, just as the cube generated the spherical tiling above. Thus, by antipodal identification, each of the Platonic solids will generate a tiling of the Elliptic plane. Match each Platonic solid with the elliptic tiling it will generate.*

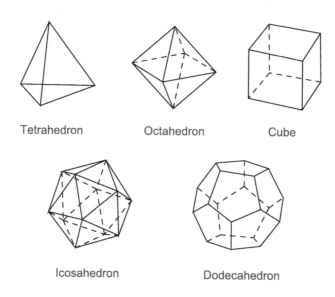

Figure 8.15 Platonic solids

Exercise 8.6.2. *The Platonic Solids were the basis of the Greeks' view of the structure of the universe. Each solid was associated with some form of matter. For example, air was associated with the octahedron. Research this aspect of Greek philosophy and write a short report of their philosophical use of the Platonic solids.*

Exercise 8.6.3. *Describe and illustrate the steps you would take to construct a (3,5) tiling of the Elliptic Plane.*

For your report give a careful and complete summary of your work on this project.

CHAPTER 9

Projective Geometry

Metrical geometry is thus a part of descriptive geometry, and descriptive geometry is all geometry.

– Arthur Cayley (1821–1895)

9.1   UNIVERSAL THEMES

We have explored three different geometries so far in this text —Euclidean, Hyperbolic, and Elliptic. In all cases we have started from a basis of the first four Euclidean axioms:

1. Between any two distinct points, a segment can be constructed.

2. Segments can be extended indefinitely.

3. Given a point and a distance, a circle can be constructed with the point as center and the distance as radius.

4. All right angles are congruent.

We have frequently pointed out that these four axioms are not, by themselves, sufficiently strong to completely develop any of the geometries we have considered. In particular, variations on the fifth axiom, the *Parallel Axiom* are needed for each geometry. Also, we need to add axioms such as Pasch's axiom and axioms concerning continuity and betweenness to fill in various axiomatic gaps. That being said, the four original Euclidean axioms serve as a simple basis from which we can compare and contrast the geometries we have considered.

In our discussion of Hyperbolic and Elliptic geometry, in Chapters 7 and 8, we carefully developed those geometries from the first four

Euclidean axioms plus modifications of the fifth axiom. A careful reading of the proofs and theorems of those two chapters yields a surprisingly similar set of theorems and proof techniques. Many of the proofs for results in Hyperbolic geometry can work (with some slight modification) in Elliptic geometry. (For example, compare the proof of Theorem 7.15 to Theorem 8.9.)

In the nineteenth century, not long after the discovery of Hyperbolic and Elliptic geometry, this similarity became a focus of study. Arthur Cayley in [8] and Felix Klein in [27] showed that all three geometries (Euclidean, Hyperbolic, and Elliptic) could be considered as sub-geometries of another, more abstract, geometry called *Projective geometry*.

Before we dive into the axiomatic details of Projective geometry, let's consider the commonalities between the three geometries we have studied so far. Referring to Euclid's axioms, we see that all three geometries have lines and points, with two points determining a unique line, and line segments extendable (perhaps not indefinitely, but they can always be extended). Linear properties of *incidence* are thus universally shared by all three geometries and will turn out to be a defining property of Projective geometry.

The third and fourth axioms deal with *metric* properties of each geometry. Metric properties deal with measurement, whether that be radii of circles (third axiom) or the congruence of right angles (fourth axiom). While all three geometries satisfy the third and fourth axioms, the nature of the measurement of distance and angle is vastly different in each geometry. Thus, a *universal* geometry that encompasses Euclidean, Hyperbolic, and Elliptic geometry as sub-geometries, must avoid specification of metric properties in its axiomatic basis.

Likewise, it is not possible to *universalize* the fifth axiom, or its hyperbolic and elliptic counterparts, as they are mutually exclusive statements.

A common theme for for Hyperbolic and Elliptic geometry was that both had circle models. A natural question to ask is whether this theme is universal —whether there is a circle model for Euclidean geometry. The answer is, surprisingly, yes! We will take a short excursion into the development of this model, as it will provide a concrete framework for our development of Projective geometry.

9.1.1 Central Projection Model of Euclidean Geometry

In the *central projection model* of Euclidean geometry, we project the Euclidean plane to points on the lower hemisphere of the unit sphere (Figure 9.1). Let C be the center of the sphere and let P be a point on the plane. Assume the sphere is situated so that the south pole (point O in Figure 9.1) is on the plane. Construct $\overleftrightarrow{CP}$ and let P' be the intersection of this line with the lower hemisphere. Clearly, for every point in the plane there is a unique point associated on the hemisphere. Also, the line x shown passing through O will project to a great circle arc x' on the sphere and any other line l that is parallel to x will project to a great circle arc l' with the same diameter endpoints as x'.

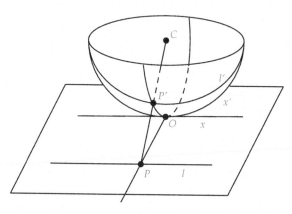

Figure 9.1

Let Σ be the set of all projected points on the sphere. Since no point in the plane will project to the two diameter endpoints, then arcs x' and l' will not intersect in Σ. Arcs such as x' and l' will then serve as the lines in the lower sphere model for Euclidean geometry. Projecting the sphere *orthographically* into the x-y plane (drop the z coordinate) yields a unit disk model for Euclidean geometry. Suitable definitions for distance and angle can be provided that make this model metrically equivalent to the standard Euclidean plane. (For details see [15], pp.212-ff).

In Figure 9.2 we see a circle, a line, and a triangle in this circle model of Euclidean geometry. Note that the circle looks like a distorted ellipse. This is due to the fact that central projection maps circles into ellipses and orthographic projection into the x-y plane distorts those elliptical shapes.

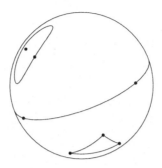

Figure 9.2

9.1.2 Ideal Points at Infinity

We have seen that all three of our geometries —Euclidean, Hyperbolic, and Elliptic —share incidence properties for lines and points and all have unit disk models. In Euclidean and Hyperbolic geometry, the points on the boundary of the disk are excluded from the geometry. However, in Hyperbolic geometry we introduced *ideal* or *omega* points as being where two parallels meet at infinity. In the central projection model of Euclidean geometry, we see clearly that two Euclidean parallel lines meet at similar ideal points on the boundary. If we expand our geometry to include these boundary points, we could say that in Euclidean geometry two lines always meet at a point or an ideal point.

From our discussion so far, it would seem reasonable, then, for a *universal* geometry to have axioms that guarantee that two points determine a unique line and that two lines always meet. This geometry would encompass both Euclidean (with ideal points) and Elliptic geometries. Since Hyperbolic geometry can be built inside Euclidean geometry, it would also encompass Hyperbolic geometry.

Before we set down the axiomatic basis for this universal geometry (Projective geometry) we will look at the historical impetus for the development of Projective geometry. We will also explore several of the main ideas of Projective geometry in an intuitive setting, to give context and motivation for the more abstract axiomatic development coming later in the chapter.

9.2   PROJECT 14 - PERSPECTIVE AND PROJECTION

In the preceding section, we saw that a unifying theme of the three geometries we have studied so far is that all have unit disk models. Another unifying theme is that of *ideal points*, or *points at infinity*.

These two ideas, central projection and points at infinity, did not originate in the study of abstract mathematics, but came from the quest for realism in the visual arts. An artist, when viewing a scene in the three-dimensional world, has the problem of depicting that scene on a flat two-dimensional canvas. To create more realistic depictions of three dimensional figures, painters begin experimenting with the notion of *linear perspective* to give their artwork a feeling of depth.

The development of linear perspective is most often credited to Filippo Brunelleschi (1337–1446). He realized that there was a vanishing point, a point at infinity, to which parallel lines in a picture would converge.

Imagine you are looking at a long, straight railroad track that goes off to distant mountains. There are trees and a small building along the side of the tracks. A parallel projection of all the objects in this scene into a flat viewing plane would give a picture like the one at the left. But, in fact, our eyes perceive an image like the one at the right, where the parallel tracks appear to converge at the distant mountain range.

The introduction of vanishing points in perspective drawing gave artists a precise way to find the angle of projection for lines in a scene and also provided a way to calculate the length of projected lines and shapes in the flat canvas of a painting. Brunelleschi used all of these ideas to create lifelike renderings of scenes, but he did not write down the relationships he discovered. One of his pupils, Leon Battista Alberti did record Brunelleschi's ideas in the book *Della Pittura (On Painting)*, first published in 1435 [1].

In this book Alberti describes the procedure one should use to correctly draw a figure using linear perspective. He defines a painting as an object that uses this perspective system:

> A painting is the intersection of a visual pyramid at a given distance, with a fixed centre and a defined position of light, represented by art with lines and colours on a given surface. ([1] pp.32–33)

To illustrate Alberti's definition, imagine the artist's eye is at position O in Figure 9.3 and the artist is looking at triangle $\triangle ABC$ on a horizontal surface, say the floor of the artist's studio. This is plane Π in the figure. If we connect the vertices of the triangle to the eye position O, we generate a visual *pyramid* as Alberti described. The intersection of this pyramid with a vertical planar surface Π' (e.g. the canvas of a painting) yields triangle $\triangle A'B'C'$.

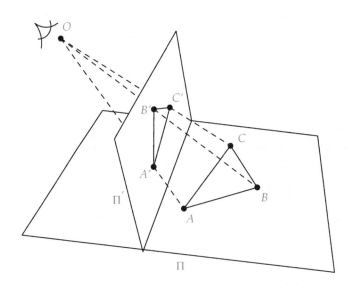

Figure 9.3

Exercise 9.2.1. *Construct a figure like that shown in Figure 9.3 with a square in the horizontal plane. Show by your drawing that lengths and right* *angles will not be preserved under the perspective projection onto a vertical plane. This is a general property of projective geometry, as we will see later.*

While Brunelleschi and Alberti are rightfully credited with the creation of the rules for linear perspective, their work is more algorithmic than mathematical. The first thoroughly mathematical development of perspective came with the work of Girard Desargues (1591–1661).

Desargues was a contemporary of Descartes and worked as an architect, engineer, and mathematician. He is credited as being the "founder of projective geometry" [40]. Desargues was the first mathematician to stipulate that projection could produce points at infinity, and that the set of all points at infinity creates a line at infinity.

We can visualize a point at infinity by considering points O, B, and B' in Figure 9.3. In perspective drawing, we start with B and draw a line to O to get the perspective point B'. But, we could also imagine that there is a light source at O and this light is projecting B' onto point B in the horizontal plane. In this *projective* view of the scene, if we move O so that it is at the same height as B' (relative to plane Π), the projected point B would move infinitely far away —it would become a point at infinity.

Desargues was the first person to rigorously explore this projective view of perspective drawing. Unfortunately, his writings on the subject used rather obscure terms, such as "bough" and "trunk" to refer to mathematical quantities and he did not provide proofs of many of his results. Also, the analytic geometry of Descartes was considered more accessible and applicable to everyday problems. Desargues' work was largely ignored during his lifetime. It wasn't until the 19th century, when projective geometry was reinvented by Gaspard Monge (1746–1818) and his pupil Victor Poncelet (1788–1867), that Desargues' work was fully recognized.

One of Desargues' most important results has become known as "Desargues' Theorem." This is one of the fundamental ideas in projective geometry. To get some intuition about Desargues' Theorem, we will explore how perspective projection works in more detail.

In Figure 9.4 we have two triangles ABC and $A'B'C'$ that are seen in perspective from point O.

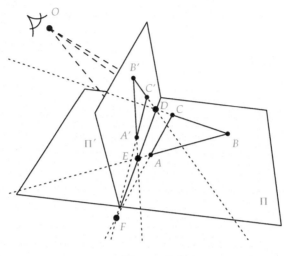

Figure 9.4

The dotted lines in the figure are extensions of corresponding pairs of sides from these two triangles. For example, one pair of sides ($\overline{AB}$ and $\overline{A'B'}$), when extended will intersect at a point E. Likewise, the other two pairs of sides intersect at points D and F.

Exercise 9.2.2. *Explain why the three intersection points D, E, and F must be collinear.*

Now, imagine that we rotate plane Π' and point O toward plane Π as in Figure 9.5. Clearly, the collinearity property for D, E, and F will persist as long as the two planes are different.

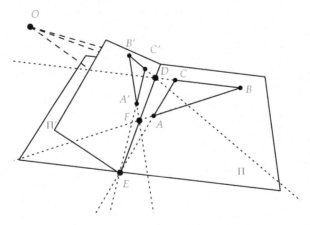

Figure 9.5

But, what happens in the "limit" as the planes become identical and point O lies in the plane of both triangles? Does the collinearity property still hold true? What happens in this case is the crux of Desargues' Theorem.

To investigate this "limit" case, we will use dynamic geometry software. Notes on how to use software for this project can be found at http://www.gac.edu/~hvidsten/geom-text.

Start up your geometry software and construct a point O that will serve as the perspective point. Then, construct three rays emanating from O as shown.

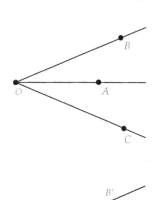

Now, construct $\triangle ABC$ and attach three new points A', B', and C' somewhere along $\overrightarrow{OA}$, $\overrightarrow{OB}$, and $\overrightarrow{OC}$. Construct $\triangle A'B'C'$.

We have constructed two triangles that are perspective from O. Next, construct the lines for each pair of corresponding sides, as shown, and find the intersection points for each pair of sides. These are points D, E, and F. Note: the intersection points might lie outside of the window. You can zoom out to see more of the diagram.

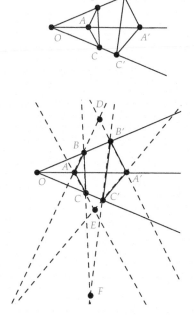

Now, the big question: will D, E, and F be collinear? Construct line $\overleftrightarrow{DE}$. Move point O and the triangle vertices around. It appears that the three points are collinear – most of the time!

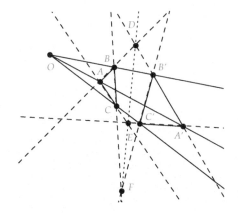

Exercise 9.2.3. *What happens when two pairs of sides, for example $\overline{BC}$ and $\overline{B'C'}$, are parallel? Are the three points still collinear? Do they still exist? Why does it seem reasonable to talk about points at infinity in this case?*

Exercise 9.2.4. *Suppose sides $\overline{BC}$ and $\overline{B'C'}$ are parallel and sides $\overline{AC}$ and $\overline{A'C'}$ are parallel. Show that sides $\overline{AB}$ and $\overline{A'B'}$ must be parallel (using theorems from Euclidean geometry). Explain why it makes sense to say that the three pairs of sides are collinear along the* line at infinity.

In Projective geometry we expand the set of allowable points to include the points at infinity, with the set of all such points comprising the line at infinity. From the discussion so far, the following result that Desargues discovered in the 17th century seems entirely reasonable and natural.

Theorem 9.1. *(Desargues' Theorem) Given two triangles $\triangle ABC$ and $\triangle A'B'C'$, if the two triangles are perspective from a point O, then corresponding sides, when extended, intersect at three points which are collinear.*

For your report give a careful and complete summary of your work on this project.

9.3 FOUNDATIONS OF PROJECTIVE GEOMETRY

Now that we have gained some historical perspective on the development of Projective geometry, and have explored how a mathematical analysis of perspective projection naturally leads to the notion of points and lines at infinity, we are ready to formally define the axiomatic basis of this new geometry.

As we saw in Chapter 1, axiom systems do not arise from thin air, but are abstractions of intuitive notions of space that are developed by our experience with the world. As mentioned in the beginning of this chapter, the goal of Projective geometry is to provide a unifying framework for studying commonalities between the three main geometries we have covered in this text: Euclidean, Hyperbolic, and Elliptic. Such a unifying framework must focus primarily on properties of *incidence*, whether that be two points being incident on (defining) a line or two lines incident (meeting) at a point. Metric notions such as distance and angle must be avoided.

To understand the basic axioms of Projective geometry, it will be useful to consider the axioms for an abstraction of Euclidean geometry that avoids metric properties. This geometry is called *Affine geometry*.

9.3.1 Affine Geometry

If we remove all notions of metric properties from Euclidean geometry, we are left with just two basic properties: the incidence property that two points determine a unique line and the Euclidean parallel property. We also need to guarantee that the geometry is truly two-dimensional. Thus, the following three axioms are natural for Affine geometry.

A1 Given two distinct points P and Q, there is a unique line l incident on these points.

A2 Given a line l and a point P not on l, there is a unique line m parallel to l through P.

A3 There exist three non-collinear points.

In this axiomatic framework, which follows the development presented by Hartshorne in [20], *point*, *line*, and *incident* are undefined terms, while *parallel*, *collinear*, and *concurrent* have their standard meanings. We call a set of points and lines that satisfies these three axioms an *Affine Plane*.

Clearly, planar Euclidean geometry satisfies the three axioms of Affine geometry, as it is our intuitive basis for this axiomatic system. In designing an axiomatic basis for Projective geometry, let's consider what happens to the Affine geometry axioms when we extend the Euclidean plane by adding points at infinity. First of all, two lines will always have at least one intersection point. Thus, the parallel axiom of Affine geometry must be replaced by a new incidence axiom. Also, the addition of points at infinity will mean that lines, which are typically defined by two non-infinite points, will necessarily have a third point– a point at infinity.

9.3.2 Axioms of Projective Geometry

The axioms for Projective geometry are as follows:

P1 Given two distinct points there is a unique line incident on these points.

P2 Given two distinct lines, there is at least one point incident on these lines.

P3 There exist three non-collinear points.

P4 Every line has at least three distinct points incident on it.

We call any set of points and lines that satisfy these four axioms a *Projective Plane*.

Axioms P1 and P2 have a remarkable similarity. By switching the terms "point" and "line," the two statements are nearly interchangeable. In fact, we can prove that they are exactly interchangeable.

> **Theorem 9.2.** *Given two distinct lines l and m, there is a unique point P incident on these lines.*

Proof: Suppose P and Q were distinct points incident on lines l and m. Axiom P1 then implies that l and m have to be the same line, which contradicts the assumption that they are distinct. □

9.3.3 Duality

> **Definition 9.1.** *The dual of an axiom or theorem is created by switching all occurrences of the term "point" with the term "line" and all occurrences of the term "line" with the term "point." An axiom system satisfies the principle of duality if all duals of axioms or theorems are again axioms or theorems.*

The dual of axioms P3 and P4 also hold true. The proofs will be left as exercises. Thus, Projective geometry satisfies the principle of duality. This is very handy when investigating properties of the geometry —if you can prove one result, you get the dual result for free!

Let's consider what the minimum number of points could be for a projective plane (assuming that the plane could have a finite number of points).

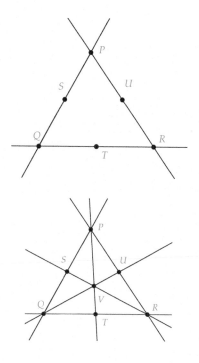

By axiom P3 we know we have at least three points, call them P, Q, and R. These three points define three non-concurrent lines. There must be at least three points on each line, so we get three new points S, T, and U.

We then must have lines $\overleftrightarrow{PT}$, $\overleftrightarrow{RS}$, and $\overleftrightarrow{QU}$. By Theorem 9.2 we know that these lines cannot match any of the three already defined. Each of these new lines must have at least three points, so we need to add at least one more point, but this point can also serve as the intersection of the three lines.

From this analysis, it would appear that the minimum number of points for a projective plane is 7. If we look at the figure carefully, we see that we have only 6 lines. The principle of duality requires that the number of lines matches the number of points. Where is the missing line? If we consider all pairs of points, it seems that we are missing a line for (S, T), (T, U), and (S, U).

If we allow circles to be lines (an idea we have seen before in this text!), we can add one more "line" to completely satisfy the requirements for a projective plane.

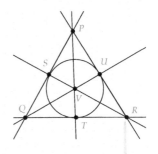

9.3.4 Triangles and Quadrangles

In the example given in the last section with 7 points and 7 lines, we note that there are many three-vertex shapes (QSV, RTP, etc) and many four-vertex shapes ($QSVT$, $SVUT$, etc). Such shapes were important figures of study in Euclidean and other geometries and they are equally as important in Projective geometry.

In defining a shape like a triangle, we have to be careful, however, as there is no notion of "betweenness" in Projective geometry. Thus, segments cannot be defined based solely on the four projective axioms. A triangle can only be defined by using points and lines.

> **Definition 9.2.** *A* triangle *consists of three non-collinear points (or* vertices*) and the three lines (or* sides*) that are defined by these points.*

In the example above, there are triangles such as $\triangle PQR$ that look to us like triangles, and others like $\triangle SVT$ that look a bit strange.

There is another strange property of this finite projective plane that has to do with four-sided shapes. Consider the shape with points S, V, T, and Q, and the six lines that are determined by pairs of these points.

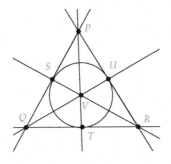

Figure 9.6

Opposite sides $\overleftrightarrow{QS}$ and $\overleftrightarrow{TV}$ intersect at P, sides $\overleftrightarrow{SV}$ and $\overleftrightarrow{QT}$ intersect at R and the two "diagonals" $\overleftrightarrow{QV}$ and $\overleftrightarrow{ST}$ intersect at U. The three intersection points P, U, and R are collinear. In the Projective geometry model that is based on Euclidean geometry, these three points are never collinear. (This will be proved later in this chapter when we consider models of Projective geometry.) If we want to ensure that Projective geometry *axiomatically* has the property that P, U, and R are not collinear, we would have to add another axiom that addresses these four-sided shapes. To state this axiom precisely, we need the following definition:

Definition 9.3. *A* complete quadrangle *consists of four distinct points (or* vertices*), no three of which are collinear, and the six lines (or* sides*) defined by these four points. If A, B, C, and D are the vertices of the quadrangle, then sides $\overleftrightarrow{AB}$ and $\overleftrightarrow{CD}$, $\overleftrightarrow{AC}$ and $\overleftrightarrow{BD}$, and $\overleftrightarrow{AD}$ and $\overleftrightarrow{BC}$ are called* opposite sides *of the quadrangle. The points at which opposite sides intersect are called* diagonal points *of the quadrangle.*

For example, in Figure 9.7 we have the complete quadrangle $ABCD$, as depicted in Euclidean Geometry. The diagonal points are E, F, and G.

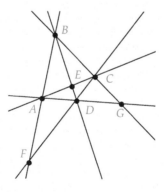

Figure 9.7

Let's consider again the 7 point figure from the discussion above (Figure 9.8). Quadrangle $QSVT$ has diagonal points P, U, and R which are collinear.

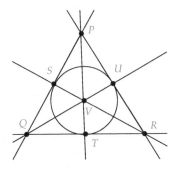

Figure 9.8

To create a Projective geometry where the diagonal points are not collinear, we would need a new axiom:

Axiom P5: (Fano's Axiom) The three diagonal points of a complete quadrangle are not collinear.

This axiom is named for the Italian mathematician Gino Fano (1871–1952). Fano discovered the 7 point figure above, which is now called the *Fano plane*. The Fano plane is the simplest figure which satisfies Axioms P1–P4, but which has a quadrangle with collinear diagonal points.

What about the dual to Fano's axiom? First, we need a dual definition to the definition for a complete quadrangle. For this definition, it will be useful to have a shorthand notation for the intersection point of two lines or the line defined by two points.

Definition 9.4. *Given two distinct lines a and b, $a \cdot b$ is the unique point of intersection of these lines. Given two distinct points A and B, AB is the unique line defined by these two points.*

Definition 9.5. *A* complete quadrilateral *consist of four distinct lines (or sides), no three of which are concurrent, and the six points (or vertices) defined by these four lines (intersections). If a, b, c, and d are the sides of the quadrilateral, then the pairs of vertices $E = a \cdot b$ and $F = c \cdot d$, $G = a \cdot c$ and $H = b \cdot d$, and $I = a \cdot d$ and $J = b \cdot c$ are called* opposite vertices *of the quadrilateral. The sides defined by opposite vertices are called* diagonal sides *of the quadrilateral.*

For example, in Figure 9.9 we have the complete quadrilateral $abcd$, as depicted in Euclidean Geometry. The diagonal sides are e, f, and g.

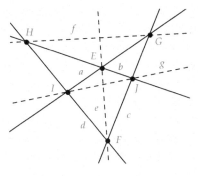

Figure 9.9

The dual to axiom P5 would then read:

Theorem 9.3. *The three diagonal sides of a complete quadrilateral are not concurrent.*

Proof: Exercise □

9.3.5 Desargues' Theorem

Let's recall Desargues' Theorem from Project 9.1.2.

Theorem 9.4. *(Desargues' Theorem) Given two triangles $\triangle ABC$ and $\triangle A'B'C'$, if the two triangles are perspective from a point O, then corresponding sides, when extended, intersect at three points which are collinear.*

In Project 9.1.2 we saw that Desargues' Theorem was a natural result in the Projective geometry developed by extending the Euclidean plane with points at infinity. A natural question to ask would be whether Desargues' Theorem is true for *any* Projective geometry with axioms P1–P4 as its basis. It turns out that Desargues' Theorem is true for projective planes that can be embedded in three-dimensional projective spaces, but is not true for every projective plane. One example of a projective plane that does not satisfy Desargues' Theorem is the *Moulton Plane*, which is explored in the exercises at the end of this section. Other examples can be found in Hartshorne [20, pages 16–18]. If we want to guarantee that Desargues' Theorem holds, we need a new axiom:

Axiom P6: Given two triangles $\triangle ABC$ and $\triangle A'B'C'$, if the two triangles are perspective from a point O, then corresponding sides, when extended, intersect at three points which are collinear.

Let's investigate the dual to this axiom. As was the case when we investigated the dual to Axiom P5 above, it will be useful to have some shorthand notation to discuss the configurations of triangles, lines, and points in Desargues' Theorem.

> **Definition 9.6.** *Triangles $\triangle ABC$ and $\triangle A'B'C'$ are perspective from point O if the lines joining corresponding vertices (lines AA', BB', and CC') are concurrent at O. The triangles are perspective from line l if corresponding pairs of sides of the triangles (AB and $A'B'$, BC and $B'C'$, AC and $A'C'$) intersect at three points on l.*

With this definition, we can restate Axiom P6 as:

Axiom P6: If two triangles $\triangle ABC$ and $\triangle A'B'C'$ are perspective from a point, then they are perspective from a line.

The dual to Axiom P6 would then be:

> **Theorem 9.5.** *If two triangles $\triangle ABC$ and $\triangle A'B'C'$ are perspective from a line, then they are perspective from a point.*

Proof:

Let $P = AB \cdot A'B'$ (i.e., P is the intersection of line $\overleftrightarrow{AB}$ and line $\overleftrightarrow{A'B'}$) and let $Q = BC \cdot B'C'$ and $R = AC \cdot A'C'$. Since the triangles are perspective from a line l, then P, Q, and R are three points on l. We need to prove that lines AA', BB', and CC' are concurrent.

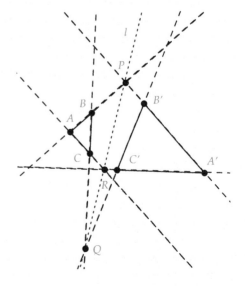

Let O be the intersection of AA' and BB'. Consider shapes RAA' and QBB'. If these are both triangles, then these two triangles are perspective from P. By Axiom P6 we have that corresponding sides are collinear. That is, $BQ \cdot AR = C$, $B'Q \cdot A'R = C'$, and $BB' \cdot AA' = O$ are collinear. Thus, CC' passes through O and the two triangles are perspective from O.

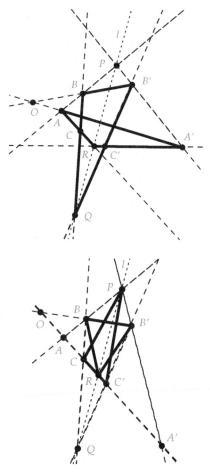

If one of RAA' and QBB' is not a triangle, say R, A, and A' are collinear, then R, A, C, and C' are all collinear, which means that CC' passes through O and the two triangles are perspective from O. A similar argument can be used if QBB' is not a triangle. □

Exercise 9.3.1. *In Affine geometry, show that two distinct lines have at most one point in common.*

Exercise 9.3.2. *Show that an affine plane must contain at least four distinct points. (Hint: By A3 there are three non-collinear points, say P, Q, and R. By A2 there is a line l through P parallel to $\overleftrightarrow{QR}$ and another line m through R parallel to $\overleftrightarrow{PQ}$. Show that l cannot be parallel to m and that the point of intersection of these two lines cannot be one of P, Q, or R.)*

Exercise 9.3.3. *Show that there is an affine plane with exactly four distinct points and exactly six distinct lines. (Hint: Start from the results of the last exercise)*

Exercise 9.3.4. *Show that the dual to axiom P3 is true. That is, in Projective geometry, show that there exist three non-concurrent lines, where we say that two lines are* concurrent *if they have a point in common.*

Exercise 9.3.5. *Show that the dual to axiom P4 is true. That is, in Projective geometry, show that every point has at least three distinct lines incident on it.*

Exercise 9.3.6. *Given a projective plane, let u be a line in the plane. Let A be the set of all points and lines from the projective plane minus the line u and all points incident on u. Show that A is an affine plane. (This implies that the "line at infinity" is not a special line in the projective plane. Any line could serve that role.)*

Exercise 9.3.7. *Verify that the example of a finite projective plane of 7 lines and 7 points satisfies all of the axioms for projective geometry.*

Exercise 9.3.8. *Prove Theorem 9.3 (Hint: Let abcd be a complete quadrilateral. Show that EFGH is a complete quadrangle. Assuming the three diagonal sides of abcd are concurrent, show that the diagonal points of EFGH are collinear.)*

In the exercises at the end of section 1.5 we defined a *Projective Plane of Order n*. The axioms for this geometry are as follows:

P(n)1 Given two distinct points, there is exactly one line incident with them both.

P(n)2 Given two distinct lines, there is at least one point that is incident with both lines.

P(n)3 There are at least four points, no three of which are collinear.

P(n)4 There exists at least one line with exactly $n+1(n > 1)$ distinct points incident with it.

Exercise 9.3.9. *Show that a Projective Plane of Order n satisfies axioms P1-P4 for a (general) Projective plane.*

Exercise 9.3.10. *Show that our example of a finite projective plane of 7 lines and 7 points is a Projective Plane of Order 2.*

In the *Moulton Plane*, the points are the standard points of the Euclidean plane, plus all of the extended points at infinity. The lines will consist of all standard lines in the Euclidean plane that have zero, infinite, or negative slope, together with a set of *broken lines* that consist of two half-lines of positive slope defined as follows:

$$y = \frac{1}{2}mx + b \qquad\qquad x \leq 0$$
$$y = mx + b \qquad\qquad x > 0$$

The points at infinity for these broken lines will be the extended points at infinity for the part of the line defined by $y = mx + b$.

Here are some examples of lines in the Moulton Plane:

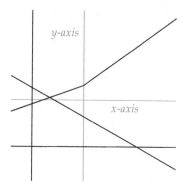

Exercise 9.3.11. *Prove that the Moulton Plane is a Projective Plane.*

[Hint: The tricky Axiom will be P1. Let A and B be distinct points. The only case to prove is where A is to the left and below B. Consider

$$\frac{slope(\overleftrightarrow{BY})}{slope(\overleftrightarrow{AY})}$$

as Y ranges from Y_0 to Y_1.]

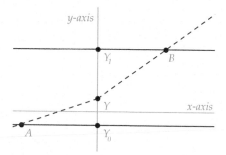

Exercise 9.3.12. *In the figure below, triangles ABC and A'B'C' are perspective from point O in the Moulton Plane. Explain why Desargues' Theorem does not hold for this configuration.*

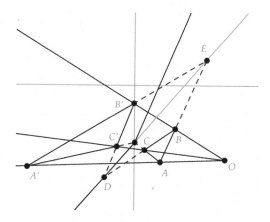

9.4 TRANSFORMATIONS AND PAPPUS'S THEOREM

In our analysis of Euclidean, Hyperbolic, and Elliptic Geometries, we utilized the notion of *transformations* extensively. In particular, we focused on *isometries*, transformations that preserved length. In Projective Geometry metric properties such as length are not defined, so there is no natural analog to isometries. However, in our development of properties of isometries, we proved that they had two important properties that do translate to Projective geometry —they mapped lines to lines and they preserved incidence (intersections of lines).

We have already considered a type of projective transformation that has these two properties —the perspective transformation based on artistic renderings of three-dimensional scenes from Project 9.1.2.

In Figure 9.10 we have two triangles ABC and $A'B'C'$ that are seen in perspective from point O. It is clear that this type of perspective transformation from three dimensions to two-dimensions maps lines to lines and intersections of lines to intersections of lines.

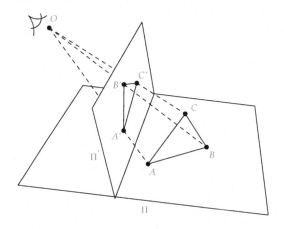

Figure 9.10

Now, consider rotating the vertical plane (and point O) toward the horizontal plane until the two planes coincide and point O is incident with the planes. The two triangles are now *perspective* from point O.

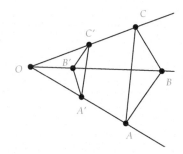

If we consider the effect of this transformation on the points on a single line l, we see that this transformation maps the points on l in a 1–1 and onto fashion onto the points of another line l'.

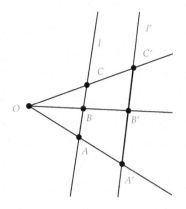

This perspective transformation maps sets of points on a line to sets of points on another line. For ease of description, we create a name for these sets of points.

> **Definition 9.7.** *The* pencil of points with axis l *is the set of all points on* l. *In a dual sense, the* pencil of lines with center O *is the set of all lines through point* O. *(Figure 9.11)*

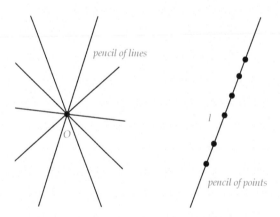

Figure 9.11

9.4.1 Perspectivities and Projectivities

We now carefully define what we mean by being *perspective from O* in terms of pencils of points. We also note the dual definition for pencils of lines.

> **Definition 9.8.** *A perspectivity with center O is a 1–1 mapping of a pencil of points with axis l to a pencil of points with axis l' such that if A on l is mapped to A' on l', then line AA' passes through O. In a dual sense, the perspectivity with axis l is a 1–1 mapping of a pencil of lines with center O to a pencil of lines with center O' such that if line a through O is mapped to line a' through O', then the intersection of a and a' lies on l.*

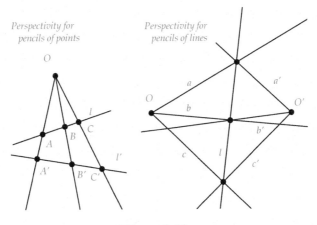

Figure 9.12

A perspectivity with center O is denoted by the symbol $\overset{O}{\barwedge}$. A perspectivity with axis l is denoted by the symbol $\overset{l}{\barwedge}$. Thus, in Figure 9.12 we have $ABC \overset{O}{\barwedge} A'B'C'$ and $abc \overset{l}{\barwedge} a'b'c'$.

It should be noted that our definition of a perspectivity with center O is only defined for pairs of lines —it is a transformation between two one-dimensional geometries. Later in this chapter we will expand our notion of projective transformations to include functions that map the entire Projective plane to itself.

The definition of perspectivity given above (hopefully) seems natural, given its intuitive roots in the artistic projection of three dimensions into a plane. It is interesting to compare the set of perspectivities in Projective geometry to the set of isometries in Euclidean geometry. One fundamental property of the set of Euclidean isometries is that it forms a *group*. In particular, if one composes two isometries, one gets another isometry. Interestingly, this is not the case in general for perspectivities.

In Figure 9.13 we have that $ABC \overset{O}{\wedge} A'B'C'$ and $A'B'C' \overset{O'}{\wedge} A''B''C''$.

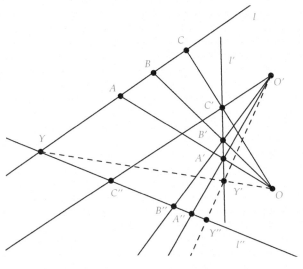

Figure 9.13

The composition of these two perspectivities maps line l to line l''. If this composition is again a perspectivity, then the intersection point, Y, of lines l and line l'' would be fixed under the perspectivity. But, this is clearly not the case, as Y is mapped to point Y''.

So, the composition of two perspectivities need not be a perspectivity, but it certainly is still a composition. Also, compositions of compositions of perspectivities will again be a composition of perspectivities. Thus, it is these compositions that seem the likely candidates to form the group of transformations in Projective geometry. This suggests the following definition.

> **Definition 9.9.** *A mapping of one pencil into another is a pro-jectivity if the mapping can be expressed as a finite composition of perspectivities.*

A projectivity is denoted by the symbol $\overline{\wedge}$. Thus, if $ABC \overline{\wedge} A'B'C'$ then there is a finite sequence of perspectivities that maps the collinear points ABC to the collinear points $A'B'C'$.

It will be left as an exercise to verify that the set of projectivities of a line into itself does indeed form a group.

9.4.2 Projectivity Constructions and Pappus's Theorem

We will now consider some basic results on projectivities.

> **Theorem 9.6.** *Let A,B,C be three distinct points on line l and A',
> B', and C' be three distinct points on line l', with l ≠ l'. Then, there
> is a projectivity taking A,B,C to A',B',C'.*

Proof: In the proof of this result we will be careful to use just the axioms
P1–P4 in the preceding section, plus any dual results based on those
axioms.

Since l and l' are distinct, we can assume that one of the pairs of
mapped points (AA', BB', CC'), consists of two different points. We
assume that A is not on l' and A' is not on l.

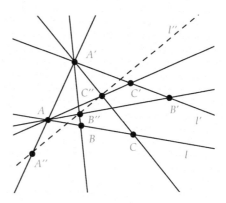

By axiom P1, lines AB', $A'B$,
AC', and $A'C$ exist. By Theorem
9.2 the points $B'' = AB' \cdot A'B$ and
$C'' = AC' \cdot A'C$ exist. ($AB' \cdot A'B$ is
the intersection of the two lines)
At least one of these two points
must be non-collinear with l and
l', as A and A' cannot both be on
l and l'.

The line $B''C''$ will intersect AA' at a point A'' by axiom P2. Also,
$A'' \neq A$, for if $A'' = A$, then points A, B'', and C'' must be collinear and
thus points A, B', and C' must be collinear, which is impossible as A is
not on l'. Similarly, we have that $A'' \neq A'$. Thus, $ABC \overset{A'}{\wedge} A''B''C''$ and
$A''B''C'' \overset{A}{\wedge} A'B'C'$ □

Using this result, we can show the following:

> **Theorem 9.7.** *Let A,B,C and A', B', and C' be two sets of dis-
> tinct points on line l. Then, there is a projectivity taking A,B,C to
> A',B',C'.*

The proof of this result is left as an exercise.

One interesting result of this theorem is that there exist *non-identity* projectivities that map two points of a line back to themselves, i.e., that leave two points of a line invariant. This is quite different than what we observed with isometries in Euclidean and Hyperbolic geometry. If an isometry fixed two points, then it was a reflection across the line through those two points, and it left every point on the line invariant. The isometry acted as the identity transformation on the line.

A natural question to ask, then, is whether a projectivity that fixes *three* points on a line must be the identity on that line. We will see in the next section that this is true for the *Real Projective Plane*, the Projective plane that is derived from adding points at infinity to the Euclidean plane. However, it is not provable by using just axioms P1–P4, or even if we add P5 and P6 from the last section (Fano's axiom and Desargues' Theorem). If we want this property of projectivities to be true in Projective geometry, we would have to add this property as an additional *axiom*.

Axiom P7: If a projectivity leaves three distinct points on a line invariant, then the projectivity must be the identity on the pencil of points on that line.

It turns out that this axiom is logically equivalent to a classical result in Geometry that was first discovered by Pappus of Alexandria (c. 290–350 AD).

Theorem 9.8. Theorem of Pappus *Let A,B,C be three distinct points on line l and A', B', and C' be three distinct points on line l', with $l \neq l'$. Then, the intersection points $X = AB' \cdot A'B$, $Y = AC' \cdot A'C$, and $Z = BC' \cdot B'C$ are collinear.*

An alternate statement of this theorem involves pencils of points and *cross joins*.

Definition 9.10. *Given two pencils of points with axes p and p', suppose there is a projectivity mapping the pencil at p to the pencil at p'. Let A and B be points on p and A', B' the corresponding points on p' under the projectivity. Then, the* cross *joins of these pairs of points are the lines AB' and BA'.*

Since a projectivity is uniquely defined by three pairs of points from one line to another (Exercise), we get the following equivalent statement for Pappus's Theorem.

> **Theorem 9.9. Theorem of Pappus** *A projectivity between two different pencils of points determines a unique line called the* axis of homology, *which contains the intersections of the cross joins of all pairs of corresponding points.*

We illustrate here two possible configurations of points in Pappus's theorem. In the one at the left, none of the points on the two lines are the point of intersection of the lines. In the figure on the right, one of the points, A, is the intersection point of the lines. Pappus's Theorem can be proven in this case, using results solely based on axioms P1–P4 (Exercise).

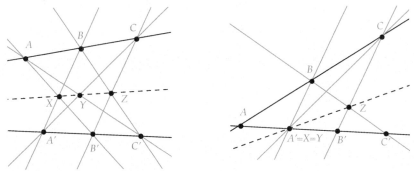

There are many applications of Pappus's Theorem. Here is a result that will be useful when we discuss conic sections:

> **Lemma 9.10.** *Let l, m, and n be three lines with $l \neq n$. Suppose there is a projectivity π given by $l \overset{O}{\wedge} m \overset{P}{\wedge} n$. If $X = l \cdot n$ is invariant under the projectivity, then l is perspective to n. That is, there is a point R such that $l \overset{R}{\wedge} n$ and the projectivity is equivalent to this perspectivity.*

Proof: We first note that if $m = n$ or $l = m$ then one of the perspectivities in the projectivity is the identity and the result is true. Thus, we only need to consider the case where all three lines are distinct.

We can also assume the three lines are non-concurrent, for if they were concurrent at X, then for points $A \neq B \neq X$ on l, and $C = \pi(A)$ and $D = \pi(b)$ on l', we could form the perspectivity with center $AC \cdot BD$. Since a perspectivity is uniquely defined by three distinct points on a line, this perspectivity must be equivalent to π.

Let $X = l \cdot n$, $Y = l \cdot m$, and $Z = m \cdot n$. Also, let $Q = m \cdot OX$ and $R = PY \cdot OZ$. Under the perspectivity from O, X goes to Q. Since X must be invariant, then $Q \overset{P}{\wedge} X$, which implies that O,P, and X must be collinear.

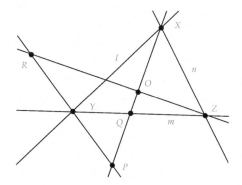

Let A be any point on l and let $A \overset{O}{\wedge} A'$, $A' \overset{P}{\wedge} A''$. Then, Y, A', and Z are three distinct points on m, while O, X, and P are three distinct points on XP. Applying Pappus's Theorem to these two triples, we get that $A = YX \cdot OA'$, $R = YP \cdot OZ$, and $A'' = A'P \cdot XZ$ are collinear.

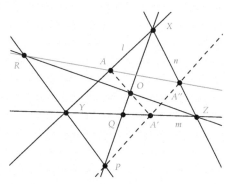

Now, R is defined independent of the point A. So, for any A on l we get that the line through R and A will yield the point A'', which is the result of the projectivity applied to A. Thus, the projectivity is the same as the perspectivity from R. □

In the next section, we will see that Pappus's Theorem can be *proven* if we are using the Real Projective Plane model for Projective geometry. In the more abstract setting of this section, what we can say is that Pappus's Theorem and axiom P7 are *logically equivalent*.

> **Theorem 9.11.** *If one assumes axioms P1–P4 and axiom P7, then Pappus's Theorem is true. Conversely, if one assumes axioms P1–P4 and Pappus's Theorem, then axiom P7 can be proven as a theorem.*

We will not take the time at this point in the text to go through the proof of this equivalence. The proof can be found in the web resources for this text, which are located at http://www.gac.edu/~hvidsten/geom-text. A complete discussion of the axiomatic intricacies for Projective geometry can be found in Hartshorne [20, Chapter 5].

We point out that in the course of the proof of Theorem 9.11, it is necessary to use the following result which says that a projectivity is essentially two perspectivities.

> **Theorem 9.12.** *If we assume Axioms P1–P4 and Pappus's Theorem, then a projectivity between two distinct lines can be written as the composition of at most two perspectivities.*

The proof can be found on the web resource page.

Before we end our discussion of projective transformations, we need to make sure that we have covered all possible duality considerations. Here is the dual to Axiom P7.

> **Theorem 9.13.** *Assume axioms P1–P4 and Pappus's Theorem. Then, if a projectivity leaves three distinct lines through a point invariant, then the projectivity must be the identity on the pencil of lines through that point.*

The dual to Pappus's Theorem involves a second type of cross-join.

> **Definition 9.11.** *Given two pencils of lines with centers P and P', suppose there is a projectivity mapping the pencil at P to the pencil at P'. Let a and b be lines incident on P and a', b' the corresponding lines incident on P' under the projectivity. Then, the cross joins of these pairs of lines are the points $a \cdot b'$ and $b \cdot a'$.*

> **Theorem 9.14. Dual to Theorem of Pappus** *Assume Axioms P1–P4 and Pappus's Theorem. Then, a projectivity between two pencils of lines determines a unique point called the* center of homology, *which is incident on the lines containing all cross joins of pairs of corresponding lines.*

The proofs of the preceding theorems can be found on the web resource page.

Exercise 9.4.1. *Show that the set of projectivities of a line l into itself forms a group. Refer to Exercise 5.6.5 for the definition of a group.*

Exercise 9.4.2. *Prove Theorem 9.7. [Hint: Use a result from the preceding section of this chapter to show there must be a line $l' \neq l$. Then use Theorem 9.6.]*

Exercise 9.4.3. *The duals to Theorem 9.6 and Theorem 9.7 must be automatically true by the principle of duality. Write down the duals to these two results.*

Exercise 9.4.4. *Prove that Pappus's theorem is true in the case where one of the points, say A, is the intersection point of the two lines l and l'. Your proof should only use results based on axioms P1–P4.*

Exercise 9.4.5. *Assume axioms P1–P4, and axiom P7 are true. Prove that if two projectivities between pencils of points with axes l and l' both have the same values on three distinct points A, B, and C on l, then the projectivities must be the same on l. Note: This result is often called the* Fundamental Theorem of Projective Geometry.

Exercise 9.4.6. *Assume axioms P1–P4, and axiom P7 are true. Prove that if two projectivities between pencils of lines with centers O and O' both have the same values on three distinct lines a, b, and c through O, then the projectivities must be the same on the pencil of lines at O. [Hint: Use the dual to axiom P7 (Theorem 9.13).]*

Exercise 9.4.7. *Using just Axioms P1–P4, prove that every projectivity can be written as the composition of at most three perspectivities. [Hint: there are two cases —either the projectivity maps a line to a different line or it maps a line to itself. Use the construction from the proof of Theorem 9.6.]*

In the next three exercises we will show that axioms P1–P4 and Pappus's Theorem imply axiom P6 – Desargues' Theorem.

We start by assuming axioms P1–P4 and Pappus's Theorem. Then, we assume the hypotheses of Desargues' Theorem, that triangles ABC and $A'B'C'$ are perspective from O, as shown, with corresponding sides intersecting at points P, Q, and R.

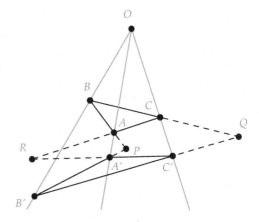

Exercise 9.4.8. *Let $S = A'B' \cdot BC$. Note that S cannot be B, B', or C' (No two points on a side of a triangle are collinear with the center of perspective). Assume S is not equal to C. Also, assume that $A \neq A'$, $B \neq B'$, and $C \neq C'$. Fill in the* Why? *parts of the argument below:*

Consider OAA' and CSB (Figure 9.14). We have the points

$$T = OS \cdot AC$$
$$U = OB \cdot A'C$$
$$P = AB \cdot A'S$$

Then, points T, U, and P are collinear (Why?).

For OCC' and $A'B'S$, we have $U = OB' \cdot A'C, V = OS \cdot A'C'$, and $Q = CS \cdot B'C'$. Then, points U, V, and Q are collinear (Why?)

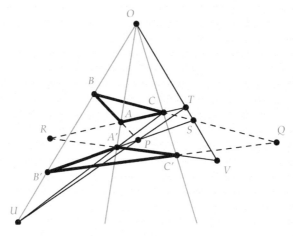

Figure 9.14

Consider the two triples of points $CA'U$ and VTS. Then, $R = CT \cdot A'V$, $Q = CS \cdot UV$ and $P = A'S \cdot TU$ (Why?) Thus, R, P, and Q are collinear. (Why?)

Exercise 9.4.9. *Assume axioms P1–P4 and Pappus's Theorem and assume that triangles ABC and $A'B'C'$ are perspective from O. If $A = A'$ or $B = B'$ or $C = C'$ show that Desargues' Theorem is true.*

Exercise 9.4.10. *The only case left to consider to finish the proof that axioms P1–P4 and Pappus's Theorem imply axiom P6 – Desargues' Theorem is the case where we have the same assumptions as in Exercise 9.4.8, but we suppose that the point $S = A'B' \cdot BC$ is equal to point C. Let $S_{A'C'} = A'C' \cdot AB$, $S_{B'C'} = B'C' \cdot AC$, $S_{AB} = AB \cdot B'C'$, $S_{AC} = AC \cdot A'B'$, and $S_{BC} = BC \cdot A'C'$. Show that $S_{A',C'} = B$, $S_{B',C'} = A$, $S_{A,B} = C'$, $S_{A,C} = B'$, and $S_{B,C} = A'$. Then, show that this produces an impossible case for Desargues' Theorem.*

9.5 MODELS OF PROJECTIVE GEOMETRY

Recall that a *model* for an axiomatic system was an interpretation of the undefined terms of the system such that the axioms held true. In this section we consider several models for Projective geometry. We start with the model alluded to at the end of the first section of this chapter. This model is based on the standard Euclidean plane.

9.5.1 The Real Projective Plane

We start by defining points at infinity.

Definition 9.12. *Let l be a Euclidean line and let $[l]$ represent the set of all Euclidean lines that are parallel to l. Then, we define $[l]$ to be an* ideal point *or a* point at infinity. *The set of all ideal points is called the* line at infinity. *Given a Euclidean line l, we define the* extended line *through l to be the set consisting of the points of l plus $[l]$.*

Definition 9.13. *The points of the* Real Projective Plane *consist of all ordinary Euclidean points plus all ideal points. The lines consist of all extended Euclidean lines plus the line at infinity.*

For this interpretation of points and lines to be a model for Projective geometry, we must show that it satisfies all of the axioms for Projective geometry.

Theorem 9.15. *The Real Projective Plane satisfies axioms P1–P4 of Projective geometry.*

Proof: **Axiom P1**: Let P and Q be two points in the Real Projective Plane. If the two points are Euclidean points, then from Euclidean geometry we know there is a unique line incident on them.

Suppose P is Euclidean and Q is an ideal point. Then, $Q = [l]$ from some Euclidean line l. Any line incident on P and Q must be parallel to l and pass through P. We know there is a unique Euclidean line m through P that is parallel to l (Playfair's Postulate), and thus is incident on P and Q.

Suppose P and Q are both ideal points. Then, they belong to the line at infinity and no other line.

Axiom P2: Let l and m be lines in the Real Projective Plane with $l \neq m$. If they are both Euclidean lines, then there are two cases to consider. If they are not parallel, then they intersect. If they are parallel, then the point $[l] = [m]$ is on both.

Suppose l is Euclidean and m is the line at infinity. Then, the point $[l]$ is on m and on l.

Axiom P3: The Real Projective Plane has an infinite number of Euclidean points.

Axiom P4: Every Euclidean line clearly has at least three points on it. Also, there are at least three distinct Euclidean lines that are pair-wise non-parallel (take the lines extending the three sides of an equilateral triangle). These three lines define three distinct points on the line at infinity. □

From the work we did in the last section, we know that we can now define perspectivities and projectivities, that is, projective transformations. There is an alternate, but equivalent, model for the Real Projective Plane that will allow us to easily analyze such projective transformations.

The 3D Model of the Real Projective Plane

In the three dimensional model of the Real Projective Plane, we interpret the two dimensional points of the Euclidean plane as three dimensional points with z-coordinate equal to 1, much as we did in section 5.7.1, where we discussed the matrix form of Euclidean isometries.

Given a Euclidean point (x, y) we identify this point with the point $(x, y, 1)$. An ordinary Euclidean line will then be identified with the corresponding line at height 1.

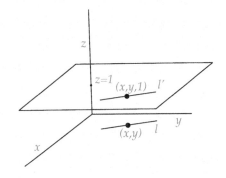

We note that there is a 1–1 relationship between the points $(x, y, 1)$ and the lines through the origin and these points. Also, there is a similar relationship between lines l' in the $z = 1$ plane (labeled π in the figure) and *planes* through the origin and l'.

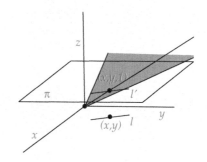

Thus, we can identify the ordinary points of the Real Projective Plane with points $(x, y, 1)$, or equivalently with any point along the line $k(x, y, 1)$. What about ideal points? Clearly, as a point moves farther and farther from the z-axis, the slope of the line through the origin and that point will decrease. In the "limit," this line will approach a line in the x-y plane. Thus, points at infinity can be identified with lines through the origin with z-coordinate equal to zero.

We summarize this discussion by introducing the notion of *homogeneous coordinates* .

Definition 9.14. Homogeneous coordinates (x_1, x_2, x_3) *for an ordinary Euclidean point* $P = (x, y)$ *are a choice of* x_1, x_2, x_3 *such that* $\dfrac{x_1}{x_3} = x$ *and* $\dfrac{x_2}{x_3} = y$. *One such representation is* $(x, y, 1)$. *Homogeneous coordinates of the form* $(x_1, x_2, 0)$ *represent points at infinity.*

Let P be the Euclidean point (x, y) and v the Euclidean 3-D line through $(x, y, 1)$ and the origin. Homogeneous coordinates give us a way to represent P (equivalently v) as a single analytical quantity, and they also give a simple analytical description of points at infinity. Points at infinity, with coordinates $(x_1, x_2, 0)$, represent the set of lines parallel to a given line having slope $\dfrac{x_2}{x_1}$. Homogeneous coordinates thus give us a natural way to "coordinatize" the Real Projective Plane.

A natural question to ask is whether there is a simple way to describe lines in the Real Projective Plane. From the discussion above, a line can be considered equivalent to a plane through the origin. The line at infinity includes all ideal points, thus all points of the form $(x, y, 0)$. The line at infinity is then the x-y plane. Now, any plane through the origin must satisfy an equation of the form $ax + by + cz = 0$. The point (x, y, z)

is on the plane and (a, b, c) is the normal vector, which uniquely defines the plane. We see that vectors can define both points and lines in the Real Projective Plane. To distinguish the two uses for vectors, we will use square brackets "[ ]" to denote planar coordinates (those that define lines).

We can now define the 3-D model of the Real Projective Plane.

Definition 9.15. *The points of the 3-D model include all non-zero homogeneous coordinate vectors $v = (x, y, z)$. The lines of the 3-D model include all non-zero homogeneous planar vectors $[a, b, c]$. A point $v = (x, y, z)$ is on a line $u = [a, b, c]$ iff $ax + by + cz = 0$. That is, if the dot product $u \cdot v = 0$.*

For the 3-D model we will think of vectors for lines $[a, b, c]$ as row vectors and vectors for points (x, y, z) as column vectors. Then, $u \cdot v$ can be thought of as a matrix multiplication.

It will be left as an exercise to verify that Axioms P1–P4 hold in this new model. The points and lines in the original Real Projective Plane can be identified in a 1–1 and onto fashion with the points and lines of the 3-D model. In fact, the 3-D model is essentially just a re-interpretation of the points and lines of the original model —the two models are essentially the same model. Thus, any statement about lines, points, and incidence that is proven in one of the models is automatically true in the other.

Here are a couple of useful results in homogeneous coordinates.

Theorem 9.16. *Three points $X = (x_1, x_2, x_3)$, $Y = (y_1, y_2, y_3)$ and $Z = (z_1, z_1, z_3)$ are collinear if and only if the determinant*
$$\begin{vmatrix} x_1 & y_1 & z_1 \\ x_2 & y_2 & z_2 \\ x_3 & y_3 & z_3 \end{vmatrix}$$ *is equal to 0.*

Proof: From linear algebra, we know that the determinant of the square matrix is 0 if and only if one of the columns (points) is a linear combination of the other two. This is equivalent to one of the points being on the same line as the other two. □

Theorem 9.17. *Three lines* $u = (u_1, u_2, u_3)$, $v = (v_1, v_2, v_3)$ *and* $w = (w_1, w_2, w_3)$ *are concurrent (meet at a single point) if and only if the determinant* $\begin{vmatrix} u_1 & u_2 & u_3 \\ v_1 & v_2 & v_3 \\ w_1 & w_2 & w_3 \end{vmatrix}$ *is equal to 0.*

Proof: From linear algebra, we know that the determinant of a square matrix is 0 if and only if one of the rows (lines) is a linear combination of the other two. This is equivalent to one of the lines having a normal vector which is in the same plane as the other two normals. This means the three planes intersect along a common line, i.e., a common projective point. □

9.5.2 Transformations in the Real Projective Plane

In section 9.4 we considered perspectivities and projectivities of one line to another. In this section we show that we can create coordinate representations of projectivities by the use of *matrix* transformations. We start by developing a coordinate parametrization for points on a line.

Theorem 9.18. *Let* $P = (p_1, p_2, p_3)$ *and* $Q = (q_1, q_2, q_3)$ *be distinct points. Let* $X = (x_1, x_2, x_3)$ *be a point on the line* $P \cdot Q$. *Then,* $X = \alpha P + \beta Q = (\alpha p_1 + \beta q_1, \alpha p_2 + \beta q_2, \alpha p_3 + \beta q_3)$, *with* α *and* β *real constants with at least one being non-zero. Conversely, if* $X = \alpha P + \beta Q$, *with at least one of* α *and* β *not zero, then* X *is on the line through* P *and* Q.

Proof: We start by assuming that X is on the line through P and Q. By Theorem 9.16 the vectors representing P, Q, and X are linearly dependent. That is, there are constants a, b, and c such that $aP + bQ + cX = (0, 0, 0)$. Since P and Q are distinct homogeneous points, then $aP + bQ \neq (0, 0, 0)$. Thus, at least one of a and b is non-zero. Also, $c \neq 0$. Thus, $X = \frac{a}{-c}P + \frac{b}{-c}Q = \alpha P + \beta Q$.

Conversely, suppose we have a point $X = \alpha P + \beta Q$. Then, clearly X is a linear combination of P and Q and Theorem 9.16 can be used to show that X is on the line through P and Q. □

> **Definition 9.16.** *The coordinates* (α, β) *are called* homogeneous parameters *or* parametric homogeneous coordinates *of the point X with respect to the* base points P *and* Q.

Homogeneous parameters of a point are not unique. From the proof of the preceding theorem, any multiple of a given homogeneous representation, (α, β) is equivalent to any $(\lambda\alpha, \lambda\beta)$, for $\lambda \neq 0$. What is unique is the *ratio* of the two parameters $\dfrac{\alpha}{\beta}$.

> **Definition 9.17.** *Given homogeneous parameters* (α, β) *of the point X with respect to the* base points P *and* Q, *the ratio* $\dfrac{\alpha}{\beta}$ *is called the* parameter *of the point.*

We automatically have the following dual theorem to Theorem 9.18:

> **Theorem 9.19.** *Let* $u = [u_1, u_2, u_3]$ *and* $v = [v_1, v_2, v_3]$ *be two distinct lines with homogeneous coordinates in the Real Projective Plane. Let* $w = [w_1, w_2, w_3]$ *be any line concurrent with* $u \cdot v$. *Then, w can be represented as* $w = Au + Bv$, *with A and B real constants, with at least one being non-zero. Conversely, if* $w = Au + Bv$, *with at least one of A and B not zero, then w is concurrent with* $u \cdot v$.

We now can provide a coordinate description of a perspectivity.

> **Theorem 9.20.** *Let l and l' be distinct lines perspective from point* O. *Let $P \neq Q$ be on l with* $PQ \overset{O}{\wedge} P'Q'$. *Let X be on l with parametric homogeneous coordinates* (α, β). *Let X' be on l' with* $PQX \overset{O}{\wedge} P'Q'X'$ *and parametric homogeneous coordinates* (α', β') *(with respect to points P' and Q'). Then, the perspectivity mapping X to X' can be represented in parametric coordinates as:*
>
> $$\lambda \begin{bmatrix} \alpha' \\ \beta' \end{bmatrix} = \begin{bmatrix} a & b \\ c & d \end{bmatrix} \begin{bmatrix} \alpha \\ \beta \end{bmatrix} \quad \text{with } ad - bc \neq 0, \lambda \neq 0$$

Proof: Let $u = P \cdot P'$, $v = Q \cdot Q'$, and $w = X \cdot X'$ (Figure 9.15). By Theorem 9.19 we know that w has parametric homogeneous coordinates $w = A\,u + B\,v$.

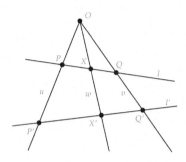

Figure 9.15

X is on the line w if the dot product of w and X is zero, or $[A\,u + B\,v] \cdot (\alpha P + \beta Q) = 0$. Expanding this out we get:

$$[Au_1 + Bv_1, Au_2 + Bv_2, Au_3 + Bv_3] \cdot (\alpha p_1 + \beta q_1, \alpha p_2 + \beta q_2, \alpha p_3 + \beta q_3) = 0$$

Re-arranging, we get

$$(u_1p_1 + u_2p_2 + u_3p_3)A\alpha + (u_1q_1 + u_2q_2 + u_3q_3)A\beta +$$
$$(v_1p_1 + v_2p_2 + v_3p_3)B\alpha + (v_1q_1 + v_2q_2 + v_3q_3)B\beta = 0$$

Let a_{21}, a_{22}, $-a_{11}$, and $-a_{12}$ be the multipliers of $A\alpha$, $A\beta$, $B\alpha$ and $B\beta$ respectively. Then,

$$a_{21}A\alpha + a_{22}A\beta = a_{11}B\alpha + a_{12}B\beta$$
$$\frac{A}{B} = \frac{a_{11}\alpha + a_{12}\beta}{a_{21}\alpha + a_{22}\beta}$$

In matrix form this can be written as

$$\lambda_1 \begin{bmatrix} A \\ B \end{bmatrix} = \begin{bmatrix} a_{11} & a_{12} \\ a_{21} & a_{22} \end{bmatrix} \begin{bmatrix} \alpha \\ \beta \end{bmatrix} \tag{9.1}$$

for some $\lambda_1 \neq 0$. We note that the matrix $\begin{bmatrix} a_{11} & a_{12} \\ a_{21} & a_{22} \end{bmatrix}$ is non-singular. If it were singular, then there would be a set of parameters (α, β) such that Equation 9.1 has A and B both equal to zero. But, we know that one of A or B must always be non-zero, which means that the matrix must be non-singular, and thus invertible.

We could carry out a similar argument for the point X' to be on line w and get

$$\lambda_2 \begin{bmatrix} A \\ B \end{bmatrix} = \begin{bmatrix} b_{11} & b_{12} \\ b_{21} & b_{22} \end{bmatrix} \begin{bmatrix} \alpha' \\ \beta' \end{bmatrix}$$

for some $\lambda_2 \neq 0$, with $\begin{bmatrix} b_{11} & b_{12} \\ b_{21} & b_{22} \end{bmatrix}$ again non-singular. Let $\begin{bmatrix} c_{11} & c_{12} \\ c_{21} & c_{22} \end{bmatrix}$ be the inverse matrix to $\begin{bmatrix} b_{11} & b_{12} \\ b_{21} & b_{22} \end{bmatrix}$. Then,

$$\frac{1}{\lambda_2} \begin{bmatrix} \alpha' \\ \beta' \end{bmatrix} = \begin{bmatrix} c_{11} & c_{12} \\ c_{21} & c_{22} \end{bmatrix} \begin{bmatrix} A \\ B \end{bmatrix} \tag{9.2}$$

Combining equations 9.1 and 9.2 we get

$$\frac{\lambda_1}{\lambda_2} \begin{bmatrix} \alpha' \\ \beta' \end{bmatrix} = \begin{bmatrix} c_{11} & c_{12} \\ c_{21} & c_{22} \end{bmatrix} \begin{bmatrix} a_{11} & a_{12} \\ a_{21} & a_{22} \end{bmatrix} \begin{bmatrix} \alpha \\ \beta \end{bmatrix}$$

The result stated in the theorem follows by letting $\lambda = \frac{\lambda_1}{\lambda_2}$ and letting

$$\begin{bmatrix} a & b \\ c & d \end{bmatrix} = \begin{bmatrix} c_{11} & c_{12} \\ c_{21} & c_{22} \end{bmatrix} \begin{bmatrix} a_{11} & a_{12} \\ a_{21} & a_{22} \end{bmatrix}$$

□

Since a projectivity is a finite sequence of perspectivities, we have:

Corollary 9.21. *Given a projectivity $l \barwedge l'$, with l and l' distinct. Let $P \neq Q$ be on l with $PQ \barwedge P'Q'$. Let X be on l with parametric homogeneous coordinates (α, β). Let X' be on l' with $PQX \barwedge P'Q'X'$, and parametric homogeneous coordinates (α', β') (with respect to P' and Q'). Then, the projectivity, as a mapping of points on l, can be represented by a matrix equation of the form:*

$$\lambda \begin{bmatrix} \alpha' \\ \beta' \end{bmatrix} = \begin{bmatrix} a & b \\ c & d \end{bmatrix} \begin{bmatrix} \alpha \\ \beta \end{bmatrix} \text{ with } ad - bc \neq 0, \lambda \neq 0$$

The converse to the above result also holds —every non-singular 2×2 matrix generates a projectivity with respect to local homogeneous coordinates.

Theorem 9.22. *Let* $P \neq Q$ *be on* l. *Let* X *be a point on* l *with parametric homogeneous coordinates* (α, β) *with respect to* P *and* Q.

Let $\begin{bmatrix} a & b \\ c & d \end{bmatrix}$ *be a matrix with* $ad - bc \neq 0$. *Then, the transformation of* l *into* l' *defined by the matrix equation:*

$$\begin{bmatrix} \alpha' \\ \beta' \end{bmatrix} = \begin{bmatrix} a & b \\ c & d \end{bmatrix} \begin{bmatrix} \alpha \\ \beta \end{bmatrix}$$

defines a projectivity from l *to* l'.

Proof: Consider the three points P, Q, and R on l, where R has homogeneous coordinates $(1,1)$. Since $P = (1,0)$ and $Q = (0,1)$, then the matrix transformation maps P, Q, R to S, T, U on l' with homogeneous coordinates (a, c), (b, d) and $(a+b, c+d)$.

Now, we know that there is a projectivity that takes P, Q, and R on l to S, T, and U on l'. By the preceding corollary, this projectivity can be represented as a matrix by

$$\lambda \begin{bmatrix} \alpha' \\ \beta' \end{bmatrix} = \begin{bmatrix} a_{11} & a_{12} \\ a_{21} & a_{22} \end{bmatrix} \begin{bmatrix} \alpha \\ \beta \end{bmatrix}$$

where the determinant $ad - bc \neq 0$ and $\lambda \neq 0$. Thus,

$$\lambda_1 \begin{bmatrix} a \\ c \end{bmatrix} = \begin{bmatrix} a_{11} & a_{12} \\ a_{21} & a_{22} \end{bmatrix} \begin{bmatrix} 1 \\ 0 \end{bmatrix}$$

$$\lambda_2 \begin{bmatrix} b \\ d \end{bmatrix} = \begin{bmatrix} a_{11} & a_{12} \\ a_{21} & a_{22} \end{bmatrix} \begin{bmatrix} 0 \\ 1 \end{bmatrix}$$

$$\lambda_3 \begin{bmatrix} a+b \\ c+d \end{bmatrix} = \begin{bmatrix} a_{11} & a_{12} \\ a_{21} & a_{22} \end{bmatrix} \begin{bmatrix} 1 \\ 1 \end{bmatrix}$$

We can assume that one of the three scalar multiples, say λ_3 is 1. Then,

$$\lambda_1 a = a_{11} \quad \lambda_2 b = a_{12} \quad a+b = a_{11} + a_{12}$$
$$\lambda_1 c = a_{21} \quad \lambda_2 d = a_{22} \quad c+d = a_{21} + a_{22}$$

Then,

$$a + b = \lambda_1 a + \lambda_2 b$$
$$c + d = \lambda_1 c + \lambda_2 d$$

Or,

$$a(1 - \lambda_1) + b(1 - \lambda_2) = 0$$
$$c(1 - \lambda_1) + d(1 - \lambda_2) = 0$$

This can be written as $\begin{bmatrix} a & b \\ c & d \end{bmatrix} \left(\begin{matrix} (1 - \lambda_1) \\ (1 - \lambda_2) \end{matrix} \right) = 0$. Since $ad - bc \neq 0$ the only solution to this equation is $\lambda_1 = 1, \lambda_2 = 1$.

Thus, $\begin{bmatrix} a & b \\ c & d \end{bmatrix} = \begin{bmatrix} a_{11} & a_{12} \\ a_{21} & a_{22} \end{bmatrix}$ □

We note that the duals to the above theorems are automatically true.

Recall our discussion in section 9.4.2 concerning projectivities. In that section we realized that we needed an axiom (P7) to ensure the uniqueness of projectivities for a projective geometry satisfying axioms P1–P4. Axiom P7 stated that if a projectivity leaves three distinct points on a line invariant, then the projectivity must be the identity. It will be left as an exercise to prove that projectivities in the 3-D model of the Real Projective Plane satisfy Axiom P7.

Since the Real Projective Plane model satisfies P7, then all of the results from 9.4.2 hold. In particular, we know that Pappus's Theorem and Desargues' Theorem are true in this model.

The only remaining axiom that we have yet to prove in the Real Projective Plane is Axiom P5 (Fano's Axiom) on complete quadrangles. To consider this axiom, we will have to expand our notion of transformations from that of projectivities (from lines to lines) to general transformations of the entire projective plane.

9.5.3 Collineations

We know that a projectivity maps a line to another line in a 1–1 fashion. If we want to extend the one-dimensional notion of projectivities to mappings of the entire projective plane to itself, it makes sense to look at transformations that are 1–1 and onto, that map lines to lines, and that preserve incidence.

Definition 9.18. *A* collineation *is a 1–1 and onto transformation of the Real Projective Plane to itself that maps lines to lines and preserves intersections of lines.*

We know from linear algebra that the requirement for a collineation to take lines to lines means that the analytic (coordinate) form of the mapping will be a matrix transformation from Euclidean three-dimensional space ($\mathbb{R}^3$) to itself. That is,

Theorem 9.23. *A collineation can be represented by a matrix equation of the form $\lambda X' = AX$ where A is a 3×3 matrix with non-zero determinant and $\lambda \neq 0$.*

Collineations have an interesting connection with complete quadrangles. Recall that a complete quadrangle consist of four distinct points, no three of which are collinear. Perhaps the easiest quadrangle to define is the set $X_1 = (1, 0, 0)$, $X_2 = (0, 1, 0)$, $X_3 = (0, 0, 1)$, and $X_4 = (1, 1, 1)$. It will be left as an exercise to show that these four points form a complete quadrangle.

Theorem 9.24. *Given a complete quadrangle consisting of points P, Q, R, and S, there is a unique collineation that takes P to X_1, Q to X_2, R to X_3, and S to X_4.*

Proof: Let $b_{11}x_1 + b_{12}x_2 + b_{13}x_3 = 0$ be the equation of the line $Q \cdot R$, $b_{21}x_1 + b_{22}x_2 + b_{23}x_3 = 0$ be the equation of the line $P \cdot R$, and $b_{31}x_1 + b_{32}x_2 + b_{33}x_3 = 0$ be the equation of the line $P \cdot Q$, Consider the matrix

$$B = \begin{bmatrix} b_{11} & b_{12} & b_{13} \\ b_{21} & b_{22} & b_{23} \\ b_{31} & b_{32} & b_{33} \end{bmatrix}$$

Then, $BP = (\alpha, 0, 0) = (1, 0, 0) = X_1$, $BQ = (0, \beta, 0) = X_2$, and $BR = (0, 0, \gamma) = X_3$. (two points are equivalent if their homogeneous coordinates are non-zero multiples of one another.)

Let

$$A = \begin{bmatrix} k_1 b_{11} & k_1 b_{12} & k_1 b_{13} \\ k_2 b_{21} & k_2 b_{22} & k_2 b_{23} \\ k_3 b_{31} & k_3 b_{32} & k_3 b_{33} \end{bmatrix}$$

for non-zero constants k_1, k_2, and k_3. Then, $AP = (k_1\alpha, 0, 0) = X_1$, $AQ = (0, k_2\beta, 0) = X_2$, and $AR = (0, 0, k_3\gamma) = X_3$. So, A and B have the same effect on P, Q, and R.

Consider the effect of B on the fourth point in the quadrangle. $BS = (s_1, s_2, s_3) = s_1(1, 0, 0) + s_2(0, 1, 0) + s_3(0, 0, 1)$. Since S is not collinear with any pair from P, Q, and R, then BS is not collinear with any pair from $(1, 0, 0)$, $(0, 1, 0)$, and $(0, 0, 1)$. This means that s_1, s_2, and s_3 are all non-zero. Let $k_1 = \frac{1}{s_1}$, $k_2 = \frac{1}{s_2}$, and $k_3 = \frac{1}{s_3}$. Then, $AS = (1, 1, 1)$. Also, if we use any other homogeneous coordinates for BS, say $BS = (\lambda s_1, \lambda s_2, \lambda s_3)$, $\lambda \neq 0$, then $AS = (\lambda, \lambda, \lambda) = (1, 1, 1) = X_4$. Thus, A maps P, Q, R, and S to X_1, X_2, X_3, and X_4.

To show that A is unique, suppose that there is another collineation given by a matrix A' that takes P to X_1, Q to X_2, R to X_3, and S to X_4. Let the homogeneous coordinates for points P, Q, R, and S be fixed. Let $\begin{bmatrix} P & Q & R \end{bmatrix}$ be the matrix with column vectors being the homogeneous coordinates for P, Q, and R. From our work above we have that

$$A \begin{bmatrix} P & Q & R \end{bmatrix} = \begin{bmatrix} k_1\alpha & 0 & 0 \\ 0 & k_2\beta & 0 \\ 0 & 0 & k_3\gamma \end{bmatrix}$$

Or,

$$A = \begin{bmatrix} k_1\alpha & 0 & 0 \\ 0 & k_2\beta & 0 \\ 0 & 0 & k_3\gamma \end{bmatrix} \begin{bmatrix} P & Q & R \end{bmatrix}^{-1}$$

For this last equation, we use the fact that P, Q, and R are non-collinear, which means that the three homogeneous column vectors for these points are linearly independent. Thus the matrix formed from these three vectors is invertible.

For A', we have that

$$A' \begin{bmatrix} P & Q & R \end{bmatrix} = \begin{bmatrix} \lambda_1 & 0 & 0 \\ 0 & \lambda_2 & 0 \\ 0 & 0 & \lambda_3 \end{bmatrix}$$

Or,

$$A' = \begin{bmatrix} \lambda_1 & 0 & 0 \\ 0 & \lambda_2 & 0 \\ 0 & 0 & \lambda_3 \end{bmatrix} \begin{bmatrix} P & Q & R \end{bmatrix}^{-1}$$

Let $(\sigma_1, \sigma_2, \sigma_3) = \begin{bmatrix} P & Q & R \end{bmatrix}^{-1}(S)$. Then, $AS = (k_1 \alpha \sigma_1, k_2 \beta \sigma_2, k_3 \gamma \sigma_3)$ and $A'S = (\lambda_1 \sigma_1, \lambda_2 \sigma_2, \lambda_3 \sigma_3)$. Since $AS = cA'S$, then

$$k_1 \alpha \sigma_1 = c\lambda_1 \sigma_1$$
$$k_2 \alpha \sigma_2 = c\lambda_2 \sigma_2$$
$$k_3 \alpha \sigma_3 = c\lambda_3 \sigma_3$$

Or,

$$\begin{pmatrix} k_1 \alpha \\ k_2 \alpha \\ k_3 \alpha \end{pmatrix} = c \begin{pmatrix} \lambda_1 \\ \lambda_2 \\ \lambda_3 \end{pmatrix}$$

Thus,

$$A = \begin{bmatrix} k_1 \alpha & 0 & 0 \\ 0 & k_2 \beta & 0 \\ 0 & 0 & k_3 \gamma \end{bmatrix} \begin{bmatrix} P & Q & R \end{bmatrix}^{-1}$$

$$= \begin{bmatrix} c\lambda_1 & 0 & 0 \\ 0 & c\lambda_2 & 0 \\ 0 & 0 & c\lambda_3 \end{bmatrix} \begin{bmatrix} P & Q & R \end{bmatrix}^{-1} = cA'$$

We conclude that, in homogeneous coordinates, A and A' are the same transformation. It is also clear that A is invertible. Thus, A is the unique collineation as described in the theorem. □

The next theorem is often called the *Fundamental Theorem of Projective Geometry*. We have already discussed another result (Exercise 9.4.5) which also had this moniker. The previous fundamental result dealt with triples of points on a single line, while the following fundamental result deals with non-collinear sets of points. This second Fundamental Theorem follows almost immediately from the preceding result.

Theorem 9.25. *There exists a unique collineation that takes the four points of a complete quadrangle to any other four points of another complete quadrangle.*

The proof will be left as an exercise.

While the proof of the Fundamental Theorem of Projective Geometry presented here is based on using homogeneous coordinates in the Real Projective Plane, the result actually holds for any abstract Projective geometry that satisfies Fano's axiom and Pappus's Theorem. A proof of this result can be found in [19][Chapter 8].

9.5.4 Homogeneous Coordinates and Perspectivities

So far we have considered perspectivities of two types —those from a pencil of points on one line to a pencil of points on another, or from a pencil of lines on one point to a pencil of lines on another. (Figure 9.16)

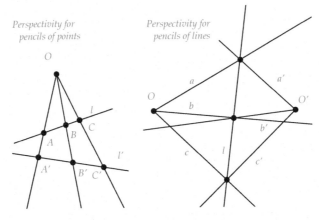

Figure 9.16

We have seen that in both cases, the perspectivity can be represented in homogeneous coordinates by a 2×2 matrix. There is a third type of perspectivity that will come in handy later in this chapter —that of a pencil of points to a pencil of lines, or vice-versa.

> **Definition 9.19.** *A perspectivity with center O and axis l is a 1–1 mapping of a pencil of points with axis l to a pencil of lines with center O such that if P on l is mapped to p through O, then p passes through P.*

Here we have an illustration of a perspectivity of this third type. Lines p, q, and r are mapped to points P, Q, and R.

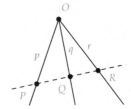

If we are careful with the choice of homogeneous coordinates for lines and points, we can synchronize the coordinates for a perspectivity from a pencil of points to a pencil of lines. The proof of the next result can be found on the web resource page.

> **Theorem 9.26.** *Given a perspectivity with center O and axis l, we can choose homogeneous coordinates for the pencil of points on l and the pencil of lines through O so that the same coordinates can represent both sets.*

9.5.5 Elliptic Model

Before we leave the subject of models for Projective geometry, we will note the connection between the 3-D Real Projective Plane model and the model for Elliptic geometry developed in section 8.3.3. In the Real Projective Plane, points are represented as lines through the origin. Each such line will intersect the unit sphere in a pair of antipodal points. Since a line represents a single point in the Projective plane, we can identify a pair of antipodal spherical points as representing a single projective point. Also, a projective line is defined as a plane through the origin. This intersects the sphere in a great circle. We can identify great circles as representing projective lines.

This representation of projective point and line was precisely the definition we used in section 8.3.3 to define elliptic points and lines in the Single Elliptic model of Elliptic Geometry. Thus, we immediately see that Single Elliptic geometry, or Elliptic geometry in general, is a model for Projective geometry. Elliptic geometry is distinguished from Projective geometry only in that it adds the metric notions of distance and angle measure.

Exercise 9.5.1. *Show that the 3-D model of the Real Projective Plane, as defined in 9.15 satisfies axioms P1–P4 for a Projective Plane.*

Exercise 9.5.2. *Use the matrix form of a projectivity to prove that the 3-D model of the Real Projective Plane satisfies Axiom P7: If a projectivity leaves three different points —P, Q, and R —on a line invariant, then the projectivity is the identity. [Hint: Use Corollary 9.21 and facts about eigenvectors.]*

Exercise 9.5.3. *What are the homogeneous coordinates of the line at infinity?*

Exercise 9.5.4. *Let P and Q be distinct points on a projective line l and let X be a point on l with parametric homogeneous coordinates (α, β) with respect to P and Q. Suppose R and S are also distinct points on l. Then, $R =$*

$\lambda_1 P + \lambda_2 Q$ and $S = \mu_1 P + \mu_2 Q$. Let (α', β') be the homogeneous coordinates for X with respect to R and S. Show that

$$\begin{pmatrix} \alpha \\ \beta \end{pmatrix} = \begin{bmatrix} \lambda_1 & \mu_1 \\ \lambda_2 & \mu_2 \end{bmatrix} \begin{pmatrix} \alpha' \\ \beta' \end{pmatrix}$$

Exercise 9.5.5. *Let $u = [u_1, u_2, u_3]$ be a line in the Real Projective Plane that is not the line at infinity. Let P and Q be any two non-ideal points (points not at infinity). Find the parametric homogeneous coordinates of the ideal point X of u.*

Exercise 9.5.6. *Prove that the four points $A = (0, 0, 1)$, $B = (1, 0, 0)$, $C = (0, 0, 1)$, and $D = (1, 1, 1)$ form a complete quadrangle. [Hint: use Theorem 9.16]*

Exercise 9.5.7. *Prove Theorem 9.25*

Exercise 9.5.8. *Suppose a collineation fixes the four points of a complete quadrangle. Prove that the collineation is the identity.*

Exercise 9.5.9. *Show that a collineation, when restricted to the points on a line l, is actually a* projectivity *on the points of l. [Hint: Let l be a line and $l' = Al$. Let P and Q be base points for homogeneous coordinates for l and P', Q' homogeneous coordinates on l'. Given X on l, use homogeneous coordinates to show that AX can be written as a 2×2 matrix equation.]*

Define a collineation to be an *affine* collineation if the collineation maps parallel lines to parallel lines. That is, the transformation preserves the property of parallelism. Two lines are parallel if they intersect at an ideal point.

Exercise 9.5.10. *Prove that a collineation is affine if and only if the collineation maps ideal points to other ideal points. Thus, the collineation maps the line at infinity to itself.*

Exercise 9.5.11. *Show that if a collineation is affine then the matrix for the collineation has its third row equivalent to $[0\ 0\ 1]$. [Hint: use homogeneous coordinates for ideal points.]*

9.6 PROJECT 15 - RATIOS AND HARMONICS

We saw in section 1.2.2 that ratios of distances were of particular importance to early Euclidean Geometry. The *golden ratio* was defined to be a splitting of a segment into two parts so that the ratio of the whole to the larger part was equal to the ratio of the larger part to the smaller. Such a splitting was called a *golden section*. It is clear that any Euclidean isometry will preserve a golden section, as isometries preserve Euclidean distance and length.

In this project we will explore whether we can extend this notion of *invariant* ratios to Projective geometry. It is not obvious that this would even be possible, as Projective geometry has no axioms that define metric properties such as distance. However, we can discuss distance in the 3-D model of Projective geometry. We will see that there is a ratio of distances that is preserved under projective *collineations*.

In 9.3.2 we discussed how a projective geometry could be built from an affine geometry by adding points at infinity. Before we look at ratios in Projective geometry, we will take a slight side trip to look at ratios in Affine geometry.

9.6.1 Ratios in Affine Geometry

Affine geometry is an abstraction of Euclidean geometry where we are concerned only with the incidence of lines and points and with the parallel properties of lines. Parallelism has the same meaning and properties in Affine geometry as it has in Euclidean geometry.

In Exercises 9.5.10 and 9.5.11 we studied the nature of affine collineations in the 3-D model of the Real Projective Plane. Affine collineations preserve the Euclidean notion of parallelism.

In Exercise 9.5.10 we showed that an affine collineation maps the line at infinity to itself. This means that an affine collineation, when restricted to ordinary Euclidean points, maps these points to other Euclidean points. Thus, we can think of an affine collineation as a map on points whose homogeneous representation has the form $(x, y, 1)$.

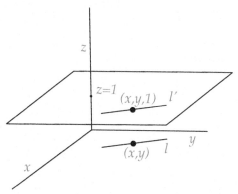

In Exercise 9.5.11 we showed that the matrix for an affine collineation has its third row equivalent to $[0\ 0\ 1]$. An affine collineation has the form

$$\begin{bmatrix} a & b & e \\ c & d & f \\ 0 & 0 & 1 \end{bmatrix}$$

In the first part of this project, we will discover a ratio that is preserved under transformation by affine collineations. We will use dynamic geometry software to carry out our investigation. Notes on how to use software for this project can be found at http://www.gac.edu/~hvidsten/geom-text.

Start up your geometry software and construct a line $\overleftrightarrow{AB}$. Attach a point C to this line and measure the distance from A to B and the distance from A to C. Then, compute the ratio of these two distances.

Dist(A,B) = 3.21
Dist(A,C) = 4.38
Dist(A,C) /Dist(A,B) = 1.36

Next, create an affine collineation. A suggestion might be:

$$\begin{bmatrix} 1 & -1 & 1 \\ 1.5 & 0.5 & 1 \\ 0 & 0 & 1 \end{bmatrix}$$

Transform the line and the three points using this collineation. The transformed line and transformed points $(D, E, \text{and } F)$ will appear in the window. You may need to zoom out to ensure that all of the new points are visible in the window.

Dist(A,B) = 3.21
Dist(A,C) = 4.38
Dist(A,C) /Dist(A,B) = 1.36

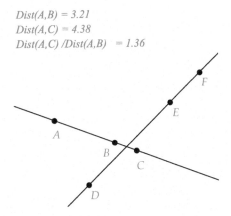

As we did before measure the distance from D to E and the distance from D to F and calculate the ratio as shown. We note that the two ratios are the same, even though the distances are certainly not the same. If we move points A, B and C around the distances might change, but the ratio defined by A, B, and C is always equal to the ratio defined by D, E, and F.

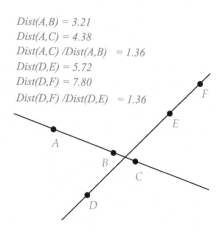

Dist(A,B) = 3.21
Dist(A,C) = 4.38
Dist(A,C) /Dist(A,B) = 1.36
Dist(D,E) = 5.72
Dist(D,F) = 7.80
Dist(D,F) /Dist(D,E) = 1.36

Our exploration suggests a Theorem!

Theorem 9.27. *Let A, B, and C be three distinct points in the Real Projective Plane and let M be an affine collineation. Let $D = M(A)$, $E = M(B)$, and $F = M(C)$. Then,*

$$\frac{dist(A, C)}{dist(A, B)} = \frac{dist(D, F)}{dist(D, E)}$$

The proof will be carried out in the following exercises:

Exercise 9.6.1. *The line $\overleftrightarrow{AB}$ can be parameterized as $P + tv$, where P is a point on the line and v is the direction vector. Then, $A = P + t_1v$, $B = P + t_2v$, and $C = P + t_3v$. Show that*

$$\frac{dist(A, C)}{dist(A, B)} = \frac{|t_3 - t_1|}{|t_2 - t_1|}$$

Exercise 9.6.2. *Use the preceding exercise, and the fact that $dist(D, F) = \|F - D\| = \|MC - MA\| = \|M(C - A)\|$ to finish the proof.*

9.6.2 Cross-Ratio

Now, we will consider what ratios are invariant in the Real Projective Plane. As background, we note that in Euclidean geometry, the distance between two points is always invariant under a Euclidean isometry. We might call this the trivial ratio (denominator is 1).

For a general affine collineation, the distance between two points

is not necessarily invariant. For example, the affine collineation
$\begin{bmatrix} 2 & 0 & 0 \\ 0 & 2 & 0 \\ 0 & 0 & 1 \end{bmatrix}$ will clearly double the distance between two points. In our
exploration above, we have seen that three collinear points suffice to define a ratio of distances that is invariant under an affine collineation. A logical conjecture would be that 4 points might suffice for an invariant ratio in Projective geometry that is invariant under a perspectivity.

Axiomatic Projective geometry does not have a defined distance function. However, in our 3-D model of the Real Projective Plane, we do have a notion of distance for points of the form $(x, y, 1)$. We have the usual Euclidean distance function. Let's explore ratios of distances for perspectivities of four points.

Clear your geometry software window, construct a line $\overleftrightarrow{AB}$ and attach two points C and D to the line. Construct another line $\overleftrightarrow{EF}$. Construct a point O off of both of these lines. Then draw lines from O to each of A, B, C, and D. Create the intersections A', B', C', and D' as shown from point O. We have created a *perspectivity* from O.

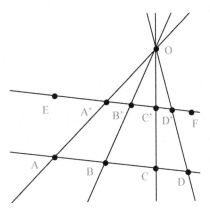

We now look for a ratio defined on four points that is invariant under the perspectivity. Let's see if there is a pattern in the ratios we have considered so far. In Euclidean geometry, the distance between two points was invariant. In Affine geometry, the ratio that was invariant was the ratio of two distances from point A, that is the ratio of two Euclidean ratios. If we consider points A and B as base points for the triples $\{A, C, D\}$ and $\{B, C, D\}$ we would have "affine" ratios $\dfrac{dist(A, C)}{dist(A, D)}$ and $\dfrac{dist(B, C)}{dist(B, D)}$. Perhaps the ratio of these "affine" ratios is invariant?

Find the four distance measurements and calculate the two ratios $\dfrac{dist(A,C)}{dist(A,D)}$ and $\dfrac{dist(B,C)}{dist(B,D)}$. Then, find the ratio of these two ratios (Figure 9.17).

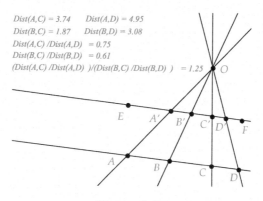

Figure 9.17

Hide some of the measurements to unclutter the screen and then do the same sequence of measurements and calculations to find $\dfrac{dist(A',C')}{dist(A',D')}$ and $\dfrac{dist(B',C')}{dist(B',D')}$. Calculate the ratio of these two ratios (Figure 9.18).

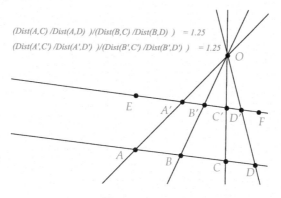

Figure 9.18

It looks like our guess was pretty good! Move points A, B, C, and D around and also move the perspective point O. The "double" ratio stays preserved. Our evidence suggests that this new ratio is invariant under perspectivities, at least perspectivities from points that are not at infinity.

For many applications of this new ratio, it will be useful to compute the ratio in terms of *directed* distances.

Definition 9.20. *Let l be a Euclidean line l defined by points A and B. Given two points C and D on l, let d be the Euclidean distance between C and D. Then, the directed distance from C to D, symbolized by dd(C, D), will be equal to d if the vector D − C is in the same direction as B − A. Otherwise, the directed distance will be equal to −d.*

The double ratio defined by directed distances with points A, B, C, and D on line l will be called the *cross-ratio* of these four points.

Definition 9.21. *Given four points A, B, C, and D on a line l, the* cross-ratio *of these four points is defined as*

$$R(A, B; C, D) = \frac{\frac{dd(A,C)}{dd(A,D)}}{\frac{dd(B,C)}{dd(B,D)}} = \frac{dd(A, C)\, dd(B, D)}{dd(A, D)\, dd(B, C)}$$

The cross-ratio is also denoted by some authors as $R(A, B, C, D)$. We use the semi-colon notation to designate that A and B are base points for the distance measurements in the calculation of this ratio.

One natural question to ask is what happens when you move A, B, C, and D around in the cross-ratio formula.

Theorem 9.28. *Suppose* $r = R(A, B; C, D)$. *Then,*

$$R(A, C; B, D) = R(D, B; C, A) = 1 - r$$

Proof: For shorthand notation, let $dd(X, Y) = XY$. Then,

$$R(A, C; B, D) = \frac{AB\ CD}{AD\ CB}.$$

Now,

$$
\begin{aligned}
1 - r &= 1 - \frac{AC \ BD}{AD \ BC} \\
&= \frac{(AD \ BC) - (AC \ BD)}{AD \ BC} \\
&= \frac{[(AD \ BC) - (BD \ BC)] + [(BD \ BC) - (AC \ BD)]}{AD \ BC} \\
&= \frac{(AD - BD)BC + BD(BC - AC)}{AD \ BC}
\end{aligned}
$$

Using directed distances, we have $(AD - BD) = AB$ and $(BC - AC) = BA = -AB$. (Verify this on a separate piece of paper). Thus,

$$
\begin{aligned}
1 - r &= \frac{(AD - BD)BC + BD(BC - AC)}{AD \ BC} \\
&= \frac{AB \ BC - BD \ AB}{AD \ BC} \\
&= \frac{AB(BC - BD)}{AD \ BC} \\
&= \frac{AB \ CD}{AD \ BC} \\
&= R(A, C; B, D)
\end{aligned}
$$

The proof for $R(D, B; C, A)$ follows from an analogous argument. □

Exercise 9.6.3. *Show, by algebraic manipulation, that the cross-ratio $R(A,B;C,D)$ is unchanged if any two pairs of points are interchanged. That is, $R(A, B; C, D) = R(B, A; D, C) = R(C, D; A, B) = R(D, C; B, A)$*

Exercise 9.6.4. *Suppose $r = R(A, B; C, D)$. Show that $R(B, A; C, D) = R(A, B; D, C) = \frac{1}{r}$. Also, show that $R(A, D; C, B) = \frac{r}{1-r}$.*

9.6.3 Harmonious Ratios

We now look at a fascinating connection between the cross-ratio and musical harmony. We will first review the connection between musical tones and mathematical ratios.

The Pythagoreans were the first to systematically analyze the mathematical properties of musical pitches. They discovered that we perceive two pitches as sounding the same if one pitch is doubled in frequency from the other. In musical terminology, the higher pitch is an *octave* above the lower. The frequency of a pitch, say 440 herz (cycles per second), is thus in a 2:1 ratio with the pitch an octave below (220 herz).

The Pythagoreans thought the most harmonious pitches were those with ratios consisting of small integer values. The simplest such ratio is the octave at 2:1. The next simplest would be a ratio of 3:2. This is the *perfect fifth*. For a base pitch of 220 herz, the perfect fifth above would be 330 herz. Or, in our modern labeling of notes, for a base pitch equal to a middle C, the perfect fifth would be G.

The Pythagoreans tuned their instruments based on this notion of perfect fifths, and then went on to build a musical scale of 12 notes. They did this by successively moving up by perfect fifths six times to get six main notes. Then, from the new top note, they moved down six times to get a scale of 12 notes. Unfortunately, there is no way to preserve the 2:1 ratio of an octave by moving by perfect fifths. Also, in this system, if the C to G is a perfect fifth, then, it is impossible to have other inter-scale fifths (e.g. D to A) be perfect. For a nice treatment of this mathematical dilemma, look at Chapters 2 and 3 of [18].

Just Tuning was created as an attempt to solve this problem of trying to merge a perfect fifth system with an octave system. It was widely used in the Renaissance period. In Just tuning, the intervals between notes of the 12 tone scale are chosen to have ratios as small as possible, with the goal of making as many inter-scale fifths be as perfect as possible. The notes of the C major scale in Just tuning are:

base - C	1:1
second - D	9:8
third - E	5:4
fourth - F	4:3
fifth - G	3:2
sixth - A	5:3
seventh - B	15:8
octave - C	2:1

The fifth, fourth, and third form a nice progression of 3:2, 4:3, 5:4. The *major triad* C-E-G incorporates the ratios 3:2 and 5:4. Given a base pitch of frequency f, the major triad has frequencies of $\frac{3}{2}f$ and $\frac{5}{4}f$ respectively.

It is interesting to compare the pitch of a note to the length of a string that produces that pitch. Mersenne's Law (Marin Mersenne 1588–1648) states that the frequency of a pitch is inversely proportional to the length of the string producing the pitch.

For a major triad, assume that L is the length of the string producing the base tone. Then, the lengths for the other strings will be $\frac{2}{3}L$ and $\frac{4}{5}L$. We will now explore the relationship between these length ratios and the cross-ratio.

Clear your geometry software window and create $\overline{C1\,C2}$. This segment will represent the length of a string vibrating at a base pitch.

Dilate point $C2$ by a ratio of $\frac{2}{3}$ towards $C1$. Point G will be created as the dilated point with the defined ratio.

Similarly, define a dilation of $\frac{4}{5}$ and dilate point $C2$ toward $C1$ to get new point E as shown.

Now, use the distance measuring property of your software to calculate the cross-ratio $R(C1, E; G, C2)$. If your software does not have directed distances, the cross-ratio will turn out to equal 1. A quick calculation shows that it must actually be -1, if we use directed distances.

Exercise 9.6.5. *Use the ratios defined above, and the properties of directed distances, to prove that the cross-ratio $R(C1, E; G, C2)$ is equal to -1.*

We have seen that a harmonious set of intervals (the fifth and third) turns out to have cross-ratio of -1 when considering string lengths. The following definition is not that surprising.

Definition 9.22. *Let A, B, C, and D be collinear points. We say that (A, B, C, D) forms a* **harmonic set** *if $R(A, B; C, D) = -1$. We call points C and D* **harmonic conjugates** *of each other. We write $H(A, B; C, D)$ if A, B, C, and D form a harmonic set.*

For your report give a careful and complete summary of your work on this project.

9.7 HARMONIC SETS

9.7.1 Harmonic Sets of Points

Recall from earlier in this chapter that a *complete quadrangle* consisted of four distinct points (vertices), no three of which are collinear, and the six lines (sides) defined by the four points.

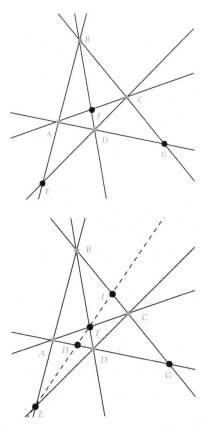

If A, B, C, and D are the points of the quadrangle, then sides AB and CD, AC and BD, and AD and BC are called *opposite sides* of the quadrangle. The points at which opposite sides intersect are called *diagonal points* of the quadrangle. Here the diagonal points are E, F, and G.

Now, given any pair of diagonal points there are two sides of the quadrangle that do not pass through these points. For example, AD and BC do not pass through E or F. It will be left as an exercise to show that the line through any pair of diagonal points will not pass through a vertex. For example EF will intersect sides AD and BC at points H and I which are different than the vertices of the quadrangle.

Definition 9.23. *Let ABCD be a complete quadrangle and let E and F be diagonal points. Let AD be a side of the quadrangle such that E and F do not lie on AD. Let H be the intersection point of EF with AD. Then, H is called a* harmonic point *with respect to E and F.*

Since we can form three pairs of points from the three diagonal points of a complete quadrangle, then there are six harmonic points defined for any complete quadrangle. Taking these in pairs, we get the following definition.

> **Definition 9.24.** *Let ABCD be a complete quadrangle and let E and F be diagonal points. Let H and I be the harmonic points on EF. Then, (E, F, H, I) are said to form a* harmonic set *(or harmonic tetrad). We denote a harmonic set by H(E, F; H, I). Points H and I are called* harmonic conjugates *of each other.*

We note here that at the end of Project 9.5.5, we had a very different definition of harmonic sets and conjugates. We will see later that these two seemingly different definitions of harmonic sets are equivalent. Before we get to that result, we will explore more fully the present definition of harmonic sets.

> **Theorem 9.29.** *Given any three distinct, collinear points E, F, and H on a line l, there is a unique point I ≠ H on l such that H(E, F; H, I).*

Proof: We start by constructing a complete quadrangle $ABCD$ that has the points E and F as diagonal points.

Let A be some point that is not on EF. We can construct a line $m \neq l$ through F that does not pass through A.

Let $B = m \cdot AE$, $C = m \cdot AH$, and $D = CE \cdot AF$. Then, $I = l \cdot BD$ is the harmonic conjugate of H.

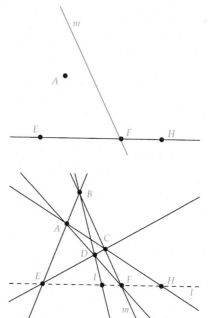

Now, suppose that we start with a different point A' not on EF and different line $m' \neq l$ through F that does not pass through A'. We do the same construction as above to get point I'. We will show that $I = I'$.

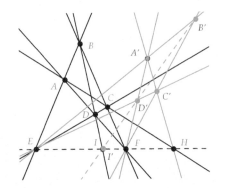

For the rest of the proof, we will use the results on Desargues Theorem and its dual as discussed in 9.3.5.

In the final figure above, we see that triangles ABC and $A'B'C'$ are both perspective from line EF. Thus, lines AA', BB', and CC' are concurrent. Likewise, triangles ACD and $A'C'D'$ are perspective from EF, so lines AA', CC', and DD' are concurrent. Thus, all four lines AA', BB', CC', and DD' are concurrent at one point. This means that triangles ABD and $A'B'D'$ are perspective from this point of concurrency. This means that points $E = AB \cdot A'B'$, $F = AD \cdot A'D'$ and $BD \cdot B'D'$ are collinear. This can only happen if $I = I'$ □

9.7.2   Harmonic Sets of Lines

We will now consider the dual results to those discussed above. Note that the duality principle guarantees that dual results will hold true without the need for proof. We begin with the dual to the complete quadrangle.

A *complete quadrilateral* consists of four distinct lines (sides), no three of which are concurrent, and the six points (vertices) defined by these four lines (as intersections). If a, b, c, and d are the sides of the quadrilateral, then the pairs of vertices $A_1 = a \cdot b$ and $A_3 = c \cdot d$, $A_2 = b \cdot c$ and $A_4 = a \cdot d$, and $A_5 = b \cdot d$ and $A_6 = a \cdot c$ are opposite vertices of the quadrilateral. The sides defined by opposite vertices are called *diagonal sides* of the quadrilateral.

In Figure 9.19 we have the complete quadrilateral $abcd$. The diagonal sides are e, f, and g.

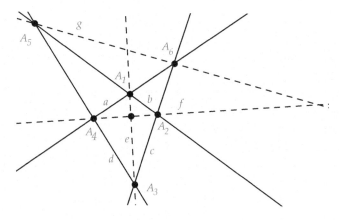

Figure 9.19

Given any pair of diagonal sides there are two points of the quadrilateral that do not lie on these lines. For example, A_5 and A_6 do not lie on e or f. It will be left as an exercise to show that the point of intersection of any pair of diagonal sides will not lie on any side of the quadrilateral. Thus, the line h through $e \cdot f$ and A_5 will be different from any side of the quadrilateral, as will the line i through $e \cdot f$ and A_6 (Figure 9.20).

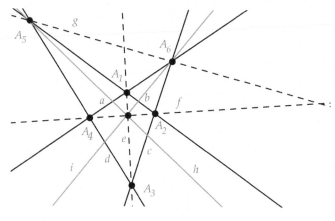

Figure 9.20

Definition 9.25. *Let abcd be a complete quadrilateral and let e and f be diagonal lines. Let A be a vertex of the quadrilateral such that e and f do not contain A. Let h be the line through e · f and A. Then, h is called a* harmonic line *with respect to e and f.*

Since we can form three pairs of lines from the three diagonal lines of a complete quadrilateral, then there are six harmonic lines defined for any complete quadrilateral. Taking these in pairs, we get the following definition.

> **Definition 9.26.** *Let abcd be a complete quadrilateral and let e and f be diagonal lines. Let h and i be the harmonic lines on e·f. Then, (e, f, h, i) are said to form a* harmonic set *(or harmonic tetrad). We denote a harmonic set by H(e, f; h, i). Lines h and i are called* harmonic conjugates *of each other.*

There is a fundamental connection between harmonic sets of lines and harmonic sets of points. In Figure 9.21 points H and I are the intersections of diagonal lines e and f with g. From the construction, we can see that $A_1 A_2 A_3 A_4$ will be a complete quadrangle and points A_5, A_6, H, and I will form a harmonic set of points.

Conversely, if we had started with harmonic points A_5, A_6, H, and I we could choose a point G not on $A_5 A_6$ and construct four lines through G and each of these four points. We could create a complete quadrilateral by choosing an arbitrary point for A_1 and then computing intersections to get $abcd$. Then, the four lines through G would form a harmonic set.

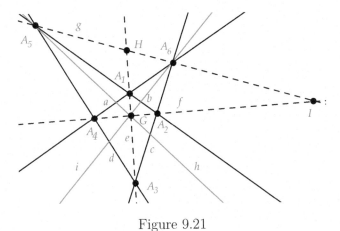

Figure 9.21

We will use this observation to prove the following result:

Theorem 9.30. *Let E, F, H, and I be a harmonic set on a line l. That is, $H(E, F; H, I)$. Suppose that l' is perspective to l through G and let $EFHI \overset{G}{\wedge} E'F'H'I'$. Then, $H(E', F'; H', I')$. That is, the image of a harmonic set under a perspectivity is again a harmonic set.*

Proof: From our discussion above, we know that if we construct EG, FG, HG, and IG, then this will form a harmonic set of lines, and these lines will intersect l' at E', F', H', and I'. Now, we can construct the points A_1, etc, for these points on l', as we did above, and get a complete quadrangle, guaranteeing that $H(E', F'; H', I')$. □

We immediately get the following result on projectivities:

Corollary 9.31. *Suppose that $H(E, F; H, I)$ for points on a line l and there is a projectivity taking l to l'. Let $EFHI \overline{\wedge} E'F'H'I'$. Then, $H(E', F'; H', I')$.*

The next theorem says that harmonic sets are always related by a projective transformation.

Theorem 9.32. *Let E,F,H, and I be distinct collinear points such that $H(E, F; H, I)$. Let E',F',H', and I' be distinct collinear points such that $H(E', F'; H', I')$. Then, there exists a projectivity mapping E,F,H, and I to E',F',H', and I'.*

Proof: By Theorem 9.7 we know there is a projectivity taking E,F and H to E',F' and H'. Let I'' be the image of I under this projectivity. By the previous Corollary, we know that $H(E', F'; H', I'')$. But, by Theorem 9.29 we know that the fourth point in a harmonic set is uniquely defined, based on the first three points. Thus, $I' = I''$. □

9.7.3 Harmonic Sets and the Cross-Ratio

We will now consider the relationship between harmonic sets and the cross-ratio in the Real Projective Plane model of Projective geometry. At the end of Project 9.5.5, we defined harmonic sets in terms of the cross-ratio. Given collinear points A, B, C, and D, we called (A, B, C, D) a harmonic set if the cross-ratio $R(A, B; C, D)$ was equal to -1. In this section we will show that this is equivalent to the definition of harmonic sets in terms of complete quadrangles. Before we do that, we need to analyze the cross-ratio in more detail.

Let A, B, C, and D be distinct collinear points on a line l and let P and Q be points distinct from these four on l. Then, A, B, C, and D have parametric homogeneous coordinates (α_1, α_2), (β_1, β_2), (γ_1, γ_2) and (δ_1, δ_2). (Definition 9.16) We have $A = \alpha_1 P + \alpha_2 Q$, $B = \beta_1 P + \beta_2 Q$, etc. Since P and Q are different than A, B, C, and D, then none of the parameters are zero and we can assume that the homogeneous parameters for the four points are $(\alpha, 1)$, $(\beta, 1)$, $(\gamma, 1)$ and $(\delta, 1)$, where $\alpha = \frac{\alpha_1}{\alpha_2}$, $\beta = \frac{\beta_1}{\beta_2}$, $\gamma = \frac{\gamma_1}{\gamma_2}$, and $\delta = \frac{\delta_1}{\delta_2}$. In this parametrization the cross-ratio has a very simple form:

Theorem 9.33. *If A, B, C, and D are distinct collinear points with parametric homogeneous coordinates $(\alpha, 1)$, $(\beta, 1)$, $(\gamma, 1)$ and $(\delta, 1)$ with respect to points P and Q. Then*

$$R(A, B; C, D) = \frac{dd(A, C)\, dd(B, D)}{dd(B, C)\, dd(A, D)}$$
$$= \frac{(\gamma - \alpha)(\delta - \beta)}{(\gamma - \beta)(\delta - \alpha)}$$

The notation $dd(A, C)$ stands for the *directed* distance between A and C. The proof of this theorem is left as an exercise.

Corollary 9.34. *Let A, B, C, and D be distinct collinear points on a line l and let P and Q be points distinct from these four on l. Let A, B, C, and D have parametric homogeneous coordinates (α_1, α_2), (β_1, β_2), (γ_1, γ_2) and (δ_1, δ_2) with respect to points P and Q. Then,*

$$R(A, B; C, D) = \frac{\det\left(\begin{bmatrix} \gamma_1 & \alpha_1 \\ \gamma_2 & \alpha_2 \end{bmatrix}\right) \det\left(\begin{bmatrix} \delta_1 & \beta_1 \\ \delta_2 & \beta_2 \end{bmatrix}\right)}{\det\left(\begin{bmatrix} \gamma_1 & \beta_1 \\ \gamma_2 & \beta_2 \end{bmatrix}\right) \det\left(\begin{bmatrix} \delta_1 & \alpha_1 \\ \delta_2 & \alpha_2 \end{bmatrix}\right)}$$

The proof is left as an exercise. We note here that this representation of the cross-ratio would be well-defined even if one or more of the base points for the homogeneous parametric coordinates was one of the four points A,B,C, or D. In fact, this determinant expression for the cross-ratio is how many authors define the cross-ratio. We started with the directed distance definition to be better understand its geometric significance.

For future reference, we note that the determinant method for defining the cross-ratio can be easily extended, in a dual fashion, to a cross-ratio for lines through a common point.

Definition 9.27. *Let a, b, c, and d be be distinct lines through a point P and let l and m be lines distinct from these four that pass through P. Let a, b, c, and d have parametric homogeneous coordinates (α_1, α_2), (β_1, β_2), (γ_1, γ_2) and (δ_1, δ_2) with respect to lines l and m. Then, the cross-ratio of a, b, c, and d is defined as*

$$R(a,b;c,d) = \frac{det\left(\begin{bmatrix} \gamma_1 & \alpha_1 \\ \gamma_2 & \alpha_2 \end{bmatrix}\right) det\left(\begin{bmatrix} \delta_1 & \beta_1 \\ \delta_2 & \beta_2 \end{bmatrix}\right)}{det\left(\begin{bmatrix} \gamma_1 & \beta_1 \\ \gamma_2 & \beta_2 \end{bmatrix}\right) det\left(\begin{bmatrix} \delta_1 & \alpha_1 \\ \delta_2 & \alpha_2 \end{bmatrix}\right)}$$

Using the results from Theorem 9.26 one can prove the following result. The proof is left as an exercise.

Corollary 9.35. *Given a perspectivity with center O and axis l, with l not passing through O, let a, b, c, and d be lines in the pencil of lines at O, and A, B, C, and D the corresponding points on l defined by the perspectivity. Then, $R(a,b;c,d) = R(A,B;C,D)$*

The most important corollary to Theorem 9.33 is the following:

Corollary 9.36. *The cross-ratio is invariant under projectivities.*

Proof: Let A,B,C, and D be distinct collinear points on a line l with parametric homogeneous coordinates (with respect to points P and Q) (α_1, α_2), (β_1, β_2), (γ_1, γ_2) and (δ_1, δ_2). Let T be a projectivity from l to l' and let A',B',C', and D' be the images of A,B,C, and D under T. These four image points have parametric homogeneous coordinates with respect to $P' = T(P)$ and $Q' = T(Q)$. Let these parametric coordinates be (α_1', α_2'), (β_1', β_2'), (γ_1', γ_2') and (δ_1', δ_2').

Then,

$$R(A', B'; C', D') = \frac{det\left(\begin{bmatrix} \gamma_1' & \alpha_1' \\ \gamma_2' & \alpha_2' \end{bmatrix}\right)det\left(\begin{bmatrix} \delta_1' & \beta_1' \\ \delta_2' & \beta_2' \end{bmatrix}\right)}{det\left(\begin{bmatrix} \gamma_1' & \beta_1' \\ \gamma_2' & \beta_2' \end{bmatrix}\right)det\left(\begin{bmatrix} \delta_1' & \alpha_1' \\ \delta_2' & \alpha_2' \end{bmatrix}\right)}$$

By Corollary 9.21 we know that the projectivity T can be represented by a matrix equation. For example, for points A and A' we have

$$\lambda_A \begin{bmatrix} \alpha_1' \\ \alpha_2' \end{bmatrix} = \begin{bmatrix} a & b \\ c & d \end{bmatrix}\begin{bmatrix} \alpha_1 \\ \alpha_2 \end{bmatrix}$$

where the determinant $ad - bc \neq 0$ and $\lambda \neq 0$.

Thus, we can write $R(A', B'; C', D')$ as

$$\frac{\lambda_C\lambda_A det\left(\begin{bmatrix} a & b \\ c & d \end{bmatrix}\begin{bmatrix} \gamma_1 & \alpha_1 \\ \gamma_2 & \alpha_2 \end{bmatrix}\right)\lambda_D\lambda_B det\left(\begin{bmatrix} a & b \\ c & d \end{bmatrix}\begin{bmatrix} \delta_1 & \beta_1 \\ \delta_2 & \beta_2 \end{bmatrix}\right)}{\lambda_C\lambda_B det\left(\begin{bmatrix} a & b \\ c & d \end{bmatrix}\begin{bmatrix} \gamma_1 & \beta_1 \\ \gamma_2 & \beta_2 \end{bmatrix}\right)\lambda_D\lambda_A det\left(\begin{bmatrix} a & b \\ c & d \end{bmatrix}\begin{bmatrix} \delta_1 & \alpha_1 \\ \delta_2 & \alpha_2 \end{bmatrix}\right)}$$

Now we use the fact that $det(MN) = det(M)det(N)$ for square matrices M and N, and the fact that $ad - bc \neq 0$, to cancel the terms involving $det\left(\begin{bmatrix} a & b \\ c & d \end{bmatrix}\right)$. Also, since all of the λ terms are non-zero, we can cancel them as well, and the result holds. □

Now we can show that the definition for harmonic sets given at the end of Project 9.5.5 is equivalent to the definition of harmonic sets in terms of complete quadrangles.

> **Theorem 9.37.** *If A, B, C, and D are distinct collinear points, then the four points form a harmonic set ($H(A, B; C, D)$, based on the complete quadrangle definition) if and only if $R(A, B; C, D) = -1$.*

Proof: If $H(A, B; C, D)$ then, clearly $H(A, B; D, C)$, as the order of C and D is independent of the definition of a harmonic set. By Theorem 9.32 there is a projectivity taking A, B, C, and D to A, B, D, and C. By Corollary 9.36 we know that $R(A, B; C, D) = R(A, B; D, C)$. Let

$R(A, B; C, D) = r$. By exercise 9.6.4 we know that $R(A, B; D, C) = \frac{1}{r}$.
Thus, $r = \frac{1}{r}$ or $r^2 = 1$. Since $r \neq 1$ (Exercise) we have that $r = R(A, B; C, D) = -1$.

Now, assume that $R(A, B; C, D) = -1$. We know that given points
A, B, and C there is a unique point D' for which A, B, C and D'
form a harmonic set. By the first part of this theorem we know that
$R(A, B; C, D') = -1$. Thus, $R(A, B; C, D) = R(A, B; C, D')$. But, there
is a unique point D for which $R(A, B; C, D) = -1$. (Exercise). So, $D = D'$ and A, B, C and D form a harmonic set. □

Exercise 9.7.1. *Let $ABCD$ be a complete quadrangle and let E, F, and G
be its diagonal points. Show that any line through two of the diagonal points
will not pass through one of the vertices of the quadrangle.*

Exercise 9.7.2. *Let $abcd$ be a complete quadrilateral and let e, f, and g
be its diagonal lines. Show that a point of intersection of two of the diagonal
lines will not lie on one of the sides of the quadrangle.*

Exercise 9.7.3. *Suppose $H(E, F; H, I)$ for points on a line l and that
$H(E, F'; H', I')$ for points on a line l'. (So, E is on both lines) Show that
there is a perspectivity that takes $\{E, F, H, I\}$ to $\{E, F', H', I'\}$.*

Exercise 9.7.4. *Prove Theorem 9.33 . (Hint: The proof is quite similar to
the work for exercise 9.6.1).*

Exercise 9.7.5. *Prove Corollary 9.34.*

Exercise 9.7.6. *Prove Corollary 9.35.*

Exercise 9.7.7. *Let A, B, and C be three distinct collinear points in the Real
Projective Plane. Show that if $R(A, B; C, D) = R(A, B; C, D')$, then $D = D'$.
(Hint: Use the parametric form of the cross-ratio from Theorem 9.33.)*

Exercise 9.7.8. *Show that the cross-ratio of four distinct collinear points
A,B,C, and D in the Real Projective Plane cannot equal 0, 1, or infinity.
(Hint: Use the parametric form of the cross-ratio from Theorem 9.33. You
may also want to use Exercise 9.6.4)*

Exercise 9.7.9. *Suppose A, B, C, D, and E are five distinct collinear
points in the Real Projective Plane. Show that $R(A, B; C, D)R(A, B; D, E) = R(A, B; C, E)$*

Exercise 9.7.10. *Suppose that distinct collinear points A, B, C, and D in
the Real Projective Plane form a harmonic set (based on the complete quadrangle definition). That is $H(A, B; C, D)$. Show that $H(C, D; A, B)$. (Hint:
Use Theorem 9.37)*

Exercise 9.7.11. *Let A, B, and C be distinct collinear points on a line l in the Real Projective Plane with B the midpoint of segment $\overline{AC}$. Explain why it makes sense to describe the harmonic conjugate of B to be the point at infinity of the line l. [Hint: Let D be the point at infinity on l. Consider the coordinate forms of the points A, B, C, and D and show that the cross ratio of these four points will be equal to −1.]*

9.8 CONICS AND COORDINATES

To this point in our review of Projective geometry we have been exclusively concerned with points, lines, and linear figures such as triangles and quadrangles. Noticeably missing from our discussion has been any mention of circles, which were a fundamental (axiomatic!) topic in our study of Euclidean, Hyperbolic, and Elliptic geometries. This is not that surprising given that circles are defined in terms of *metric* properties - as the set of points equally distant from a center point.

The analog of circles in Projective geometry is the idea of *conic sections*. These include all of the traditional sections of the cone —the circle, ellipse, hyperbola, and parabola. In Projective geometry, these are all equivalent figures under the appropriate projective transformation. Conic sections in projective geometry will be defined using properties of projective transformations. This might seem a bit strange at first, but there is a nice analog in Euclidean geometry where we can construct conic sections via Euclidean transformations.

9.8.1 Conic Sections Generated by Euclidean Transformations

Recall that the *pencil of points with axis l* is the set of all points on *l*. The *pencil of lines with center O* is the set of all lines through point *O*. (Figure 9.22)

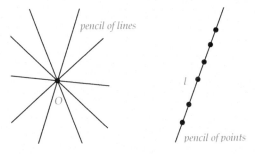

Figure 9.22

We have used this terminology frequently in our study of Projective geometry, but in this section we will assume that pencils of points and lines consist solely of Euclidean points and lines.

We will be concerned with how pencils of points and lines are transformed under Euclidean transformations, i.e., isometries. Consider how the pencil of lines with center O is transformed under the composition of two Euclidean isometries. We know from our work in Chapter 5 that the composition of two isometries is always another isometry —either a reflection, a translation, a rotation, or a glide reflection.

For example, let r be the reflection isometry across line $\overleftrightarrow{AB}$ as shown. Consider the pencil of lines with center O. Under the isometry r, one of these lines, say $\overleftrightarrow{OP}$, will map to $\overleftrightarrow{O'P}$, where P is the intersection point of the lines on $\overleftrightarrow{AB}$ and $r(O) = O'$. We say that $\overleftrightarrow{OP}$ and $\overleftrightarrow{O'P}$ are *corresponding* lines under the reflection r.

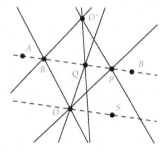

Now, consider the set of all intersection points of corresponding lines. These would include points P, Q, and R as shown above. This set of points is the *locus* of the points of intersection of corresponding lines of two pencils (the one at O and the one at O') that are related by the reflection r.

> **Definition 9.28.** *A set of points is called a* locus of points *if each point in the set satisfies some geometric condition. A point is a member of the locus of points if and only if it satisfies the condition.*

In the previous example, a point is in the locus of points if it satisfies the condition that it is a point of intersection of corresponding lines of the two pencils. Clearly, this set of points is the line $\overleftrightarrow{AB}$. There is one unique line which does not generate an element in this locus —the line $\overleftrightarrow{OS}$ which is parallel to $\overleftrightarrow{AB}$ at O. However, if we consider this example in the extended Euclidean plane, with points at infinity attached, then $\overleftrightarrow{OS}$ and $r(\overleftrightarrow{OS})$ will intersect at a point at infinity, which we would then have to add to the locus of points.

Now, let's consider the locus of points that are generated from two pencils that are related by the composition of *two* reflections.

Let $\overleftrightarrow{AB}$ and $\overleftrightarrow{CD}$ serve as two lines of reflection for reflections r_{AB} and r_{CD}. Let O be a center for a pencil of lines and let $O' = r_{AB}(O)$ and $O'' = r_{CD}(O')$. For a given line $l = \overleftrightarrow{OP}$ in the pencil at O, let $m = r_{AB}(l)$ and $n = r_{CD}(m)$.

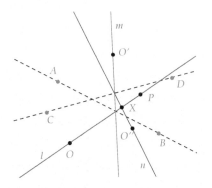

The intersection point, X, of l and n will be a point on the locus of points that are generated from the two pencils of lines at O and O'' that are related by the composition of r_{AB} and r_{CD}.

The next theorem is a non-obvious result about the set of all locus points X constructed as above.

Theorem 9.38. *Let $\overleftrightarrow{AB}$ and $\overleftrightarrow{CD}$ be intersecting lines in the plane. Let r_{AB} and r_{CD} be reflections across these lines. Let $O'' = r_{CD}(r_{AD}(O))$. Then, the locus of points of intersection of corresponding lines of the two pencils at O and O'' forms a circle.*

Proof: Let E be the intersection of $\overleftrightarrow{AB}$ and $\overleftrightarrow{CD}$. This point is on the locus, as it is fixed by both reflections. We will show that the locus of points generated by the reflections is the circle through E, O, and O''.

The circle through E, O, and O'' is the circumscribed circle through these points, as we saw in Project 2.2. This center of this circle is constructed by finding the intersection point H of the perpendicular bisectors of $\overline{EO}$ and $\overline{EO''}$. (Midpoints F and G)

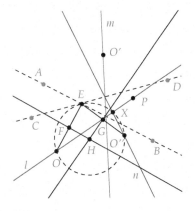

We know from Chapter 5 that if the two lines of reflection in the construction above are not parallel, then the composition f of the two reflections across these lines will be a *rotation* about the point where the lines intersect —in our case, a rotation about point E. Also, we know the angle of rotation is twice the angle made by the two lines. Thus, the measure of $\angle OEO''$ is twice the measure of $\angle AEC$. From our work on circles in 2.6 we know that inscribed angle $\angle OEO''$ has a measure that is one-half the measure of central angle $\angle OHO''$.

From Exercise 5.4.7 in Chapter 5 we know that the measure of the vertical angle at X made by $\overleftrightarrow{OP}$ and n must equal the angle of rotation. Thus, the measure of $\angle OXO''$ equals the measure of $\angle OEO''$, and thus the measure of $\angle OXO''$ is one-half of the measure of central angle $\angle OHO''$. By Theorem 2.42 the point X must be on the circle. □

9.8.2 Point Conics in Projective Geometry

As mentioned earlier, conics defined by metric properties are not feasible in Projective geometry. However, the methods used in the last section to construct a Euclidean circle is generalizeable to Projective geometry. Instead of constructing a locus of points by using Euclidean isometries, we will instead use Projective transformations, i.e., *projectivities*. Our development follows closely the work of W. T. Fishback in [13].

> **Definition 9.29.** *A* point conic *is the locus of points of intersection of corresponding lines of two pencils of lines, where the first pencil is transformed to the second by a projectivity. If the projectivity is equivalent to a single perspectivity, or if the centers of the pencils are the same point, the point conic will be called* singular. *Otherwise, the point conic is called* non-singular.

Here we illustrate a non-degenerate point conic that is generated by some projectivity between pencils of lines at O and O''. A point X on the conic is defined as the intersection of two lines l and l'' that are projectively related.

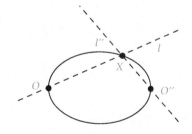

From the figure, it appears that the centers of the two pencils lie on the point conic. This is always the case.

> **Theorem 9.39.** *The centers of the pencils that define a point conic always lie on the point conic.*

Proof: We know that the line $n = OO''$ is a member of the pencil of lines at O and also the pencil of lines at O'' (Figure 9.23). The projectivity defining the point conic will map n to some other line n'' through O''. Clearly, n and n'' intersect at O''. Consider $m'' = n$ as a line through O''. The inverse of the projectivity will map m'' to a line m through O. Then, m and m'' intersect at O.

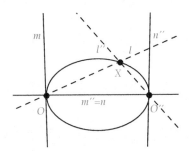

Figure 9.23

□

Our definition of point conics splits the family of possible conics into two groups, the singular point conics and the non-singular point conics. We begin our analysis with the singular point conics.

> **Theorem 9.40.** *The possible singular point conics include the following: the entire Projective plane, the set of points on two distinct lines, the set of points on a single line, and a single point.*

Proof: A singular point conic is defined by two cases: either its defining projectivity is equivalent to a single perspectivity or the centers of the pencils of lines of the point conic are the same point.

Suppose the point conic's defining projectivity is equivalent to a single perspectivity with axis l, with two distinct centers $O \neq O'$ of pencils of lines. Clearly, the corresponding lines of the pencils will meet on the axis l and so the point conic consists of the points on l.

Now, suppose the centers of the pencils are the same point $O = O'$. We have to consider all possible types of projectivities. One possible projectivity would be the identity. In this case, every line corresponds to itself, and all points on all lines would be on the point conic. The point conic is thus the entire projective plane.

If the projectivity is not the identity, then the projectivity maps all lines through O back to lines through O. So, corresponding lines would either be the same line (a fixed line under the projectivity) or different lines. If they were the same line, the entire line would form part of the point conic. If they were different, only the point O would be common.

Consider the set of fixed lines under the projectivity. By Theorem 9.13 the projectivity has at most two fixed lines, because if it had three, it would be the identity. If the projectivity has two fixed lines, then these together form the conic. If there is just one fixed line, that will be the point conic. If there are no fixed lines, only the point O will be on the conic. □

By the preceding theorem, singular point conics can consist of subsets that are lines. This is not the case for non-singular point conics.

Theorem 9.41. *There are at most two distinct points of a non-singular point conic that lie on a given line.*

Proof: Suppose line l has three distinct points A, B, and C that are on the conic. These three points are the intersections of corresponding lines from the pencils of lines at points O and O''. Let a, b, and c be the lines through O and a'', b'', and c'' the corresponding lines through O''.

The three pairs of lines $\{a, a''\}$, $\{b, b''\}$, and $\{c, c''\}$ will define a perspectivity with axis l. By Exercise 9.4.6 we know that the projectivity defining the conic must be the same as the perspectivity. But, this contradicts the fact that the conic is non-singular.

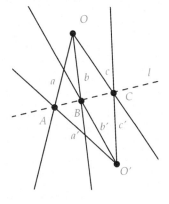

□

9.8.3 Non-singular Conics and Pascal's Theorem

We now investigate the structure of non-singular conics. The next theorem tells us how to recognize when a set of points lies on a non-singular point conic.

> **Theorem 9.42.** *Let A, B, C, D, E, and F be distinct points such that no subset of three of the six points is collinear. Let P be the intersection of AE with CF, Q the intersection of AD with CB, and R the intersection of BE with DF. Then C, D, E, and F are on a non-singular point conic determined by projectively related pencils of lines at A and B if and only if P, Q, and R are collinear.*

Proof: For the first half of the proof, assume that C, D, E, and F are on a non-singular point conic determined by pencils of lines at A and B. We need to show that P, Q, and R are collinear. For this proof, we will use the cross-ratio from the Real Projective model of Projective geometry. We do this for clarity of exposition. The result does hold in abstract, axiomatic Projective geometry as well. For a proof see Section 6.5 of [42].

Let L, M, and N be the intersections of BC, AC, and AE with DF. Since there is a projectivity that takes the pencil of lines at A to the pencil of lines at B, and since the cross-ratio is invariant under projectivities, we have that $R(AC, AD; AE, AF) = R(BC, BD; BE, BF)$.

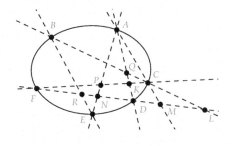

By Corollary 9.35 we have that $R(AC, AD; AE, AF) = R(M, D, N, F)$ and $R(BC, BD; BE, BF) = R(L, D, R, F)$. Thus, $R(M, D, N, F) = R(L, D, R, F)$.

By Theorem 9.7 there is a projectivity π that takes M, D, and F to L, D, and F. Since $R(M, D, N, F) = R(\pi(M), \pi(D), \pi(N), \pi(F)) = R(L, D, \pi(N), F)$ and $R(M, D, N, F) = R(L, D, R, F)$, then we have $R(L, D, \pi(N), F) = R(L, D, R, F)$. By Exercise 9.7.7 we have that $\pi(N) = R$.

Since C, F, and D are not collinear, there is a perspectivity π_1 taking

DF to CF with center A. Under this perspectivity M, D, and F map to C, K, and F, where K is the intersection of AD with CF. Since no subset of three of A, B, C, D, E, or F is collinear, then Q is not on CF or DF. Thus, the perspectivity π_2 from CF to DF with center Q will map C, K, and F to L, D, and F. The combination of these two perspectivities then maps M, D, and F to L, D, and F. By Exercise 9.4.5, the projectivity π must equal the composition of the two perspectivities.

Let's consider what happens to point N under the two perspectivities π_1 and π_2. π_1 will map N to P and then π_2, the perspectivity with center Q, will map P to R, as $\pi(N) = R$. This is only possible if R is on QP. This finishes the first half of the proof.

For the second part of the proof, assume P, Q and R are collinear. We can use the perspectivities π_1 and π_2 defined above to create a projectivity π that maps points M, D, N, and F to points L, D, R, and F. The cross-ratio results from above will still hold and thus we get $R(AC, AD; AE, AF) = R(BC, BD; BE, BF)$.

For the pencil of lines at A there is a projectivity S that will take the three lines AC, AD, AE to lines BC, BD, BE in the pencil at B. The cross-ratio equality then implies that $S(AF) = BF$. Thus, C, D, E, and F are on the point conic defined by the pencils at A and B. Since $A \neq B$ and the projectivity S is not a simple perspectivity, we know the conic is non-singular. $\square$

This theorem is a re-statement of one of the most famous theorems in geometry, Pascal's Theorem.

In the statement of Theorem 9.42 we looked at intersections of certain lines. If we take these lines and list them in the order where the vertices match we have a six-sided figure, a hexagon. The hexagon is defined by AE, EB, BC, CF, FD, and DA.

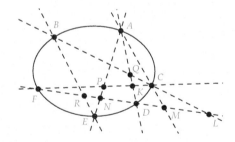

Of course, this hexagon is not made up of segments, but lines. How are the intersections chosen? Here we have "unwrapped" the hexagon into a more standard configuration. A quick check of the intersections from Theorem 9.42 shows that we are choosing *opposite sides* of the hexagon for intersections.

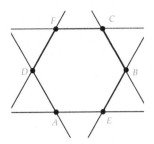

Pascal's Theorem in its classical form is as follows:

Theorem 9.43. *(Pascal's Theorem) If a hexagon is inscribed in a non-singular conic, then points of intersection of opposite sides are collinear.*

Blaise Pascal (1623–1662) proved this theorem when he was just 16 years old!

It is interesting to note that are exactly 60 different hexagons possible by connecting the six points A, B, C, D, E, and F. To see why, consider that any hexagon must pass through point A, so we could consider the hexagon as a path through the points, starting at A. From A there are 5 other points to choose as the next hexagon vertex. Once we have chosen the second vertex, there are four points to choose as the third vertex, and so on until we reach the last vertex. Thus, one would think that there are $5! = 120$ different hexagons. However, a path such as $ABCDEF$ and $AFEDBC$ represent the same hexagon, but reversing the path. Thus, we need to divide by two to get the non-directed paths that form the possible hexagons.

Pascal's Theorem leads directly to one of the key results in the theory of conics, first discovered by Jakob Steiner (1796–1963), who did pioneering work in the foundations of projective geometry. The proof will be left as an exercise.

Theorem 9.44. *(Steiner's Theorem) A non-singular point conic can be defined as the locus of points of intersection of two projectively related pencils of lines with centers at two arbitrarily chosen (distinct) points on the conic.*

Steiner's Theorem leads to the following existence result for point conics.

> **Theorem 9.45.** *Let A, B, C, D, and E be distinct points, no three of which are collinear. Then, there exists a unique non-singular point conic passing through these five points.*

Proof: Since no subset of three points is collinear, then the lines AC, AD, AE are distinct and the lines BC, BD, BE are distinct. Considering the pencil of lines at A there is a unique projectivity S that maps AC, AD, AE to BC, BD, BE in the pencil of lines at B. By Steiner's theorem, any conic through the five points can be determined by projectively related pencils at A and B. Thus, there can be only one such point conic. □

9.8.4 Line Conics

Before we continue in our investigation of point conics, we should take a break and consider the dual results to what we have shown. The dual to a point conic is a line conic.

> **Definition 9.30.** *A line conic is the envelope of lines defined by corresponding points of two pencils of points, where the first pencil is transformed to the second by a projectivity. If the projectivity is equivalent to a single perspectivity, or if the axes of the pencils are the same line, the line conic will be called* singular. *Otherwise, the line conic is called* non-singular.

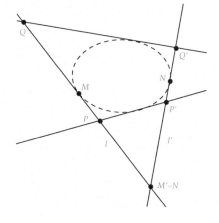

Here we have a non-degenerate line conic defined by a projectivity between pencils of points on l and l'. For example, points P and P' are projectively related, so the line PP' is on the line conic. Note that for some point M on l, there will be a corresponding point M' on l' such that M' is on l. Likewise, there is some point N' on l' that corresponds to point $N = M'$ on l.

In the example above, we have shaded in an ellipse for reference purposes to help visualize the line conic. Here are a few duals to the Theorems we have already proven for point conics.

Theorem 9.46. *(Dual to Theorem 9.39) The axes of the pencils that define a line conic always lie on the line conic.*

Theorem 9.47. *(Dual to Theorem 9.40) The possible singular line conics include the following: the entire Projective plane, the set of lines on two distinct points, the set of lines on a single point, and a single line.*

Theorem 9.48. *(Dual to Theorem 9.41) There are at most two distinct lines of a non-singular line conic that pass through a given point.*

Theorem 9.49. *(Dual to Pascal's Theorem) If a, b, c, d, e, and f are six distinct lines in a non-singular line conic, then the lines defined by joining opposite vertices are concurrent.*

A vertex is the intersection point of adjacent lines. For example, one vertex would be the intersection of a and b. The dual to Pascal's Theorem is known as Brianchon's Theorem, in honor of J. C. Brianchon (1785–1864).

9.8.5 Tangents

In Theorem 9.41 we proved that there are at most two distinct points of a non-singular point conic that lie on a given line. The case where a line intersects a non-singular point conic at a single point will be important enough to have its own definition.

Definition 9.31. *A line is a* tangent line *to a non-singular point conic if it intersects the conic in exactly one point.*

The next theorem guarantees that tangent lines exist.

Theorem 9.50. *At each point on a non-singular point conic there is a unique tangent line.*

Proof: Let P be a point on the point conic. Let Q be another point.

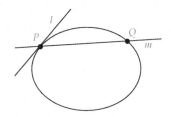

By Steiner's Theorem the point conic can be generated by pencils of lines at P and Q, under a projectivity mapping one pencil to the other. Now, the line $m = PQ$ is in the pencil at Q. Thus, there must be a line l at P that corresponds to m.

We claim that l can intersect the conic only once. For suppose l intersected the conic at another point R. Since the conic is non-singular, $R \neq Q$ (this is proved in the exercises). Then, $l = PR$ would correspond to QR. Since P, Q, and R cannot be collinear, we would have l corresponding to two different lines in the pencil at Q. This contradicts the definition of a projectivity.

Suppose another line $l' \neq l$ was tangent at P. This line has to correspond to a line $m' \neq m$ at Q. But m' and l' define a point X on the conic different than P. Then, l' intersects the conic in two places, which contradicts the assumption that it was tangent at P. □

An interesting fact about tangents and Pascal's Theorem, is that we can replace pairs of edges with tangents and still have the conclusion of that theorem.

In the statement of Pascal's Theorem (or Theorem 9.42), we have hexagon $AEBCFD$ inscribed in a point conic. Consider what happens to the lines of this hexagon as we move point A to E and point B to C. The edges rotate into what seems to be the tangent lines at $A = E$ and $B = C$.

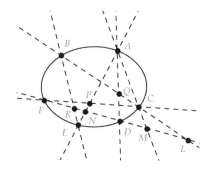

Here is a picture of what the lines and intersections look like after we replace the edges with tangent lines. That is, in the statement of the Theorem we replace AE with the tangent at A and BC with the tangent at B. If we carry this replacement of edge with tangent line throughout the proof of Theorem 9.42 we see that the proof is still correct!

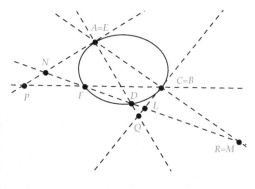

Theorem 9.51. *If $ABFD$ is a quadrangle inscribed in a non-singular point conic, let P be the intersection of the tangent at A with BF, Q the intersection of the tangent at B and AD, and R the intersection of AB with DF. Then, P, Q, and R are collinear.*

If we have a hexagon $AEBCFD$ inscribed in a conic, and let A move to D and B move to C, we get the following result:

Theorem 9.52. *If $AFBE$ is a quadrangle inscribed in a non-singular point conic, let P be the intersection of AE with BF, Q the intersection of the tangent at B with the tangent at A, and R the intersection of the tangent at E and the tangent at F. Then, P, Q, and R are collinear.*

Proof:

Let S be the intersection of AF with BE. If we replace AD with the tangent at A and BC with the tangent at B, then points P, Q, and S match points P, Q, and R in Theorem 9.42. Also, the proof will still hold for this configuration. Thus, P, Q, and S are collinear.

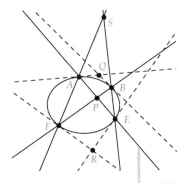

If, in this last argument, we interchange A with F and B with E, we get a configuration that will again satisfy Theorem 9.42. But now the points P, R, and S will match P, Q, and R in Theorem 9.42. Thus, P, R, and S are collinear. Since P, Q, and S are collinear we conclude that P, Q, and R are collinear. □

With what we have proven so far the following connection between line and point conics is hopefully intuitive. The proof can be found on the web resource page.

> **Theorem 9.53.** *The set of tangents to a non-singular point conic forms a line conic.*

9.8.6 Conics in Real Projective Plane

In this section, we consider conics in the Real Projective Plane model of Projective geometry. Recall that in this model a point (not at infinity) is represented by a vector $(x, y, 1)$, or equivalently any vector of the form (x, y, z) with $z \neq 0$. A point at infinity is represented by $(x, y, 0)$.

Lines are also represented as vectors of the form $u = [a, b, c]$. A point (x, y, z) is on line $[a, b, c]$ if $ax + by + cz = 0$.

Consider the pencil of lines on a point O. Let u and v be two lines in the pencil at O. Let w be another line in the pencil at O. By Theorem 9.19, we can parametrize w by $w = \alpha u + \beta v$, with α and β real constants, with at least one being non-zero.

Now, suppose we have a point conic defined by the pencils of lines on two points O and O'. Let u and v be two lines in the pencil at O and u' and v' be the projectively related lines in the pencil at O'. Then, for any line w in the pencil at O we have that $w = \alpha u + \beta v$, and for any line w' in the pencil at O' we have $w' = \alpha' u' + \beta' v'$. If $u = [u_1, u_2, u_3]$ and $v = [v_1, v_2, v_3]$ are the vectors for u and v, and $u' = [u'_1, u'_2, u'_3]$ and $v' = [v'_1, v'_2, v'_3]$ are the vectors for u' and v', then $w = [\alpha u_1 + \beta v_1, \alpha u_2 + \beta v_2, \alpha u_3 + \beta v_3]$ and $w' = [\alpha' u'_1 + \beta' v'_1, \alpha' u'_2 + \beta' v'_2, \alpha' u'_3 + \beta' v'_3]$.

If a point (x, y, z) is on the point conic, then it must be on both lines. Thus,

$$(\alpha u_1 + \beta v_1)x + (\alpha u_2 + \beta v_2)y + (\alpha u_3 + \beta v_3)z = 0$$
$$(\alpha' u'_1 + \beta' v'_1)x + (\alpha' u'_2 + \beta' v'_2)y + (\alpha' u'_3 + \beta' v'_3)z = 0$$

We can re-write these equations as follows:

$$A\alpha + B\beta = 0$$
$$C\alpha' + D\beta' = 0$$

where $A = u_1 x + u_2 y + u_3 z$, $B = v_1 x + v_2 y + v_3 z$, $C = u_1' x + u_2' y + u_3' z$, and $D = v_1' x + v_2' y + v_3' z$.

By Corollary 9.21 we have that

$$\lambda \begin{bmatrix} \alpha' \\ \beta' \end{bmatrix} = \begin{bmatrix} a & b \\ c & d \end{bmatrix} \begin{bmatrix} \alpha \\ \beta \end{bmatrix}$$

where the determinant $ad - bc \neq 0$ and $\lambda \neq 0$.

Solving for α' and β' and substituting into the previous equations, we get

$$A\alpha + B\beta = 0$$
$$\frac{1}{\lambda}(C(a\alpha + b\beta) + D(c\alpha + d\beta)) = 0$$

Or,

$$A\alpha + B\beta = 0$$
$$(Ca + Dc)\alpha + (Cb + Dd)\beta = 0$$

Writing this as a matrix equation, we get

$$\begin{bmatrix} A & B \\ Ca + Dc & Cb + Dd \end{bmatrix} \begin{bmatrix} \alpha \\ \beta \end{bmatrix} = \begin{bmatrix} 0 \\ 0 \end{bmatrix}$$

For this equation to have nonzero solutions, the determinant of the multiplying 2×2 matrix must be zero. That is, $A(Cb + Dd) - B(Ca + Dc) = 0$. Recall that $A = u_1 x + u_2 y + u_3 z$, $B = v_1 x + v_2 y + v_3 z$, $C = u_1' x + u_2' y + u_3' z$, and $D = v_1' x + v_2' y + v_3' z$. If we substitute these terms into the determinant equation, we get an equation in x, y, z of the form $b_{11}x^2 + b_{12}xy + b_{13}xz + b_{21}yx + b_{22}y^2 + b_{23}yz + b_{31}zx + b_{32}zy + b_{33}z^2 = 0$.

For example, we could have $4x^2 + 8xy - 4xz + 12yx - 3y^2 + 2yz + 2zx + 6zy + z^2 = 0$. To make this equation symmetric let $a_{ij} = \frac{b_{ij} + b_{ji}}{2}$. Then, for our example we would have $4x^2 + 10xy - xz + 10yx - 3y^2 + 4yz - zx + 4zy + z^2 = 0$. An equivalent form for this equation would be

$$\begin{bmatrix} x & y & z \end{bmatrix} \begin{bmatrix} 4 & 10 & -1 \\ 10 & -3 & 4 \\ -1 & 4 & 1 \end{bmatrix} \begin{bmatrix} x \\ y \\ z \end{bmatrix} = \begin{bmatrix} 0 \\ 0 \\ 0 \end{bmatrix}$$

Let $x_1 = x$, $x_2 = y$, and $x_3 = z$. Summarizing our work above we have the following:

Theorem 9.54. *A point conic in the Real Projective plane has an equation of the form*

$$\sum_{i,j=1}^{3} a_{ij} x_i x_j = 0$$

where $a_{ij} = a_{ji}$. In matrix form, $X^t A X = [0]$ where $X = \begin{pmatrix} x_1 \\ x_2 \\ x_3 \end{pmatrix}$.

There is an even simpler way to represent a conic in the Real Projective Plane. Since the equation for a conic is a quadratic in x, y, and z, by completing the square on each of these variables we can change coordinates and eliminate all terms such as xy, or xz. Examples of how to do this are described in the exercises. We get the following:

Theorem 9.55. *With an appropriately chosen change of coordinates, a point conic in the Real Projective plane has an equation of the form*

$$c_1 x^2 + c_2 y^2 + c_3 z^2 = 0$$

With this result we can classify conics in terms of the c_i's of the preceding theorem:

1. All $c_i \neq 0$ and all $c_1 > 0$. There is clearly no solution for this equation, so the point conic is the empty set.

2. All $c_i \neq 0$ and not all with the same sign. We can assume one is positive and the others negative. For example, $z^2 = x^2 + y^2$. Dividing by z, we get $x'^2 + y'^2 = 1$, where $x' = \frac{x}{z}$ and $y' = \frac{y}{z}$. This is the familiar equation of an ellipse from Euclidean geometry.

3. One $c_i = 0$, others non-zero, but of same sign. For example, $x^2 + y^2 = 0$. This has solution $(0, 0, 1)$. This case yields a singular conic of a single point.

4. One $c_i = 0$, others non-zero, but of differing sign. For example, $x^2 - y^2 = 0$. This has solution $x = \pm y$. This case yields a singular conic of two lines.

5. Exactly two c_i's being zero. For example, $x^2 = 0$. This case yields a singular conic of one line.

6. All $c_i = 0$. All points in the plane would satisfy this case.

We conclude that a point conic is non-singular if and only if $c_i \neq 0$ for $i = 1, 2, 3$ and not all c_i have the same sign. We also observe that all non-singular point conics are essentially equivalent in the projective plane. We do not have the separate cases of ellipses, hyperbolas, and parabolas, as we do in Euclidean geometry.

Exercise 9.8.1. *Prove that if a point conic is non-singular and is defined by pencils at points A and B, then the line AB cannot correspond to itself under the projectivity taking the pencil at A to the pencil at B. [Hint: According to Theorem 9.12 the projectivity defining the conic is equivalent to the composition of two perspectivities. Suppose that the line corresponded to itself. Use the dual to Lemma 9.10 to show the projectivity would be a perspectivity.]*

Exercise 9.8.2. *Prove that if five distinct points on a point conic have the property that no subset of three points is collinear, and the projectivity defining the conic is not the identity, then the point conic is non-singular.*

Exercise 9.8.3. *Prove Steiner's Theorem. [Hint: Use Theorem 9.42. What happens if we switch A with C, B with D, and E with F?]*

Exercise 9.8.4. *State and illustrate (provide a figure to help explain) the dual to Theorem 9.42.*

Exercise 9.8.5. *Show that two different non-singular conics can intersect in at most four points.*

Exercise 9.8.6. *Suppose the equation of a non-singular point conic in the Real Projective plane is given by $2x^2 + 8xy + 12y^2 - 4xz + 2yz + 12z^2 = 0$. Show that by completing the square on the $x - y$ terms, and using the change of coordinates $x' = x + 2y$, that the equation in the new coordinates becomes $2x'^2 + 4y^2 - 4x'z + 10yz + 12z^2 = 0$.*

Exercise 9.8.7. *Suppose the equation of a non-singular point conic in the Real Projective plane is given by $2x^2 + 4y^2 - 4xz + 10yz + 12z^2 = 0$. Show that by completing the square on the $x - z$ terms, and using the change of coordinates $x' = x - z$, that the equation in the new coordinates becomes $2x'^2 + 4y^2 + 10yz + 10z^2 = 0$.*

Exercise 9.8.8. *Suppose the equation of a non-singular point conic in the Real Projective plane is given by $2x^2 + 4y^2 + 10yz + 10z^2 = 0$. Refer to the method of the previous two exercises to show that the change of coordinates $z' = z + \frac{1}{2}y$ will transform the equation to $2x^2 + \frac{3}{2}y^2 + 10z'^2 = 0$.*

CHAPTER 10

Fractal Geometry

Why is geometry often described as "cold" and "dry"? One reason lies in its inability to describe the shape of a cloud, a mountain, a coastline, or a tree. Clouds are not spheres, mountains are not cones, coastlines are not circles, and bark is not smooth, nor does lightning travel in a straight line.

– Benoit Mandelbrot in *The Fractal Geometry of Nature* [30]

10.1 THE SEARCH FOR A "NATURAL" GEOMETRY

Classical geometry had its roots in ancient Babylonian and Egyptian calculations of land areas and architectural designs. The word *geometry* means "earth measurement." It is said that Aristotle came to believe that the earth was a sphere by watching ships disappear over the ocean's horizon.

Intuitively, we like to think our notions of geometry are the result of our interactions with nature, but is that really the case?

As mentioned in the quote above, most objects in nature are not really regular in form. A cloud may look like a lumpy ball from far away, but as we get closer, we notice little wisps of vapor jutting out in every direction. As we move even closer, we notice that the cloud has no real boundary. The solidness of the form dissolves into countless filaments of vapor. We may be tempted to replace our former notion of the cloud being three-dimensional with a new notion of the cloud being a collection of one-dimensional curves. But, if we look closer, these white curves dissolve into tiny water droplets. The cloud now appears to be a collection of tiny three-dimensional balls. As we move even closer, to the molecular level, these droplets dissolve into tiny whirling masses of hydrogen and oxygen, shapes akin to the original cloud itself.

So, a cloud in the sky cannot *really* be described as a classical geometric figure. This seems to contradict the commonly held intuition about Euclidean geometry being a product of our natural environment. How can we make geometric sense of objects like clouds, mountains, trees, atoms, planets, and so on?

We saw in Chapters 5 and 6 that there is one geometric idea that does seem to resonate with our experience of the natural world—the idea of *symmetry*. Symmetry is the idea that an object is invariant under some transformation of that object. In Chapter 5 we looked at the notion of symmetry as invariance under Euclidean isometries. Such symmetries include bilateral symmetry, rotational symmetry, translation symmetry, and glide symmetry.

In the discussion of the cloud as a geometric object, we noticed that when we viewed a section of the cloud at the molecular level, we saw a shape similar to the original cloud itself. That is, the molecular "cloud" appeared to be the same shape as the original cloud, when *scaled* up by an appropriate scale factor.

Similarly, consider the fern in Figure 10.1. Each leaf is made up of sub-leaves that look similar to the original leaf, and each sub-leaf has sub-sub-leaves similar to the sub-leaves, and so on.

Figure 10.1 Fern leaf

Thus, many natural objects are similar to parts of themselves, once you scale up the part to the size of the whole. That is, they are symmetric under a change of scale. We call such objects *self-similar* objects. Such objects will be our first example of *fractals*, a geometric class of objects that we will leave undefined for the time being.

10.2 SELF-SIMILARITY

An object will be called *self-similar* if a part of the object, when scaled by a factor $c > 0$, is equivalent to the object itself.

We can be more precise in this definition by making use of *similarity transformations*.

Definition 10.1. *A* similarity transformation S, *with* ratio $c > 0$, *is a transformation (i.e., one-to-one and onto map) from Euclidean n-dimensional space ($\mathbb{R}^n$) to itself such that*

$$|S(x) - S(y)| = c\,|x - y| \tag{10.1}$$

A self-similar set will be a set that is invariant under one or more similarity transformations.

Definition 10.2. *A* self-similar *set F in $\mathbb{R}^n$ is a set that is invariant under a finite number of non-identity similarity transformations.*

Which classical Euclidean objects are self-similar? Consider a circle C. We know that the closer we "look" at the circle, the flatter the curve of the circle becomes. Thus, a circle cannot be self-similar. Similarly, any differentiable curve in the plane will not be self-similar, with one exception—a line. Lines are perhaps the simplest self-similar figure.

Self-similarity is a concept foreign to most of the 2-dimensional geometry covered in calculus and Euclidean geometry. To develop some intuition for self-similarity, we need some examples.

10.2.1 Sierpinski's Triangle

Our first example is generated from a simple filled-in triangle, $\triangle ABC$.

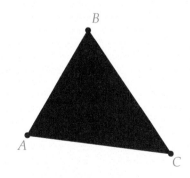

Let L, M, and N be the midpoints of the sides and remove the middle third triangle, $\triangle LMN$.

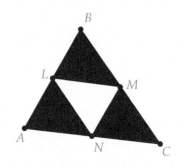

Now, each of the three smaller triangles is *almost* the same figure as the original triangle, when scaled up by a factor of three, except for the "hole" in the middle of the big triangle, which is missing in the smaller triangles. To make the smaller triangles similar to the original, let's remove the middle third of each triangle.

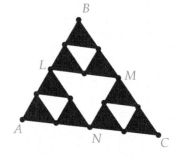

Now, we have fixed our problem, and each of the three subtriangles has a hole in the middle. But, now the big triangle has sub-triangles with holes, and so the three sub-triangles are, again, not similar to the big triangle. So, we will remove the middle third of each of the sub-sub-triangles.

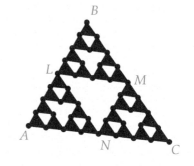

Again, we have *almost* self-similarity. To create a truly self-similar figure, we need to continue this middle third removal process to infinity! Then each sub-triangle will be *exactly* similar to the original triangle, each sub-sub-triangle will be similar to each sub-triangle, and so on, with the scale factor being 3 at each stage.

The figure that is left after carrying on the middle third removal process infinitely often is called the *Sierpinski Triangle*, or *Sierpinski Gasket*, in honor of Waclaw Sierpinski (1882–1969), a Polish mathematician who is known for his work in set theory, topology, and analysis.

Note that Sierpinski's Triangle is invariant under three basic similarities, scaling transformations by a factor of $\frac{1}{2}$ toward points A, B, and C.

Since the process of creating Sierpinski's Triangle starts with a simple 2-dimensional filled-in triangle, it is natural to try to calculate the area of the final figure.

Let's denote by *stages* the successive process of removing middle thirds from sub-triangles. At stage 0 we have the original filled-in triangle, $\triangle ABC$. At stage 1 we have removed the middle third. At stage 2 we have removed the middle third of each of the remaining triangles, and so on.

We can assume the area of the first triangle to be anything we like, so we will assume it equal to 1. At stage 1 the area remaining in the figure will be

$$Area(stage\ 1) = 1 - \frac{1}{4}$$

since all of the four sub-triangles are congruent (proved as an exercise), and thus have an area $\frac{1}{4}$ the area of the original triangle.

At stage 2 we remove three small triangles from each of the remaining sub-triangles, each of area $\frac{1}{16}$. The area left is

$$Area(stage\ 2) = 1 - \frac{1}{4} - \frac{3}{16}$$

At stage 3 we remove nine areas, each of area $\frac{1}{64}$. Thus,

$$Area(stage\ 2) = 1 - \frac{1}{4} - \frac{3}{16} - \frac{9}{64}$$

Seeing the pattern developing here, we conclude that the area left for Sierpinski's Triangle at stage n is

$$Area(stage\ n) = 1 - \frac{1}{4} \sum_{k=0}^{n} \left(\frac{3}{4}\right)^k$$

$$\lim_{n \to \infty} Area = 1 - \frac{1}{4} \frac{1}{\left(1 - \frac{3}{4}\right)}$$

$$= 0$$

This is truly an amazing result! The Sierpinski Triangle has had all of its area removed, but still exists as an infinite number of points. Also, it has all of the boundary segments of the original triangle remaining, plus all the segments of the sub-triangles. So, it must be at "least" a 1-dimensional object. We will make this idea of an "in-between" dimension more concrete in the next section.

10.2.2 Cantor Set

What kind of shape would we get if we applied the middle-third removal process to a simple line segment?

Here we have a segment $\overline{AB}$.

Remove the middle third to get two segments $\overline{AC}$ and $\overline{DB}$.

Perform this process again and again as we did for the Sierpinski Triangle. Here is a collection of the stages of the process, where we have hidden some of the points for clarity.

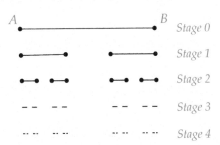

The limiting figure, which has an infinite number of middle thirds removed, is called the *Cantor Set*, in honor of Georg Cantor (1845–1918). Cantor is known for his work in set theory and in particular for his investigation of orders of infinity and denumerable sets. It is left as an exercise to show that the Cantor Set has length 0, although it is made up of an infinite number of points.

Both of our examples so far, Sierpinski's Triangle and the Cantor Set, do not seem to fit into our classical notion of 1- and 2- dimensional objects.

Sierpinski's Triangle is more than a 1-dimensional curve, yet certainly less than an area, which is 2-dimensional, while the Cantor Set lies somewhere between dimensions 0 and 1. How can this be? How can an object have *fractional* dimension? This idea of fractional dimension was the critical organizing principle for Mandelbrot in his study of natural phenomena and is why he coined the term *fractal* for objects with non-integer dimension.

10.3 SIMILARITY DIMENSION

The dimension of a fractal object turns out to be quite difficult to define precisely. This is because there are several different definitions of dimension used by mathematicians, all having their positive as well as negative aspects. In this section we will look at a fairly simple definition of dimension for self-similar sets, the *similarity dimension.*

To motivate this definition, let's look at some easy examples.

Consider our simplest self-similar object, a line segment $\overline{AB}$.

$A \bullet \!\!\!\text{———————}\!\!\! \bullet B$

It takes two segments of size $\frac{AB}{2}$ to cover this segment.

In general, it takes N sub-segments of the original segment of size $\frac{AB}{N}$ to cover the original segment. Or, if we think of the sub-segments as being similar to the original, it takes N sub-segments of *similarity ratio* $\frac{1}{N}$ to cover $\overline{AB}$.

We will define the function $r(N)$ to be the similarity ratio between a figure and its parts.

Suppose we have a rectangle $ABCD$ of side lengths $AB = a$ and $BC = b$. If we subdivide this region into N parts that are similar to the original region, then the area of each sub-part will be $\frac{ab}{N}$, and the length of each side will be scaled by the similarity ratio $r(N) = \frac{1}{\sqrt{N}} = \frac{1}{N^{\frac{1}{2}}}$. In the figure at right, we have $N = 9$.

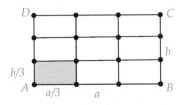

For a rectangular solid in three dimensions, the similarity ratio would be $r(N) = \frac{1}{N^{\frac{1}{3}}}$.

Note that these three similarity ratios hold no matter the size of the segment, rectangle, or solid under consideration. In higher dimensions, the similarity ratio for a d-dimensional rectangular object would be $r(N) = \frac{1}{N^{\frac{1}{d}}}$.

Equivalently, if we let $r = r(N)$, we have

$$Nr^d = 1$$
$$log(N) + d\,log(r) = 0$$
$$d = \frac{log(N)}{-log(r)} = \frac{log(N)}{log(\frac{1}{r})} \tag{10.2}$$

Since the value of d matches our Euclidean notion of dimension for simple self-similar objects, like segments and rectangles, we will define the similarity dimension to be the value of d given by equation 10.2.

Definition 10.3. *The* similarity dimension d *of a self-similar object is given by*

$$d = \frac{log(N)}{log(\frac{1}{r})} \tag{10.3}$$

where the value of $r = r(N)$ is the ratio by which a ruler measuring a side of the object will change under the assumption that each of the N sub-objects can be scaled to form the original object. The ratio r is called the similarity ratio *of the object.*

For example, consider the Cantor Set. Each sub-object is $\frac{1}{3}$ the scale of the original segment, and it takes two sub-objects to make up the whole (after stage 0). Thus, the similarity dimension of the Cantor Set is

$$\frac{log(2)}{log\left(\frac{1}{\frac{1}{3}}\right)} = \frac{log(2)}{log(3)} \approx .6309$$

This result agrees with our earlier intuition about the dimension of the Cantor Set being somewhere between 0 and 1.

As another example, it takes three sub-triangle shapes to scale up and cover the bigger triangle in Sierpinski's Triangle (again, ignoring stage 0). Each sub-triangle has side-length scaled by $\frac{1}{2}$ of the original side-length, thus the similarity ratio is $\frac{1}{2}$, and the similarity dimension is

$$\frac{log(3)}{log\left(\frac{1}{\frac{1}{2}}\right)} = \frac{log(3)}{log(2)} \approx 1.58496$$

Exercise 10.3.1. *Show that all sub-triangles at stage 1 of the Sierpinski Triangle process are congruent.*

Exercise 10.3.2. *Show that the length left after all of the stages of construction of Cantor's Set is 0.*

Exercise 10.3.3. Sierpinski's Carpet *is defined by starting with a square and removing the middle third sub-square, each side of which is $\frac{1}{3}$ the size of the original square.*

Here is stage 1 for Sierpinski's Carpet.

The Carpet is defined as the limiting process of successively removing middle-third squares. Does Sierpinski's Carpet have non-zero area? Sketch the next iteration of this figure.

Exercise 10.3.4. *Show that the similarity dimension of Sierpinski's Carpet is $\frac{log(8)}{log(3)}$. Thus, Sierpinski's Carpet is much more "area-like" than Sierpinski's Triangle.*

Exercise 10.3.5. *In the construction of the Sierpinski Carpet, instead of removing just the middle-third square, remove this square and all four squares with which it shares an edge. What is the similarity dimension of the limiting figure?*

Exercise 10.3.6. *The* Menger Sponge *is defined by starting with a cube and sub-dividing it into 27 sub-cubes. Then, we remove the center cube and all six cubes with which the central cube shares a face. Continue this process repeatedly, each time sub-dividing the remaining cubes and removing pieces. The Menger Sponge is the limiting figure of this process. Show that the dimension of the Menger Sponge is $\frac{log(20)}{log(3)}$. Is the sponge more of a solid or more of a surface? Devise a cube removal process that would produce a limiting figure with fractal dimension closer to a surface than a cube. The Menger Sponge is named for Karl Menger (1902–1985). He is known for his work in geometry and on the definition of dimension.*

Exercise 10.3.7. *Starting with a cube, is there a removal process that leads to a fractal dimension of 2, yet with a limiting figure that is fractal in nature?*

10.4   PROJECT 16 - AN ENDLESSLY BEAUTIFUL SNOWFLAKE

In this project we will use dynamic geometry software to create a self-similar fractal. Self-similar fractals are ideal for study using computational techniques, as their construction is basically a *recursive* process—one that loops back upon itself.

The self-similar fractal we will construct is "Koch's snowflake curve," named in honor of Helge von Koch (1870–1924), a Swedish mathematician who is most famous for the curve that bears his name. It is an example of a continuous curve that is not differentiable at any of its points. In the next few paragraphs we will discuss how to use a "template" curve to recursively build the Koch self-similar snowflake curve. Please read this discussion carefully (you do not need to construct anything yet).

To construct the template curve for the Koch snowflake, we start with a segment $\overline{AB}$. Just as we removed the middle third for the Cantor Set construction, we remove the middle third of $\overline{AB}$, but replace it with the two upper segments of an equilateral triangle of side length equal to the middle third we removed (Figure 10.2).

Figure 10.2

Now think of the process just described as a *replacement* process, where we take a segment and replace it with a new curve, the template curve. Each segment of the new template curve can then be replaced with a copy of the template that is scaled by a factor of $\frac{1}{3}$. If we carry out this replacement process for each of the four small segments in the template, we get the curve shown in Figure 10.3.

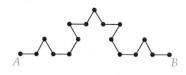

Figure 10.3

The process just outlined can be made recursive. We could take each of the new segments in the curve just described and replace them with a scaled-down copy of the template. We could then repeatedly take each new set of segments at stage n and replace them with copies of the template to get a curve at stage $n + 1$. Thus, the replacing of segments by copies of the template loops back on itself indefinitely.

At the point where we stopped the replacement process, the curve had 16 small segments, each of length $\frac{1}{9}$ of the original segment $\overline{AB}$. Replacing each of these segments with a $\frac{1}{27}$ scale copy of our template, we get a new curve with 64 segments, each a length $\frac{1}{27}$ of the original. The new curve is shown in Figure 10.4 where we have hidden all points except A and B for clarity.

Figure 10.4

The Koch curve is the curve that results from applying this template

replacement process an *infinite* number of times. The curve is self-similar in the sense that if you took a piece of the curve and magnified it by a factor of 3, you would see the same curve again.

Start up your geometry software and create a segment $\overline{AB}$ on the screen.

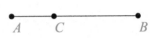

Next, we will divide segment $\overline{AB}$ into three equal parts. Set point A as a center of dilation and then dilate point B by a factor of $\frac{1}{3}$ towards A.

Similarly, carry out the steps to dilate point B by a ratio of $\frac{2}{3}$ towards point A. Then hide segment $\overline{AB}$. We have now split $\overline{AB}$ into equal thirds.

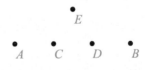

Next we create the "bump" in the middle of the template. Set C as a center of rotation and rotate point D by 60 degrees.

Finally, construct segments between A, C, E, D, and B (in that order). At this point our template curve is complete.

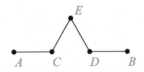

At this point in the Koch curve construction, the template can be used to replace each of the four segments that are in the template itself. Most dynamic geometry software has the capability to do this replacement process. Go to http://www.gac.edu/~hvidsten/geom-text for instructions on how to carry out this replacement process for your particular software.

Once you have defined the replacement process, carry out this process several times.

Here we have carried out the replacement process three times, each time replacing all segments with smaller copies of the template.

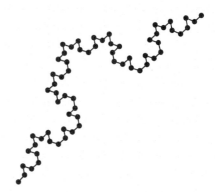

Exercise 10.4.1. *Find a formula for the length of the Koch curve at stage n of its construction, with stage 0 being the initial segment $\overline{AB}$, which you can assume has length 1. Is the Koch curve of finite or infinite length?*

The Koch Snowflake curve is the result of applying this recursive process to each of three segments of a triangle.

Create a triangle, $\triangle ABC$ and carry out the replacement process twice on each side. We can see why this curve is called the "snowflake" curve.

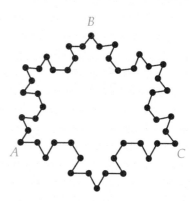

The Koch snowflake curve is the result of applying the template replacement process an infinite number of times to each of the three edges of $\triangle ABC$.

Exercise 10.4.2. *Suppose the area of $\triangle ABC$ is equal to 1. Show that the area enclosed inside the Koch snowflake curve is given by the infinite series*

$$Area = 1 + \frac{3}{9} + 4 \cdot \frac{3}{9^2} + 4^2 \cdot \frac{3}{9^3} + \cdots$$

Find the sum of this series.

Exercise 10.4.3. *Find the similarity dimension of the Koch snowflake curve.*

Let the "Koch hat curve" be the curve obtained by running the recursive process described, but replacing segments at each stage with a different template.

Instead of the triangle template, use the square template at right, where each side of the square is $\frac{1}{3}$ the length of the original segment.

Exercise 10.4.4. *Find the next stage of this fractal curve and show that the curve intersects itself. Find the dimension of the Koch hat curve.*

Exercise 10.4.5. *Design a template different from the two used above, and use it to create a fractal curve. Find the dimension of this new curve.*

10.5 CONTRACTION MAPPINGS AND THE SPACE OF FRACTALS

In the last few sections we focused our attention on the simplest kinds of self-similar sets, those that are invariant under similarities S_n, all of which share the same scaling ratio c.

For example, the Cantor Set, defined using the interval $[0, 1]$ as the initial figure, will be invariant under the similarities S_1 and S_2, defined by

$$S_1(x) = \frac{x}{3}$$

$$S_2(x) = \frac{x}{3} + \frac{2}{3}$$

Both S_1 and S_2 are special cases of more general functions called *contraction mappings.*

Definition 10.4. *A contraction mapping S is a function defined on a set D of $(\mathbb{R}^n)$ such that there is a number $0 \le c < 1$ with*

$$|S(x) - S(y)| \le c\,|x - y|$$

for all x, y in D.

One of the great insights in the subject of fractal geometry is the fact that not only are many fractals invariant under contraction mappings, but the fractal itself can be *generated* from *iterating* the contraction mappings on an initial shape.

Let's consider again the Cantor Set on the interval $[0,1]$, with the contraction mappings S_1 and S_2 as just defined. Let B_n be the result of applying these contractions repeatedly n times. That is,

$$
\begin{aligned}
B_0 &= [0,1] \\
B_1 &= S_1(B_0) \cup S_2(B_0) = [0, \tfrac{1}{3}] \cup [\tfrac{2}{3}, 1] \\
B_2 &= S_1(B_1) \cup S_2(B_1) \\
&= S_1([0, \tfrac{1}{3}]) \cup S_1([\tfrac{2}{3}, 1]) \cup S_2([0, \tfrac{1}{3}]) \cup S_2([\tfrac{2}{3}, 1]) \\
&= [0, \tfrac{1}{9}] \cup [\tfrac{2}{9}, \tfrac{1}{3}] \cup [\tfrac{2}{3}, \tfrac{7}{9}] \cup [\tfrac{8}{9}, 1] \\
&\ \cdot \\
&\ \cdot \\
&\ \cdot \\
B_n &= S_1(B_{n-1}) \cup S_2(B_{n-1})
\end{aligned}
$$

Note that B_0 is the figure from the Cantor Set construction at stage 0; B_1 is the figure at stage 1 (the middle third is gone); B_2 is the figure at stage 3; and so on. Why is this the case? Think of what the two contraction mappings are doing geometrically. The effect of S_1 is to contract everything in the interval $[0,1]$ into the first third of that interval, while S_2 also contracts by $\tfrac{1}{3}$, but then shifts by a distance of $\tfrac{2}{3}$. Thus, the effect of iterating these two maps on stage $k-1$ of the Cantor Set construction is to shrink the previous set of constructed segments by $\tfrac{1}{3}$ and then copying them to the intervals $[0, \tfrac{1}{3}]$ and $[\tfrac{2}{3}, 1]$. This has an effect equivalent to the removal of all middle-thirds from the previous segments at stage $k-1$.

Suppose we started our construction with the interval $[2, 4]$ instead of $[0, 1]$. Then

$$
\begin{aligned}
B_0 &= [2, 4] \\
B_1 &= S_1(B_0) \cup S_2(B_0) = [\frac{2}{3}, \frac{4}{3}] \cup [\frac{4}{3}, 2] = [\frac{2}{3}, 2] \\
B_2 &= S_1(B_1) \cup S_2(B_1) = [\frac{2}{9}, \frac{2}{3}] \cup [\frac{8}{9}, \frac{4}{3}] \\
B_3 &= [\frac{2}{27}, \frac{2}{9}] \cup [\frac{8}{27}, \frac{4}{9}] \cup [\frac{20}{27}, \frac{8}{9}] \cup [\frac{26}{27}, \frac{10}{9}] \\
B_4 &= [\frac{2}{81}, \frac{2}{27}] \cup \cdots [\frac{80}{81}, \frac{28}{27}]
\end{aligned}
$$

and so on.

Note what is happening to the initial and final intervals at each stage of iteration. For example, at stage 3 we have points ranging from $\frac{2}{27}$ to $\frac{10}{9}$, and at stage 4 we have points ranging from $\frac{2}{81}$ to $\frac{28}{27}$. It looks like repeatedly applying S_1 and S_2 to the interval $[2, 4]$ is closing in on the interval $[0, 1]$. This should not be surprising, as 0 and 1 are *fixed points* of S_1 and S_2. In fact, for any x we have

$$
|S_2(x) - S_2(1)| = |S_2(x) - 1| = \frac{1}{3}|x - 1|
$$

Thus,

$$
\begin{aligned}
|S_2^n(x) - 1| &= |S_2^n(x) - S_2^n(1)| = \frac{1}{3}|S_2^{n-1}(x) - S_2^{n-1}(1)| \\
&= \frac{1}{3}^2 |S_2^{n-2}(x) - S_2^{n-2}(1)| \\
& \qquad \vdots \\
&= \frac{1}{3}^n |x - 1|
\end{aligned}
$$

Clearly, as n grows without bound, $S_2^n(x)$ must approach 1. By a similar argument, we can show $S_1^n(x)$ approaches 0. On the other hand, points inside $[0, 1]$ can "survive" forever by the *combined* action of S_1 and S_2. For example, the points $\frac{1}{3}$ and $\frac{2}{3}$ are pulled back and forth by S_1 and S_2, but always by the same amount toward each of the fixed points, and thus they survive all stages of the construction process. No point outside $[0, 1]$ will have this prospect of surviving.

Thus, it appears that the points in the Cantor Set are not uniquely tied to the starting interval $[0, 1]$. We would get the same set of limiting points if we started with *any* interval. The Cantor Set is thus *attracting* the iterates of the two contraction mappings S_1 and S_2.

From this brief example one might conjecture that something like this attracting process is ubiquitous to fractals, and such a result is, in fact, the case.

To prove this, we need to develop some tools for handling the iteration of sets of functions on subsets of Euclidean space. We will first need a few definitions.

Definition 10.5. *The distance function in $\mathbb{R}^n$ will be denoted by d. Thus, $d(x, y)$ measures the Euclidean distance from x to y.*

Definition 10.6. *A set D in $\mathbb{R}^n$ is* bounded *if it is contained in some sufficiently large ball. That is, there is a point a and radius R such that for all $x \in D$, we have $d(x, a) < R$.*

Definition 10.7. *A sequence $\{x_n\}$ of points in $\mathbb{R}^n$* converges *to a point x as n goes to infinity, if $d(x_n, x)$ goes to zero. The point x is called the* limit *point of the sequence.*

Definition 10.8. *A set D in $\mathbb{R}^n$ is* closed *if it contains all of its limit points, that is, if every convergent sequence of points from D converges to a point inside D.*

Definition 10.9. *An* open ball *centered at a, of radius ϵ, is the set of points x in $\mathbb{R}^n$ such that $d(x, a) < \epsilon$.*

Definition 10.10. *A set D in $\mathbb{R}^n$ is* open *if for every x in D, there is an open ball of non-zero radius that is entirely contained in D.*

The unit disk $x^2 + y^2 \leq 1$ is closed. The interval $(0,1)$ is not. The set $x^2 + y^2 < 1$ is open, as is the interval $(0,1)$. The interval $[0,1)$ is neither open nor closed.

Definition 10.11. *A set D in $\mathbb{R}^n$ is* compact *if any collection of open sets that covers D (i.e., with the union of the open sets containing D) has a finite sub-collection of open sets that still covers D.*

It can be shown that if D is compact, then it is also closed and bounded [4, pages 20–25].

We will be working primarily with compact sets and so will define a structure to contain all such sets.

Definition 10.12. *The space $\mathcal{H}$ is defined as the set of all compact subsets of $\mathbb{R}^n$.*

The "points" of $\mathcal{H}$ will be compact subsets. In the preceding example of the Cantor Set, we can consider the stages B_0, B_1, and so forth, of the construction as a sequence B_n of compact sets, that is, a sequence of "points" in $\mathcal{H}$. It appeared that this sequence *converged* to the Cantor Set. However, to speak of a sequence converging, we need a way of measuring the distance between points in the sequence. That is, we need a way of measuring the distance between compact sets.

Definition 10.13. *Let $B \in \mathcal{H}$ and $x \in \mathbb{R}^n$. Then*

$$d(x, B) = min\{d(x,y)|y \in B\}$$

Lemma 10.1. *The function $d(x, B)$ is well defined. That is, there always exists a minimum value for $d(x, y)$ where y is any point in B.*

Proof: Let $f(y) = d(x, y)$. Since the distance function is continuous (by definition!), we have that f is a continuous function on a compact set and thus must achieve a minimum and maximum value. (For the proof of this extremal property of continuous function on a compact set, see [4, page 31].) □

Definition 10.14. *Let $A, B \in \mathcal{H}$. Then*

$$d(A, B) = max\{d(x, B)|x \in A\}$$

The value of $d(A, B)$ will be well defined by a similar continuity argument to the one given in the proof of the last lemma.

As an example, let A be a square and B a triangle as shown at the right. Given $x \in A$, it is clear that $d(x, B) = d(x, y_1)$, where y_1 is on the left edge of triangle B. Then $d(A, B) = d(x_1, y_1)$, where x_1 is on the left edge of the square. On the other hand, $d(B, A) = d(y_2, x_2) \neq d(A, B)$.

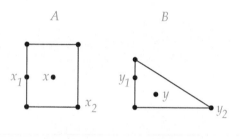

Thus, in general $d(A, B)$ need not equal $d(B, A)$. Since one of the defining conditions for a distance function (or *metric*) is that it be symmetric, we need to modify the definition a bit.

Definition 10.15. *The* Hausdorff distance *between A and B in $\mathcal{H}$ is given by*

$$d_{\mathcal{H}}(A, B) = max\{d(A, B), d(B, A)\}$$

The Hausdorff distance function satisfies all of the requirements for a metric. That is,

1. $d_{\mathcal{H}}(A, B) = d_{\mathcal{H}}(B, A)$

2. $d_{\mathcal{H}}(A, A) = 0$

3. $d_{\mathcal{H}}(A, B) > 0$ if $A \neq B$

4. $d_{\mathcal{H}}(A,B) \leq d_{\mathcal{H}}(A,C) + d_{\mathcal{H}}(C,B)$ for A, B, C in $\mathcal{H}$. This is called the *Triangle Inequality*.

Condition 1 is true by the way we defined the Hausdorff distance. Conditions 2 and 3 are left as exercises.

> **Lemma 10.2.** *The Hausdorff distance function satisfies the Triangle Inequality.*

Proof: Let a, b, and c be points in A, B, and C, respectively. Using the Triangle Inequality for Euclidean distance, we have that $d(a,b) \leq d(a,c) + d(c,b)$. Thus,

$$
\begin{aligned}
d(a,B) &= min\{d(a,b)|b \in B\} \\
&\leq min\{d(a,c) + d(c,b)|b \in B\} \; for \; all \; c \in C \\
&\leq d(a,c) + min\{d(c,b)|b \in B\} \; for \; all \; c \in C \\
&\leq min\{d(a,c)|c \in C\} + min\{min\{d(c,b)|b \in B\}|c \in C\} \\
&\leq d(a,C) + max\{min\{d(c,b)|b \in B\}|c \in C\} \\
&\leq d(a,C) + d(C,B)
\end{aligned}
$$

And so,

$$
\begin{aligned}
d(A,B) &= max\{d(a,B)|a \in A\} \\
&\leq max\{d(a,C) + d(C,B)|a \in A\} \\
&\leq d(A,C) + d(C,B)
\end{aligned}
$$

Likewise, $d(B,A) \leq d(B,C) + d(C,A)$. So,

$$
\begin{aligned}
d_{\mathcal{H}}(A,B) &= max\{d(A,B),d(B,A)\} \\
&\leq max\{d(A,C) + d(C,B),d(B,C) + d(C,A)\} \\
&\leq max\{d(A,C),d(C,A)\} + max\{d(B,C),d(C,B)\} \\
&\leq d_{\mathcal{H}}(A,C) + d_{\mathcal{H}}(C,B)
\end{aligned}
$$

□

We are now in a position to prove that contraction mappings on $\mathbb{R}^n$ generate contraction mappings on $\mathcal{H}$.

Theorem 10.3. *Let S be a contraction mapping on $\mathbb{R}^n$, with ratio $c < 1$. Then S is a contraction mapping on compact sets in $\mathcal{H}$.*

Proof: First, since a contraction map is continuous (proved as an exercise), then if D is a non-empty compact set, $S(D)$ must also be a non-empty compact set. So, S is a well-defined mapping from $\mathcal{H}$ to itself.

Now, for A and B in $\mathcal{H}$:

$$
\begin{aligned}
d(S(A), S(B)) &= max\{min\{d(S(x), S(y))|y \in B\}|x \in A\} \\
&\leq max\{min\{c\,d(x, y)|y \in B\}|x \in A\} \\
&\leq c\,max\{min\{d(x, y)|y \in B\}|x \in A\} \\
&\leq c\,d(A, B)
\end{aligned}
$$

Likewise, $d(S(B), S(A)) \leq c\,d(B, A)$, and so

$$
\begin{aligned}
d_{\mathcal{H}}(S(A), S(B)) &= max\{d(S(A), S(B)), d(S(B), S(A))\} \\
&\leq max\{c\,d(A, B), c\,d(B, A)\} \\
&\leq c\,d_{\mathcal{H}}(A, B)
\end{aligned}
$$

□

Theorem 10.4. *Let $S_1, S_2, \ldots, S_n$ be contraction mappings in $\mathbb{R}^n$ with ratios $c_1, c_2, \ldots, c_n$. Define a transformation S on $\mathcal{H}$ by*

$$
S(D) = \bigcup_{i=1}^{n} S_i(D)
$$

Then, S is a contraction mapping on $\mathcal{H}$, with contraction ratio $c = max\{c_i|i = 1, \ldots, n\}$, and S has a unique fixed set $F \in \mathcal{H}$ given by

$$
F = \lim_{k \to \infty} S^k(E)
$$

for any non-empty $E \in \mathcal{H}$.

To prove this theorem, we will use the following lemma.

Lemma 10.5. *Let A, B, C, and D be elements of $\mathcal{H}$. Then*

$$d_{\mathcal{H}}(A \cup B, C \cup D) \leq max\{d_{\mathcal{H}}(A, C), d_{\mathcal{H}}(B, D)\}$$

Proof: First, $d(A \cup B, C) = max\{d(A, C), d(B, C)\}$ (proved as an exercise). Second, $d(A, C \cup D) \leq d(A, C)$ and $d(A, C \cup D) \leq d(A, D)$ (proved as an exercise). Thus,

$$
\begin{aligned}
d_{\mathcal{H}}(A \cup B, C \cup D) &= max\{d(A \cup B, C \cup D),\ d(C \cup D, A \cup B)\} \\
&= max\{max\{d(A, C \cup D),\ d(B, C \cup D)\}, \\
&\qquad max\{d(C, A \cup B),\ d(D, A \cup B)\}\} \\
&\leq max\{max\{d(A, C), d(B, D)\}, \\
&\qquad max\{d(C, A), d(D, B)\}\} \\
&\leq max\{max\{d(A, C), d(C, A)\}, \\
&\qquad max\{d(B, D), d(D, B)\}\} \\
&\leq max\{d_{\mathcal{H}}(A, C),\ d_{\mathcal{H}}(B, D)\}
\end{aligned}
$$

$\square$

Now for the proof of Theorem 10.4. We will prove the result in the case where $n = 2$. Let A and B be in $\mathcal{H}$. Then, by the previous lemma, we have

$$
\begin{aligned}
d_{\mathcal{H}}(S(A), S(B)) &= d_{\mathcal{H}}(S_1(A) \cup S_2(A),\ S_1(B) \cup S_2(B)) \\
&\leq max\{d_{\mathcal{H}}(S_1(A), S_1(B)),\ d_{\mathcal{H}}(S_2(A), S_2(B))\} \\
&\leq max\{c_1\, d_{\mathcal{H}}(A, B),\ c_2\, d_{\mathcal{H}}(A, B)\} \\
&\leq c\, d_{\mathcal{H}}(A, B)
\end{aligned}
$$

Now let E be a non-empty set in $\mathcal{H}$. Then assuming $m \leq n$, we have

$$d_{\mathcal{H}}(S^m(E), S^n(E)) \leq c d_{\mathcal{H}}(S^{m-1}(E), S^{n-1}(E))$$

$$.$$
$$.$$
$$.$$

$$\leq c^{n-m} d_{\mathcal{H}}(E, S^{n-m}(E))$$

Also, by the Triangle Inequality, we have

$$
\begin{aligned}
d_{\mathcal{H}}(E, S^k(E)) &\leq d_{\mathcal{H}}(E, S(E)) + d_{\mathcal{H}}(S(E), S^2(E)) + \cdots \\
&\quad + d_{\mathcal{H}}(S^{k-1}(E), S^k(E)) \\
&\leq (1 + c + c^2 + \cdots + c^{k-1})\, d_{\mathcal{H}}(E, S(E)) \\
&\leq \left(\frac{c^k}{1-c}\right) d_{\mathcal{H}}(E, S(E)) \\
&\leq \left(\frac{1}{1-c}\right) d_{\mathcal{H}}(E, S(E))
\end{aligned}
$$

Thus,

$$
d_{\mathcal{H}}(S^m(E), S^n(E)) \leq \left(\frac{c^{n-m}}{1-c}\right) d_{\mathcal{H}}(E, S(E))
$$

Since $d_{\mathcal{H}}(E, S(E))$ is fixed and $0 \leq c < 1$, we can make the term $d_{\mathcal{H}}(S^m(E), S^n(E))$ as small as we want. A sequence having this property is called a *Cauchy sequence*. It is a fact from real analysis that Cauchy sequences converge in $\mathbb{R}^n$. Thus, this sequence must converge to some set F in $\mathcal{H}$.

If F' were another compact fixed set of S, then the sequence $\{S^k(F')\}$ must converge to F, which means the distance from F to F' goes to zero, and so $F = F'$. $\square$

Definition 10.16. *The set F that is the limit set of a system of contraction mappings S_k on $\mathbb{R}^n$ is called the* attractor *of the system. A system of contraction mappings that is iterated on a compact set is called an* iterated function system *or* IFS.

Note how Theorem 10.4 confirms our conjecture arising from the Cantor Set construction. This theorem guarantees that the Cantor Set is actually the attractor for the IFS consisting of the two contractions S_1 and S_2 that were used in the set's construction.

For another example, let's return to the Sierpinski Triangle, defined on an initial triangle $\triangle ABC$. This figure is fixed under three contractions: scaling by $\frac{1}{2}$ toward A, scaling by $\frac{1}{2}$ toward B, and scaling by $\frac{1}{2}$ toward C.

Suppose that points A, B, and C are at positions $(0,0)$, $(0,1)$, and $(1,1)$. Then the three contractions are defined as $S_1(x, y) = \frac{1}{2}(x, y)$, $S_2(x, y) = \frac{1}{2}(x, y) + (0, \frac{1}{2})$, and $S_3(x, y) = \frac{1}{2}(x, y) + (\frac{1}{2}, \frac{1}{2})$.

Suppose we iterate these three contractions on the unit square shown at the right.

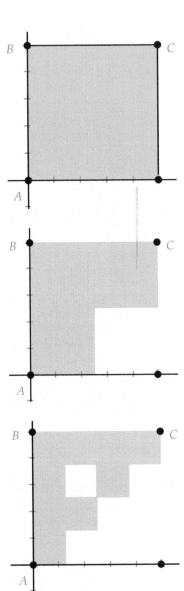

After one iteration of the three contractions, we would have three sub-squares as shown.

After two iterations, we would have nine sub-squares.

After one more iteration, we begin to see the Sierpinski Triangle take shape, as this is the attractor of the set of three contractions.

10.6 FRACTAL DIMENSION

Recall our definition of the similarity dimension of a self-similar fractal as the value of

$$d = \frac{log(N)}{log(\frac{1}{r})}$$

where $r = r(N)$ was the similarity ratio, the ratio by which a ruler measuring a side of an object at stage i of the construction would change under the assumption that N sub-objects can be scaled to exactly cover the object at stage i.

This definition assumes that the scaling of sub-parts of the fractal *exactly* matches the form of the fractal itself, the scaling factor is uniform throughout the fractal, and there is only one such scaling factor.

It would be nice to have a definition of fractal dimension that does not suffer from all of these constraints. In this section we will expand the similarity dimension concept to a more general covering-scaling concept of dimension.

First, we need to explicitly define the notion of a *covering set*.

Definition 10.17. *Let A be a compact set. Let $B(x, \epsilon) = \{y \in A | d(x, y) \leq \epsilon\}$. That is, $B(x, \epsilon)$ is a closed ball centered at x of radius ϵ. Then $\bigcup_{n=1}^{M} B(x_n, \epsilon)$, with $x_n \in A$, is an ϵ-covering of A if for every $x \in A$, we have $x \in B(x_n, \epsilon)$ for some n.*

Here we have a $\frac{1}{2}$ covering of the unit square. Note that this covering uses eight disks (2-dimensional balls) to cover the square.

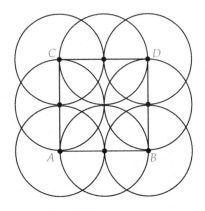

Clearly, there must be some minimal number of circles of radius $\frac{1}{2}$ that will cover the unit square. This minimal covering number will be used in our definition of fractal dimension.

Definition 10.18. *Let A be a compact set. Then the* minimal ϵ covering number *for A is*

$$\mathcal{N}(A, \epsilon) = min\{M | A \subset \bigcup_{n=1}^{M} B(x_n, \epsilon)\}$$

for some set of points $\{x_n\}$.

We note that the value of $\mathcal{N}(A, \epsilon)$ is well defined, as every set can be covered by some collection of *open* ϵ-balls. Since A is compact, then there is a finite sub-collection that still covers A, and the closure of this collection of open balls will also cover A. Thus, A has at least one finite ϵ-covering and, thus, must have one with the fewest number of elements.

Intuitively, as ϵ changes, the minimum number of balls needed to cover A should also change. For a line segment, the two should be directly related: if we decrease ϵ by a factor of $\frac{1}{2}$, the number of segments (1-dimensional balls) should go up by $2 = 2^1$. If A is a rectangle and we decrease ϵ by a factor of $\frac{1}{2}$, the number of disks should go up by $4 = 2^2$. For a cube, the increase in the number of balls should be $8 = 2^3$.

This was precisely the relationship we noticed when defining the similarity dimension, so it makes sense to define the fractal dimension to mirror the definition of similarity dimension.

Definition 10.19. *Let A be a compact set. Then if*

$$D_F = \lim_{\epsilon \to 0} \frac{log(\mathcal{N}(A, \epsilon))}{log(\frac{1}{\epsilon})}$$

exists, we call D_F the fractal dimension of A.

Using this definition, it is not hard to prove the following result.

Theorem 10.6. *(Box-Counting Theorem) Let A be compact. Cover $\mathbb{R}^n$ by just-touching square boxes of side-length $\frac{1}{2^n}$. Let $\mathcal{N}_n(A)$ be the number of boxes that intersect A. Then*

$$D_F(A) = lim_{n \to \infty} \frac{log(\mathcal{N}_n(A))}{log(2^n)}$$

The proof amounts to showing that the boxes in a covering of A can be trapped between two sequences of ϵ balls that both converge to the fractal dimension. The proof can be found in [4].

Finally, we have the following simplification of the calculation of fractal dimension in the case of an *iterated function system* (IFS).

Theorem 10.7. *(IFS Fractal Dimension Theorem) Let $\{S_n\}_{n=1}^M$ be an IFS and let A be its attractor. Assume that each S_n is a transformation with scale ratio $0 < c_n < 1$, and assume that, at each stage of its construction, portions of the fractal meet only at boundary points. Then the fractal dimension is the unique number $D = D_F(A)$ such that*

$$\sum_{n=1}^M c_n^D = 1$$

Again, the proof can be found in [4]. For example, for the Cantor Set, at each stage, the fractal consists of completely distinct segments, which do not intersect at all. Thus, we have that $\left(\frac{1}{3}\right)^D + \left(\frac{1}{3}\right)^D = 1$, and so $D = \frac{log(2)}{log(3)}$.

This theorem would also apply to Sierpinski's Triangle, as portions of the fractal at each stage in the construction are triangles that meet each other only at boundary segments.

Exercise 10.6.1. *Show that a contraction mapping must be a continuous function on its domain.*

Exercise 10.6.2. *Give an example of two compact subsets, A and B of $\mathbb{R}^2$, with $d(A, B) = d(B, A)$.*

Exercise 10.6.3. *Prove that conditions 2 and 3 in the list of properties for a metric hold for the Hausdorff distance function.*

Exercise 10.6.4. *Prove the first statement in the proof of Lemma 10.5. That is, show that $d(A \cup B, C) = max\{d(A, C), d(B, C)\}$.*

Exercise 10.6.5. *Prove the second statement in the proof of Lemma 10.5. That is, show that $d(A, C \cup D) \leq d(A, C)$ and $d(A, C \cup D) \leq d(A, D)$.*

Exercise 10.6.6. *Use the construction for Sierpinski's Triangle based on the points $A = (0, 0)$, $B = (0, 1)$, and $C = (1, 1)$ to show that $\mathcal{N}_1(S) = 3$, $\mathcal{N}_2(S) = 3^2$, $\mathcal{N}_3(S) = 3^3$, and so on, where S is Sierpinski's Triangle. Use this to find the fractal dimension. How does the fractal dimension compare to the similarity dimension?*

Exercise 10.6.7. *Use the IFS Fractal Dimension Theorem to compute the fractal dimension of Sierpinski's Triangle.*

Exercise 10.6.8. *Show that the IFS consisting of $S_1(x) = \frac{1}{2}x$ and $S_2(x) = \frac{2}{3}x + \frac{1}{3}$ does not meet the overlapping criterion of Theorem 10.7. Nevertheless, by identifying the attractor for this IFS, show that its fractal dimension can be calculated and is equal to 1.*

10.7 PROJECT 17 - IFS FERNS

We saw in the last few sections that a compact set A can be realized as the *attractor* of an iterated function system (IFS), a system of contraction mappings whose fixed set is A.

It should not be too surprising, then, that natural shapes can arise as the attractors of IFS systems. In this project we will look at an especially pretty attractor—the 2-dimensional outline of a fern.

Figure 10.5 Fern leaf

Consider the fern shown in Figure 10.5. Each leaf is made up of sub-leaves similar to the original leaf, and each sub-leaf has sub-sub-leaves similar to the sub-leaves, and so on.

How can we find the contraction mappings that lead to the fern? Michael Barnsley in [4] proved that to find an IFS whose attractor approximates a given shape F, one only needs to find *affine mappings* that when applied to F yield a union, or *collage* of shapes that approximate F. Barnsley called this the *Collage Theorem*.

> **Definition 10.20.** *An* affine mapping S *on the plane is a function of the form*
>
> $$S(x, y) = (ax + by + e, cx + dy + f) \qquad (10.4)$$
>
> *where a, b, c, d, e, and f are real constants.*

Affine mappings are generalizations of the planar isometries we studied in Chapter 5. For example, if $a = \cos(\theta)$, $b = -\sin(\theta)$, $c = \sin(\theta)$, $d = \cos(\theta)$, and $e = f = 0$, then S will be a rotation of angle θ about the origin. Note that affine mappings are not necessarily invertible. If they are invertible, we will call them *affine transformations*.

In this project we will discover a set of affine mappings that will split the fern into a collage of sub-ferns that can be reassembled into the original shape of the fern. We will use dynamic geometry software to define this set of affine mapppings. In order to accurately define the affine maps, we will need to use an electronic version of the fern image. Go to http://www.gac.edu/~hvidsten/geom-text and click on the link labeled "Projects." Then, click on "IFS Ferns." On this page, you can download the file "fern.jpg."

Start up your geometry software and paste the fern image into the main window. Create point A at the very tip of the fern, C at the second branch point, and B at the base, as shown at right.

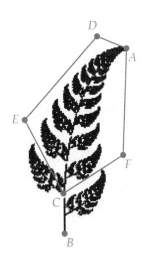

To split the fern into a collage of similar pieces, we will start with the set of branches totally enclosed in the polygon $ADECF$. Clearly, this sub-fern of branches is similar to the entire fern, if we scale the sub-fern by an appropriate scaling factor.

To calculate the scaling factor, we will measure the ratio of AC to AB. This should match the scaling factor of the sub-fern to the original fern. Measure the distance from A to C and the distance from A to B.

Then, compute the ratio of these two distances (Figure 10.6). Note that your distances may vary from the ones shown. This is okay —we are interested only in the *ratio* of these distances.

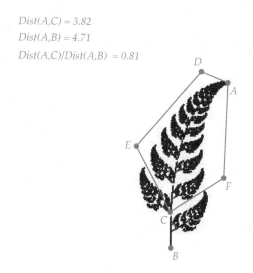

Dist(A,C) = 3.82
Dist(A,B) = 4.71
Dist(A,C)/Dist(A,B) = 0.81

Figure 10.6

It appears that the ratio of the sub-fern to the entire fern is about 0.8. This is the scaling factor we sought. However, the sub-fern is not just a scaled-down version of the bigger fern. After scaling the big fern by a scale factor of 0.8, we need to turn it slightly, about 5 degrees clockwise, and translate it upward by the length of $\overline{BC}$ to exactly match the outline of the big fern. Measure the length of $\overline{BC}$. (Again, your measurement may vary from the value shown.)

The affine mapping that will take the bigger fern to the smaller fern is then the composition of a rotation by 5 degrees, a scaling by 0.8, and a translation by the length of $\overline{BC}$, which we will denote by h. All of these functions are invertible, so the composition will be a transformation of the plane.

Exercise 10.7.1. *Let T_1 be the transformation taking the entire fern to the sub-fern enclosed by $ADECF$. Show that*

$$T_1(x, y) \approx \begin{bmatrix} 0.8 & 0.07 \\ -0.07 & 0.8 \end{bmatrix} \begin{pmatrix} x \\ y \end{pmatrix} + \begin{pmatrix} 0.0 \\ h \end{pmatrix}$$

We now have an affine transformation, T_1, that when applied to the fern will cover a major portion of the fern itself. But, we are still missing the lower two branches and the trunk.

Consider the lower branches and in particular the one enclosed by the polygon $CHGJ$. Measure appropriate distances to verify that the scaling factor for this sub-fern is about 0.3.

Exercise 10.7.2. *Let T_2 be the transformation taking the entire fern to the sub-fern enclosed by $CHGJ$. The angle of rotation for T_2 appears to be about 50 degrees. Show that*

$$T_2(x,y) \approx \begin{bmatrix} 0.19 & -0.23 \\ 0.23 & 0.19 \end{bmatrix} \begin{pmatrix} x \\ y \end{pmatrix} + \begin{pmatrix} 0.0 \\ h \end{pmatrix}$$

Exercise 10.7.3. *Let T_3 be the transformation taking the entire fern to the bottom right sub-fern. Show that the scaling factor for T_3 is about 0.3, the angle of rotation for T_3 is about -60 degrees, and T_3 includes a reflection about the y-axis. Use this information to show that*

$$T_3(x,y) \approx \begin{bmatrix} -0.15 & 0.26 \\ 0.26 & 0.15 \end{bmatrix} \begin{pmatrix} x \\ y \end{pmatrix} + \begin{pmatrix} 0.0 \\ \frac{h}{2} \end{pmatrix}$$

Using the affine transformations T_1, T_2, and T_3, we can *almost* cover the original fern with copies of itself. The only piece of the fern missing from this collage is the small piece of trunk between B and C. Since this is about $\frac{1}{5}$ of the height of the fern, we can just squash the fern down into a vertical line of length 0.2. The affine mapping that will accomplish this is

$$T_4(x,y) \approx \begin{bmatrix} 0.0 & 0.0 \\ 0.0 & 0.2 \end{bmatrix} \begin{pmatrix} x \\ y \end{pmatrix} + \begin{pmatrix} 0.0 \\ 0.0 \end{pmatrix}$$

We note here that T_4, while a valid affine mapping, is not a transformation. It is not one-to-one and, thus, not invertible.

We have now completed our splitting of the fern into a collage of four sub-pieces. According to the Collage Theorem, an IFS consisting of T_1, T_2, T_3, and T_4 should have as its attractor a shape that approximates the original fern. Let's see if that is true.

Close the window containing the fern. We will use a pre-built file to do the rest of this project. Go to http://www.gac.edu/~hvidsten/geom-text and click on the link labeled "Projects." Then, click on "IFS Ferns" to get to the website for this project. Download the file labeled "IFS" for your particular geometry software. Open this file in your geometry program to continue this project.

The affine mappings we will use in creating our fern have already been defined in this file. Each affine mapping has the general form

$$T(x, y) = \begin{bmatrix} a & b \\ c & d \end{bmatrix} \begin{pmatrix} x \\ y \end{pmatrix} + \begin{pmatrix} e \\ f \end{pmatrix}$$

For example, the transformation T_1 defined earlier has $a = 0.8$, $b = 0.07$, $c = -0.07$, $d = 0.8$, $e = 0.0$, and $f = 1.0$. (We are using $f = h = 1.0$ from the earlier measurements.)

At this point we have four affine mappings that we can apply to arbitrary geometric objects. We will apply these mappings to the point A that is in the center of your window.

Click the "Iterate" button in your window. A series of points will be created by successfully iterating each of the four affine mappings on point A, yielding a new set of points. On each of these points, all four mappings will again be carried out, and so on. The iteration is successively done four times.

The shape that appears is *something* like the fern. One problem you might have is that your points may be too large. Reducing the point size can give a more pleasing picture.

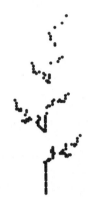

This type of iteration can create a lot of points. We started with one point A, the first iteration produced four new points. The second produced 16 points, the third 64, and the fourth 256. This exponential growth can quickly exhaust most geometry programs, so we will stop with the computer simulation at this point. If we did continue the iteration process, we would start to see the fern take shape.

Exercise 10.7.4. *The affine maps T_1, T_2, and T_3 are valid transformations (i.e., are invertible), and thus the IFS Fractal Dimension Theorem (Theorem 10.7) applies to the shape of the fern minus its trunk. Use the contraction scale factors for T_1, T_2, and T_3 to estimate the fractal dimension of the portion of the fern that does not include the main trunk.*

Exercise 10.7.5. *Find the four affine maps that will create the shape at right. You may assume that shape is bounded by the unit square (0 ≤ x ≤ 1 and 0 ≤ y ≤ 1).*

In this project we have seen how a small number of affine maps can represent or *encode* the shape of a complex natural object. In theory, any complex object can be similarly encoded once the affine maps for that object are discovered. This has important implications for the transmission of complex data, such as images over a computer network. If we can encode each image with just a small number of affine maps, we can greatly reduce the transmission time for images and video. The Collage Theorem guarantees that we can get arbitrarily close to the original image by a good choice of affine maps. Once we have these affine maps, the reproduction of the original image is completely straightforward and fast, using modern computers. The major draw-back to an encoding scheme using IFS systems is the discovery of which affine maps will produce a good approximation to an image.

10.8 ALGORITHMIC GEOMETRY

In our discussion of self-similar fractals, we saw that their construction required a looping process, whereby each level of the construction was built from specific rules using the results of previous levels. This created a *recursive* sequence of constructions and transformations to produce the fractal.

In the last section we saw how we could use a system of affine maps, an IFS system, to generate a fractal. Transformations are applied recursively to an initial object, with the fractal appearing as the *attracting set* of the IFS system.

In both cases, fractals are created using a set of *instructions* which generate a recursive procedure. Such a set of instructions is called an *algorithm*. Since the construction of many fractals requires hundreds or even thousands of calculations, fractal algorithms are most often carried out by computers.

In the construction of self-similar fractals, we used the computer as a powerful bookkeeping device to record the steps whereby edges were replaced by template curves. For IFS systems, the computer carried out the numerous iterations of a set of affine maps, leading to a fractal attractor appearing as out of a mist. In both of these fractal construction algorithms, we use the computer to make the abstract ideas of self-similarity or attracting sets a concrete reality.

10.8.1 Turtle Geometry

The notion of utilizing computing technology to make abstract mathematical ideas more "real" was the guiding principle of Seymour Papert's work on turtle geometry and the programming language *LOGO*. In his book *Mindstorms*, Papert describes turtle geometry as the "tracings made on a display screen by a computer-controlled turtle whose movements can be described by suitable computer programs" [34, page xiv].

Papert's work has inspired thousands of teachers and children to use turtle geometry in the classroom as a means of exploring geometry (and computer programming) in a way that is very accessible to young (and old) students.

Turtle geometry has also proved to be an ideal way in which to explore fractal shapes. To see how fractals can be constructed using turtle geometry, we first have to create a scheme for controlling the turtle on screen.

We will direct the behavior of the turtle with the following set of commands.

1. f Forward: The turtle moves forward a specified distance without drawing.

2. | Back: The turtle moves backward a specified distance without drawing.

3. F Draw Forward: The turtle moves forward and draws as it moves.

4. + Turn Left: The turtle rotates counterclockwise through a specified angle.

5. – Turn Right: The turtle rotates clockwise through a specified angle.

6. [Push: The current state of the turtle is pushed onto a state stack.

7.] Pop: The state of the turtle is set to the state on top of a state stack.

As an example, suppose we have specified that the turtle move 1 unit and turn at an angle of 90 degrees. Also, suppose the initial heading of the turtle is vertical. Then the set of symbols

$$F + F + F + F$$

can be considered a *program* that will generate a square of side-length 1 when interpreted, or carried out, by the turtle.

Here we have labeled the edges drawn by the turtle as a, b, c, and d, in the order that the turtle created them.

In Figure 10.7 we have an example illustrating the use of the turtle-state stack. The turtle was given the program $F[+FF + FF + FF]+$. After the first edge is drawn (edge a), the position and heading of the turtle (direction it is pointing) are stored as a *turtle-state*. This turtle-state is placed on the top of a virtual *stack* of possible turtle-states. Then the turtle interprets the symbols $+FF + FF + FF$, ending with edge g. At the symbol] the turtle "pops" the stored turtle-state off the stack and the turtle resets itself to that position and heading.

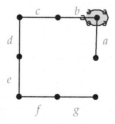

Figure 10.7

Let's see how we can use this set of symbols to represent the stages of construction of the Koch snowflake curve.

Recall that the Koch snowflake curve is a fractal that is constructed by beginning with an initial segment. This segment is then replaced by a template curve made up of four segments as shown at the right. The angles inside the peak are 60 degrees, making the triangle formed by the peak an equilateral triangle (Figure 10.8).

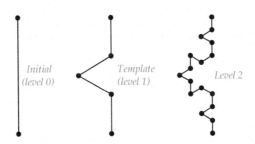

Figure 10.8

The initial segment is the Koch curve at level 0. The template is the Koch curve at level 1. If we replace each of the segments in the template with a copy of the template at a reduced scale, we get the Koch curve at level 2, and so on.

To model the Koch snowflake curve using our set of symbols, we could say that the initial segment is a Draw Forward, or the symbol F. The template is (forgetting for now the problem of scaling) a Draw Forward followed by a Turn Left of 60 degrees, then a Draw Forward followed by two Turn Rights of 60 degrees, then a Draw Forward followed by a Turn Left of 60 degrees, and finally another Draw Forward. The symbol set that describes the template is thus $F + F - -F + F$ (assuming our turns are always 60 degrees).

Thus, the Koch curve is defined by an initial symbol F and a template symbol set $F + F - -F + F$ that governs how the initial segment is replaced. At level 0 the Koch curve is F. At level 1 we replace F by $F + F - -F + F$. At level 2 we replace all segments of level 1 by the template. This is equivalent to replacing all occurrences of the symbol F in the level 1 symbol set by the template set $F + F - -F + F$. The level 2 symbol set is then $F + F - -F + F + F + F - -F + F - -F + F - -F + F + F + F - -F + F$.

10.9 GRAMMARS AND PRODUCTIONS

The set of turtle symbols described in the last section was used in the classic work by Prusinkiewicz and Lindenmayer, titled *The Algorithmic Beauty of Plants* [29]. This beautiful book describes how one can use turtle geometry to model plants and plant growth. Their method of describing natural fractal shapes has been called *Lindenmayer Systems*, or *L-systems*, in honor of Astrid Lindenmayer, who first pioneered the notion of using symbols to model plant growth.

Lindemayer's novel idea was to model plant structures through a *grammar rewriting* system. We can think of a set of symbols as a *word* in a grammar built from those symbols. The symbol set $F + F + F + F$ is a word in the grammar built on turtle command symbols.

In the Koch curve example, the curve is *grammatically* defined by an initial word F and a template word $F + F - -F + F$. At each level we *rewrite* the previous level's word by substituting in for each occurrence of the symbol F.

The Koch curve can be completely described by two words: the starter word F and template word $F + F - -F + F$. The Koch curve is then the limiting curve one gets by successively rewriting an *infinite* number of times, having the turtle interpret the final word, which is theoretically possible but physically impossible.

The formal definition of an *L-system* is as follows.

Definition 10.21. *A* Lindenmayer system *or* L-system *consists of*

- *a finite set Σ of symbols*

- *a set Ω of words over Σ*

- *a finite set P of* production rules *or* rewrite rules, *of the form $\sigma - > \omega$, where $\sigma \in \Sigma$ and $\omega \in \Omega$*

- *a symbol S in Σ that is called the* start *symbol or the* axiom *of the system*

Lindenmayer Systems are special types of *formal grammars*. A formal grammar is used in computer science to describe a formal language, a set of strings made up of symbols from an alphabet. Formal grammars are used to express the syntax of programming languages such as Pascal or

Java. These languages can be completely expressed as a set of production rules over a set of symbols and words.

For example, suppose $\Sigma = \{a, b, S\}$, and $P = \{S-> aSb, S-> ba\}$. That is, there are two production rules:

1. $S-> aSb$

2. $S-> ba$

Beginning with the start symbol S, we can rewrite using production rule 1 to get the new word aSb. Then using rule 2 on this word, we can rewrite to get $abab$. At this point, we can no longer rewrite, as there is no longer a start symbol in our word. We have reached a *terminal* word in the grammar. In fact, it is not hard to see that the set of all *producible* words (i.e., words that are the result of repeatedly rewriting the start symbol) consists of a subset of all strings containing an equal number of a's and b's. We only get a subset of such strings because the string $bbaa$, for example, is not producible. We will call the set of producible strings the *language* of the grammar.

10.9.1 Space-Filling Curves

As another example, let's consider a grammar that classifies the Draw Forward segments of a turtle into two groups: left edges, which we will denote by F_l, and right edges, which we will denote by F_r.

By artificially creating such a classification, we can control which instances of the symbol F in a word will be replaced under a production. Those edges labeled F_l will be replaced by a production rule for F_l, and those labeled F_r will be replaced by a production rule for F_r. When the turtle interprets a word that uses these two symbols, it will interpret both as a simple Draw Forward.

To see how this works, let's consider the problem of filling up a unit square by a path passing through the points in the square.

Here we have subdivided a square into 25 sub-squares. Look carefully at the path weaving through the square. It touches each of the corners of the internal sub-squares, will hit one or the other of the corners of the squares on the left or right sides, and will also do this on the top and bottom sides. Note that this path starts at the lower left corner of the square (P) and ends at the lower right corner (Q).

Consider the edge of this path in the sub-square labeled A. If we scaled down the path in the previous figure by a factor of $\frac{1}{5}$, we could replace this edge with the scaled-down copy of the original path. The original path can be considered a *template*, similar to the strategy we used for the Koch snowflake. Suppose we try to replace each edge of the original path with the template. In the figure at right, we have replaced the edges in sub-squares A and B.

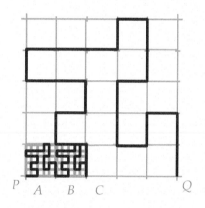

When we try to replace the edge in sub-square C, we have a problem. If we substitute our template on the left side of the edge in C, we will collide with the path we already have in B. Thus, we need to substitute to the *right* of this edge. However, our template is not oriented correctly to do this. We need to substitute a new path, one that first passes directly into square C, so that it does not "double-up" on any of the points already covered on the edge between B and C. Also, it must not "double-up" on any of the other edges it might possibly share with the previous template curve, if it is in another orientation.

Here is a picture of what the "right" template curve should be. Note that it is shown in relation to the original template curve. To see how it would appear in sub-square C, rotate the curve 90 degrees counter-clockwise. This template will not "double-up" with our old template, which we will call the "left" template. Also, if this template adjoins a left template, then all corner points on the edge where the two squares meet will be covered by one path or the other.

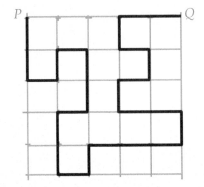

Now all we have to do to create the template replacement process is to label all edges in each of the two templates with "l" or "r" to designate which of the scaled-down templates will replace that edge. We start with the left template and label the first edge "l." Then the next edge is also "l," but the edge at C must be "r," as must be the next edge.

Continuing in this way, we get the following set of edge labels for the left template.

Note that every sub-square has a label in it. Thus, every sub-square will have a scaled-down template passing through it. Also, by the way we constructed the two templates, we are guaranteed that along an edge where two sub-squares meet, the new paths will not cross each other or intersect. A simple way to keep track of the orientations of the sub-squares is that an edge labeled "l" will have the left template constructed to the left of the edge, and vice versa for an edge labeled "r."

The two production rules for the left and right templates are, thus,

$$F_l- > F_lF_l + F_r + F_r - F_l - F_l + F_r + F_rF_l - F_r - F_lF_lF_r +$$
$$F_l - F_r - F_lF_l - F_r + F_lF_r + F_r + F_l - F_l - F_rF_r +$$
$$F_r- > -F_lF_l + F_r + F_r - F_l - F_lF_r - F_l + F_rF_r + F_l + F_r -$$
$$F_lF_rF_r + F_l + F_rF_l - F_l - F_r + F_r + F_l - F_l - F_rF_r$$

Note that for the left template, we need to add a final turn $(+)$ to ensure that any new path starts in the same direction as the original.

Now all we need is an initial starting path. Since it is customary to have the turtle heading vertically at the start, we will turn the turtle so that it heads to the right initially. Thus, the path $-F_l$ will be our starting path.

If we don't want to use subscripts for the two types of F symbols, we can instead use the following production rules with the starter word $-Fl$.

$$l- > lFl + rF + rF - Fl - Fl + rF + rFFl - rF - FlFlrF +$$
$$Fl - rF - FlFl - rF + FlrF + rF + Fl - Fl - rFrF +$$
$$r- > -FlFl + rF + rF - Fl - FlrF - Fl + rFrF + Fl + rF -$$
$$FlrFrF + Fl + rFFl - Fl - rF + rF + Fl - Fl - rFr$$

Note that we give productions only for how the symbols l and r are replaced. Thus, the initial F for the left production and the final F for the right production must be omitted from the replacement.

This set of productions will generate an equivalent rewriting system, if we interpret Fl as F_l and rF as F_r. The turtle, when interpreting a level 2 rewrite of this system, will generate the image in Figure 10.9. We can start to see how this curve will fill up the space in the square, as the rewriting level increases without bound.

Figure 10.9 Space-filling curve

Exercise 10.9.1. *Let an L-system be defined by* $\Sigma = \{a, b, 1, S\}$, *and* $P = \{S- > aSb, S- > 1\}$, *with start symbol* S. *Prove that this grammar generates the language* $\{a^n b^n | n \geq 0\}$.

Exercise 10.9.2. *Let an L-system be defined by* $\Sigma = \{F, +, -, S\}$, *and* $P = \{S- > F - F - F - F, F- > F - F + F + FF - F - F + F\}$, *with start symbol* S. *This system, when interpreted by a turtle, will generate a self-similar fractal called the* Quadratic Koch Island, *as first described in [30, page 50]. Sketch levels 1 (rewrite S once) and 2 (rewrite twice) of this curve (assume the turn angle is 90 degrees), and sketch the template for the fractal. Find the similarity dimension using the template.*

Exercise 10.9.3. *Let an L-system be defined by* $\Sigma = \{F, +, -, L, R\}$, *and* $P = \{L- > +RF - LFL - FR+, R- > -LF + RFR + FL-\}$, *with start symbol* L. *This curve is called the* Hilbert *curve. Find the level 1 and 2 rewrite words for this system and sketch them on a piece of paper using turtle geometry. Do you think this system will generate a space-filling curve?*

Exercise 10.9.4. *In our space-filling curve example, show that the right template is a simple rotation of the left template. On a 7×7 grid find a left template curve with edges labeled r and l, as we did earlier, such that all squares are visited by some edge of the curve. Also, create the curve so that a 180 degree rotation about the center of the grid will produce a right template curve with the same "doubling-up" properties as we had for the pair of curves in our example. That is, if the right and left templates would meet at an edge, then no edge of the left would intersect an edge of the right, except at the start and end of the paths.*

10.10 PROJECT 18 - WORDS INTO PLANTS

What makes many plants fractal-like is their branching structure. A branch of a tree often looks somewhat like the tree itself, and a branch's sub-branch system looks like the branch, and so forth. To model the development of branching structures, we will use the grammar rewriting ideas of the last section, plus the push and pop features of turtle geometry. This will be necessary to efficiently carry out the instructions for building a branch and then returning to the point where the branch is attached.

The grammar we will use consists of the turtle symbols described in the last section plus one new symbol "X." We can think of X as being a *virtual* node of the plant that we are creating. Initially, the start symbol for our grammar will be just the symbol X, signifying the potential growth of the plant.

For example, here is a very simplified branching system for a plant.

X

How can we represent this branching structure using our grammar rewriting system? It is clear that the plant grew in such a way that three new branch nodes were created from the original potential node X, which we represent here as a point. Thus, the start symbol X must be replaced with three new X's. Also, the branches were created at an angle to the main branch, so there needs to be some turning by the turtle. Finally, each new branch has length of two Draw Forward's, if we consider one Draw Forward to be the distance between the points on a branch. Which production rules will represent this plant? Consider the following set of productions:

$$X \;\; \to \;\; F[+X]F[-X] + X$$
$$F \;\; \to \;\; FF$$

The first production replaces a node X by the word $F[+X]F[-X] +$ X. Thus, the replacement will produce a "stalk" of length 1, a new branch node turned to the left that is *independent* of the previous symbol F, due to the push and pop symbols, a second length of main stalk, a second independent branch node coming off the stalk at an angle to the right, and finally a third branch coming off the top at an angle to the left.

The second production says that a length of stalk will grow twice as long in the next generation. Putting this all together, we have in these two productions a blueprint for the growth of the plant.

Exercise 10.10.1. *How many times was the start symbol X rewritten, using the production rules, to generate the preceding plant image?*

To generate this branching structure, we will make use of the turtle geometry capabilities of the software program *Geometry Explorer*. If you are not already using this program, go to http://www.gac.edu/~hvidsten/geom-text for instructions on how to download and use this software.

To begin, we need to define a turn angle and a heading vector for our turtle. Start *Geometry Explorer* and create segments $\overline{AB}$, $\overline{BC}$, and $\overline{DE}$ as shown.

Select A, B, and C (in that order) and choose **Turtle Turn Angle** (**Turtle** menu) to define the turn angle for our turtle. Then, multi-select D and E and choose **Turtle Heading Vector** (**Turtle** menu) to define the turtle's initial direction of motion.

Now, create a point X on the screen. Select X and choose **Create Turtle at Point** (**Turtle** menu) to create a turtle based at X.

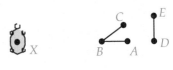

A dialog box will pop up to control the movement of the turtle. Click on the tab labeled "Grammar Turtle." In this window we can type in the start symbol (axiom) and a list of productions. Type in "X" for the axiom and then type in the two productions as shown, using "=" to designate the left and right sides of the rule. Type in "1" in the Rewrite Level box and hit the Rewrite button. The computer will rewrite the axiom, using the two production (replacement) rules.

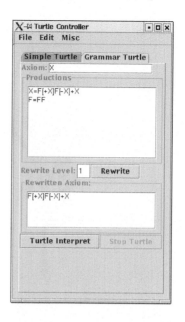

Now, click on the "Turtle Interpret" button. The turtle will carry out the commands in the rewritten word.

We do not see any actual branches yet, as the three new X's in the rewritten word are *potential* branches. Undo the turtle back to its start position by typing "Ctrl-U" repeatedly. Then, in the grammar window type in "2" for the rewrite level and hit "Rewrite." Click on "Turtle Interpret" again to see the plant starting to take shape.

The figure drawn by the turtle is clearly a branched structure, but is not really much like a plant. To more fully develop the branching pattern, we need to rewrite the axiom to a higher level.

Undo the turtle back to its start position and move E close to D, so that the turtle moves only a short distance each time it changes position. Change the rewrite level to 4 in the Turtle Controller and hit Return. Note how the new sentence has expanded.

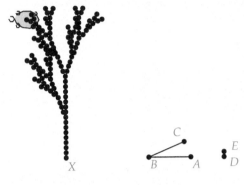

Figure 10.10

Click on "Turtle Interpret" and watch the turtle interpret the rewritten sentence (Figure 10.10). This may take awhile—anywhere from a few seconds to several minutes, depending on the speed of your computer. There are a tremendous number of actions that the turtle needs to carry out as it interprets the sentence.

You will be able to tell when the turtle is done by the state of the "Stop Turtle" button. If the turtle is still drawing, the button will be active. Once the turtle completes drawing, the button will become inactive.

The image is very "blotchy" with all of the points visible that were drawn by the turtle. Let's hide all of these points by choosing **Hide All** (**View** menu) and then choosing the **Points** submenu. Now the figure looks like the bushy branch structure of a plant (minus the leaves).

Exercise 10.10.2. *Find a set of two productions that will generate the branching pattern shown here, from an initial start symbol of X.*

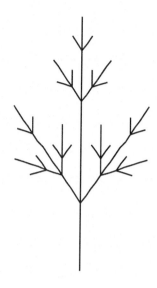

Exercise 10.10.3. *Design an L-system that will have a thick, bushy branching system.*

The analysis of complex branching patterns is quite an interesting subject, one for which we have only scratched the surface. For more information on this subject, consult the Prusinkiewicz and Lindenmayer text [29].

It is interesting to note how the use of grammar rewriting systems mirrors our discussion of axiomatic systems from Chapter 1. The start symbol and production rules can be thought of as abstract axioms or postulates. The words of the language produced by the grammar are akin to *theorems* in the axiomatic system, as they are generated using the axioms of the system. It seems that our investigation of geometry has returned full circle to where it began with the Greeks and their axiomatic system of reasoning.

A Primer on Proofs

A mathematical proof of a theorem is a logical series of statements that begin with the hypotheses of the theorem and finish with the conclusion. There is no simple recipe for writing a *good* mathematical proof. If there were, mathematics would be a very constrained and machine-like subject. In this appendix we will review a few ideas to help make the mysterious art of writing proofs less mysterious.

A.1 AXIOMATIC SYSTEMS AND PROOFS

All of the statements of a proof must be based within an axiomatic system. Statements used within a proof can consist of axioms, definitions, previously proven theorems, and general rules of logic.

The fundamental characteristics of an axiomatic system were covered in detail in Chapter 1. An axiomatic system is built from a base of concepts and statements that are used *without* proof. These are the *undefined terms* and *postulates/axioms* of the system.

Starting from a base of undefined terms and agreed upon axioms, we define other terms and use our axioms to argue the truth of other statements. These other statements are called the *theorems* of the system. An axiomatic system consists of four components:

1. Undefined Terms

2. Axioms (or Postulates)

3. Defined Terms

4. Theorems

We gave several examples of simple axiomatic systems in Chapter 1.

Proving theorems in these simple systems is good practice in that there are only a few axioms upon which to base a proof. The difficulty in developing a proof comes when we are working in more complex axiomatic systems. The variety and complexity of statements, and the numerous possible ways that one can combine previous results, can make the writing of a good mathematical proof seem like a mysterious and daunting task.

A primary difficulty students face when constructing proofs is the question of what can be assumed. There are two general ways that mathematicians approach this question.

In the first case, a complete and rigorous axiomatic foundation is laid down, with no gaps, to develop a body of theorems that can lead to a given proof. In the case of Euclidean geometry, Euclid's original five axioms were not sufficient to this task. As mentioned in the chapter on Euclidean Geometry, David Hilbert was able to devise a more rigorous set of axioms for which one can do a careful development of classical Euclidean geometry. If you wish to follow this development, consult the on-line chapter on Hilbert's geometry which can be found on the author's website. Proofs of theorems, based on first principles of Hilbert's axioms, rely on the proofs of *many* other theorems. Texts that follow this approach invariably take several chapters to prove rather elementary results.

The second approach is to briefly summarize the axiomatic basis for the system in which you are working, and then review the fundamental theorems in that system without proving the theorems. For example, in Chapter 2 we quickly reviewed the basic results of Euclidean geometry, without proof, and then went on to prove more complex theorems. In this approach, we treated standard theorems (such as the theorem that the base angles of an isosceles triangle are congruent) as fundamental results without supplying rigorous proofs. You might say that we treated these theorems like axioms. This malleability in the foundations of mathematics is an aspect of most mathematics textbooks that is very confusing to students. The complaint is that it is not clear what can be assumed and what needs to be proven. This is a very valid complaint!

In this text, the author has tried to make it clear when he is taking foundational "short-cuts" and has provided on-line resources for those students (and instructors) who wish to see all the fine details of the complete axiomatic basis for a proof. The trade-off is foundational detail versus broad coverage.

The first rule of developing a proof is to determine what can be used (assumed to be true) when constructing a proof. The second rule is to determine the type (or structure) of proof that is called for in a given setting.

A.2 DIRECT PROOFS

Most (not all) theorems are statements of the form "If .... then ...." That is, they are *conditional* statements. The part of the statement between the "if" and the "then" is called the *hypothesis*. The part after the "then" is called the *conclusion*.

Sometimes, it is not obvious that a theorem statement has a hypothesis and a conclusion. Consider the statement "The sum of the interior angles of a triangle is 180 degrees." This is not a statement of the form "If .... then ...." However, the statement does imply the existence of a triangle for which we consider an angle sum. The hypothesis is that a triangle exists. The conclusion is that the sum of its angles is 180 degrees. So, we could write the theorem statement as "If we are given a triangle, then the sum of its interior angles is 180 degrees."

Once we have identified the hypothesis and conclusion of a theorem statement, we then develop the proof of the theorem. For example, consider the statement "If two lines intersect at a point, the vertical angles created are congruent." This is Euclid's Proposition 15 of Book I of *Elements*. The hypothesis of the statement is that two lines intersect at a point. The hypothesis of a theorem statement is *assumed* to be true. To prove a theorem, one has to show that the conclusion is true. In our example, one would have to prove that the angles made by the two lines are congruent.

A *direct proof* is a proof that starts with the hypothesis as a given statement (no need to prove) and proceeds by a series of statements (axioms, definitions, or previously proved theorems), and general rules of logic, until one reaches the conclusion.

To illustrate how one can construct a direct proof under the foundational approach used in this text, let's consider one of the homework problems from Chapter 2. We will restate this exercise as a theorem:

Theorem A.1. *If a triangle has two angles congruent, then it must be isosceles.*

Before we start to write down any steps in the proof, we must first make sure we understand all of the terms in the statement. This is a critical step that is sometimes overlooked. The terms in the theorem statement are "triangle," "angle," "congruent," and "isosceles." Understanding these terms will help us understand how to organize our proof. The conclusion of the theorem statement is that the triangle is isosceles. What does this mean? By definition, an isosceles triangle is one where two sides are congruent. Thus, we need to show that the triangle has two sides that are congruent.

The hypothesis of the theorem is that the triangle has two angles congruent. As mentioned above, this can be assumed to be true when we write our proof. So, let's consider what we are given in relation to what we are to prove.

1. Given: A triangle with two angles congruent.

2. To Prove: The triangle has two congruent sides

Now that we clearly understand the hypothesis and conclusion, we consider what prior results we can use to prove the theorem. As mentioned in the previous section, this depends on the context in which we are working. In the context of exercise 2.2.7, we can use any result that comes in the preceding two sections of Chapter 2. The section in which this exercise is located is the section on Congruent Triangles and Pasch's Axiom. The hypothesis and conclusion deal with congruent sides and angles of a triangle. This would suggest that one of the triangle congruence theorems would be useful. However, the triangle congruence theorems deal with *pairs* of triangles. So, we have to think about how we can modify the given triangle so we can work with two triangles.

Up to this point, our discussion on how to think about proving a theorem has been fairly algorithmic. We first made sure we understood all the terms in the given theorem statement. Then, we identified the hypothesis and conclusion, and considered the possible prior results we could utilize.

The next step in the development of a proof is more creative. We have to play with possibilities and "massage" what we are given so that we can reach the conclusion. In our example, we need to modify a single triangle so that we can make two triangles. But, when we create the two triangles, we need to make sure that some triangle congruence theorem can apply to the new triangles.

When exploring ideas to use in a proof, it is important to label significant parts of figures so that we can refer to them without ambiguity. In our case, we have a triangle. Let's say its vertices are the points A,

B, and C. It is given that two angles are congruent, so we assume that $\angle BAC \cong \angle BCA$. Why can we just assume this? The fact that two angles are congruent does not impose restrictions on which of our labeled angles are congruent. So, we just pick two to be the congruent ones.

One way to create two triangles from a single triangle would be to construct some interior point to a side.

If we construct point D on $\overline{AB}$, then we have two triangles ADC and BDC. The problem with constructing such a point is that it messes up the given angle congruence, as we no longer have any idea about the congruence of the angles created from $\angle BCA$.

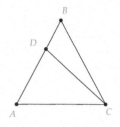

To preserve the given angle congruence, we have to construct a point on $\overline{AC}$. To be able to use triangle congruence theorems, the new point should be a midpoint.

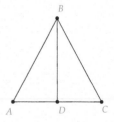

Now, we have two triangles ADB and CDB with two angles congruent and two pairs of sides congruent ($\overline{AD} \cong \overline{CD}$ and $\overline{BD}$ to itself). But, there is no triangle congruence of the form SSA, so this line of reasoning will not result in a valid proof.

However, this figure suggests a different way to split the triangle in two. Let $\overrightarrow{BE}$ be the angle bisector of the angle at B. By Pasch's axiom, this will intersect side $\overline{AC}$ at some point D. Then, we will have AAS congruence for the two triangles!

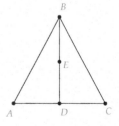

The discussion above is an illustration of the experimentation that needs to be done when working on a proof. We started with a general idea: take the given triangle and create two sub-triangles so that a triangle congruence theorem could be used. We then explored different ways to accomplish this goal.

Once we are assured that our proof plan will work, we then can write up a careful description of the proof as follows:

Proof: Let triangle $\triangle ABC$ have angles $\angle BAC$ and $\angle BCA$ congruent. Let $\overrightarrow{BE}$ be the bisector of $\angle ABC$. By Pasch's axiom, this ray will intersect $\overline{AC}$ at some point D. Then, by AAS we have that triangles $\triangle ABD$ and $\triangle CBD$ are congruent. Thus $\overline{AB} \cong \overline{CB}$. □

We summarize the important factors in writing a direct proof as follows:

1. Understood all of the terms in the given theorem statement. Check definitions.

2. Identify the hypothesis and conclusion of the theorem statement.

3. Identify the axiomatic context in which the theorem occurs. Identify valid prior results that can be used without proof.

4. Label significant parts of geometric figures. Draw some figures to aid in proof, but be careful not to reason from your figures!

5. Make a plan on how to move from hypothesis to conclusion. Use prior results to fill in the transition from hypothesis to conclusion.

6. After carefully going through above steps, write down the proof in correct logical order, using complete sentences and correct terminology.

A.3 PROOFS BY CONTRADICTION

While direct proofs make up the bulk of the proofs in this text, the second most widely used proof technique is that of proof by contradiction.

Suppose the statement of a theorem is of the form "If .... then ....". As mentioned earlier, the part of the theorem statement between the "if" and the "then" is called the *hypothesis* and the part after the "then" is called the *conclusion*. Let us symbolize the hypothesis statement by the symbol "p" and the conclusion by "q." A theorem statement can then be represented as "If p then q."

A basic logical assumption that we make when proving any theorem is that the statement we are trying to prove is either true or false, but not both. If q is the statement we are trying to prove, we assume that q and its logical negation (which we symbolize by $\sim q$) cannot simultaneously be true.

In a proof by contradiction, we start by assuming the logical negation of the conclusion we are trying to prove. That is, we assume $\sim q$ is true. Then, suppose we can develop a series of statements that follow from this assumption, ultimately resulting in a statement s that is the opposite of a statement r that we know to be true. That is, we are able to show that r and $\sim r$ are simultaneously true. Since this is logically impossible (a statement cannot be both true and false) then the assumption that $\sim q$ is true must be incorrect. That is, q must be true.

As an example, let's consider exercise 2.2.9 from Chapter 2. Let's state the first part of the exercise as a theorem.

Theorem A.2. *If two sides of a triangle are not congruent, then the angles opposite those sides are not congruent*

Let's consider how we would prove this result by the method of contradiction. The conclusion of the theorem statement is that two angles are not congruent. The logical negation of the conclusion is that the two angles *are* congruent. So, a proof by contradiction would start with this assumption, that the angles opposite the non-congruent sides of a triangle are congruent. Starting from this assumption we see what other statements this assumption would imply, with the idea in the back of our minds that we want to reach a contradictory statement to something we know is true.

Recall that in the previous section, we proved the following:

Theorem A.3. *If a triangle has two angles congruent, then it must be isosceles.*

We can use this in our current proof. The assumption that we have two angles in a triangle being congruent then implies that the triangle is isosceles. But, then the sides are congruent. This directly contradicts the hypothesis we were given in the theorem: that the two sides are *not* congruent. This contradiction then implies that our assumption about congruent angles was incorrect and thus the angles are not congruent.

We summarize the important factors in writing a proof using contradiction as follows:

1. All of the important factors used in a direct proof apply (see the previous section).

2. Identify the hypothesis (p) and conclusion (q) of the theorem statement.

3. We suppose that the logical negation of the conclusion ($\sim q$) is true.

4. We derive a statement that contradicts the hypothesis (p) or some other theorem or axiom.

5. The contradiction then implies that the conclusion (q) is true.

A.4 CONVERSE STATEMENTS

When we discussed direct proofs, we used as an example the statement found in exercise 2.2.7. This statement is the *converse* statement to Theorem 2.14:

Theorem A.4. *(Theorem 2.14) In an isosceles triangle, the two base angles are congruent.*

We restated exercise 2.2.7 as a theorem:

Theorem A.5. *If a triangle has two angles congruent, then it must be isosceles.*

Note that these two theorem statements are closely related. In fact the second statement switches the role of hypothesis and conclusion from the first.

Definition A.1. *Given a theorem statement "If p then q," where p is the hypothesis and q is the conclusion, the converse statement is the statement "If q then p." That is, the converse reverses the roles of hypothesis and conclusion.*

The hypothesis of Theorem 2.14 is that we have an isosceles triangle and the conclusion is that the base angles are congruent. The hypothesis of the converse statement is that we have a triangle with two angles congruent. Note that we do not use the term "base angles." This is because this term is defined for an isosceles triangle and we cannot assume the triangle in the hypothesis of the converse statement is isosceles. (That would be assuming the conclusion!)

The steps used in proving the converse of a statement are as follows:

1. Identify the hypothesis p and q of the original statement.

2. Write the converse statement with hypothesis (q) and conclusion (p), taking care to not assume anything from p when writing the converse hypothesis q.

3. Use the techniques of direct proof or proof by contradiction to find the proof of the converse statement.

A.5 IF AND ONLY IF STATEMENTS

Consider the statement of Theorem 2.36 in Chapter 2:

Theorem A.6. *(Theorem 2.36) Given a circle c with center O and radius $\overline{OT}$, a line l is tangent to c at T if and only if l is perpendicular to $\overleftrightarrow{OT}$ at T.*

Definition A.2. *The truth of a statement of the form "p if and only if q" is equivalent to the simultaneous truth of "If p then q" and the converse "If q then p."*

If a theorem statement is of the form "p if and only if q," then the hypothesis and conclusion follow from each other —they are essentially equivalent statements.

How does one prove an if and only if statement? The proof is essentially two different proofs. One starts by proving the statement "If p then q" followed by a proof of the statement "If q then p."

In the example of Theorem 2.36, we see that the first part of the theorem statement "Given a circle c with center O and radius $\overline{OT}$" can be assumed for either proof, as nothing is asserted in the statement other than there exists a circle. Thus, the "If p then q" statement can be stated as "If a line l is tangent to c at T then l is perpendicular to $\overleftrightarrow{OT}$ at T," and the converse as "If line l is perpendicular to $\overleftrightarrow{OT}$ at T then l is tangent to c at T."

If one checks the proof of Theorem 2.36, one can see that there are two sub-proofs clearly laid out, one for the statement "If a line l is tangent to c at T then l is perpendicular to $\overleftrightarrow{OT}$ at T" and the other for the statement "If line l is perpendicular to $\overleftrightarrow{OT}$ at T then l is tangent to c at T." The first sub-proof is a proof by contradiction and the second is a direct proof.

The steps used in proving an "if and only if" statement are as follows:

1. Identify the two statements p and q of the theorem statement.

2. Prove "If p then q".

3. Prove "If q then p".

A.6 PROOFS AND WRITING

One can think of a good proof as providing a solid logical argument, but there is also an aesthetic quality to a well-written proof. Proofs can be perfectly correct in their logical details, but till seem ugly or boring. A good proof is like telling a story —there is a beginning, middle, and end, and the writer needs to hold the reader's interest through all three parts. The following are a few hints at writing a "pretty" proof:

1. A proof should have a clearly defined beginning, middle, and end. There should be a direct line that logically moves from the hypothesis to the conclusion, without sidetracks.

2. A proof should be organized into coherent sentences and paragraphs. Each paragraph should relate to one main idea.

3. Figures and equations should relate to the proof and should flow with the text of the proof.

4. Notation and vocabulary must be used correctly.

5. A proof should be written in the first person plural. For example, "We assume that triangle ABC is isosceles." This may seem strange, but it is convention for mathematical proofs.

6. A proof should use correct English grammar and punctuation.

Book I of Euclid's *Elements*

B.1 DEFINITIONS

1. A *point* is that which has no part.

2. A *line* is breadthless length.

3. The extremities of a line are points.

4. A *straight line* is a line that lies evenly with the points on itself.

5. A *surface* is that which has length and breadth only.

6. The extremities of a surface are lines.

7. A *plane surface* is a surface that lies evenly with the straight lines on itself.

8. A *plane angle* is the inclination to one another of two lines in a plane that meet one another and do not lie in a straight line;

9. And when the lines containing the angle are straight, the angle is called *rectilinear*.

10. When a straight line set up on a straight line makes the adjacent angles equal to one another, each of the equal angles is *right*, and the straight standing on the other is called a *perpendicular* to that on which it stands.

11. An *obtuse angle* is an angle greater than a right angle.

12. An *acute angle* is an angle less than a right angle.

13. A *boundary* is that which is an extremity of anything.

14. A *figure* is that which is contained by any boundary or boundaries.

15. A *circle* is a plane figure contained by one line such that all the straight lines falling upon it from one point among those lying within the figure are equal to one another;

16. And the point is called the *center* of the circle.

17. A *diameter* of the circle is any straight line drawn through the center and terminated in both directions by the circumference of the circle, and such a straight line also bisects the circle.

18. A *semicircle* is the figure contained by the diameter and the circumference cut off by it. And the center of the semicircle is the same as that of the circle.

19. *Rectilinear figures* are those that are contained by straight lines, *trilateral* figures being those contained by three, *quadrilateral* those contained by four, and *multilateral* those contained by more than four straight lines.

20. Of trilateral figures, an *equilateral triangle* is that which has its three sides equal, an *isosceles triangle* that which has two of its sides equal, and a *scalene triangle* that which has its three sides unequal.

21. Further, of trilateral figures, a *right-angled triangle* is that which has a right angle, an *obtuse-angled triangle* that which has an obtuse angle, and an *acute-angled triangle* that which has its three angles acute.

22. Of quadrilateral figures, a *square* is that which is both equilateral and right-angled; an *oblong* that which is right-angled but not equilateral; a *rhombus* that which is equilateral but not right-angled; and a *rhomboid* that which has its opposite sides and angles equal to one another but is neither equilateral nor right-angled. And let quadrilaterals other than these be called *trapezia*.

23. *Parallel* straight lines are straight lines that, being in the same plane and being produced indefinitely in both directions, do not meet one another in either direction.

B.2 THE POSTULATES (AXIOMS)

1. To draw a straight line from any point to any point.

2. To produce a finite straight line continuously in a straight line.

3. To describe a circle with any center and distance.

4. That all right angles are equal to one another.

5. That, if a straight line falling on two straight lines make the interior angles on the same side less than two right angles, the two straight lines, if produced indefinitely, meet on that side on which are the angles less than two right angles.

B.3 COMMON NOTIONS

1. Things that are equal to the same thing are also equal to one another.

2. If equals be added to equals, the wholes are equal.

3. If equals be subtracted from equals, the remainders are equal.

4. Things that coincide with one another are equal to one another.

5. The whole is greater than the part.

B.4 PROPOSITIONS (THEOREMS)

I-1 On a given straight line to construct an equilateral triangle.

I-2 To place at a given point (as an extremity) a straight line equal to a given straight line. [Given a length and a point, we can construct a segment of that length from the point.]

I-3 Given two unequal straight lines, to cut off from the greater a straight line equal to the less. [Given two segments of different lengths, we can cut off from the larger a segment equal to the smaller.]

I-4 If two triangles have the two sides equal to two sides, respectively, and have the angles contained by the equal straight lines equal, they will also have the base equal to the base, the triangle will be equal to the triangle, and the remaining angles will be equal to the remaining angles, respectively, namely those that the equal sides subtend. [SAS Congruence]

I-5 In isosceles triangles the angles at the base are equal to one another, and if the equal straight lines be produced further, the angles under the base will be equal to one another.

I-6 If in a triangle two angles be equal to one another, the sides that subtend the equal angles will also be equal to one another.

I-7 Given two straight lines constructed on a straight line (from its extremities) and meeting in a point, there cannot be constructed on the same straight line (from its extremities) and on the side of it, two other straight lines meeting in another point and equal to the former two, respectively, namely each to that which has the extremity with it. [Given a base length and two other lengths, there is only one triangle possible on a particular side of the base.]

I-8 If two triangles have the sides equal to two sides, respectively, and have also the base equal to the base, they will also have the angles equal that are contained by the equal straight lines. [SSS Congruence]

I-9 To bisect a given rectilinear angle.

I-10 To bisect a given finite straight line.

I-11 To draw a straight line at right angles to a given straight line from a given point on it. [Given a line and a point on the line, we can construct a perpendicular to the line at the point.]

I-12 To a given infinite straight line, from a given point that is not on it, to draw a perpendicular straight line. [Given a line and a point not on the line, we can construct a perpendicular to the line through the point.]

I-13 If a straight line set up on a straight line makes angles, it will make either two right angles or angles equal to two right angles. [Supplementary angles add to 180.]

I-14 If with any straight line, and at a point on it, two straight lines not lying on the same side make the adjacent angles equal to two right angles, the two straight lines will be in a straight line with one another. [Given two angles that share a line as a common side, if the angles add to 180, then the non-shared sides of the two angles must be coincident on a line.]

I-15 If two straight lines cut one another, they make the vertical angles equal to one another. [Vertical Angle Theorem]

I-16 In any triangle, if one of the sides be produced, the exterior angle is greater than either of the interior and opposite angles. [Exterior Angle Theorem]

I-17 In any triangle two angles taken together in any manner are less than two right angles.

I-18 In any triangle the greater side subtends the greater angle. [In a triangle the larger side is opposite the larger angle.]

I-19 In any triangle the greater angle is subtended by the greater side. [In a triangle the larger angle is opposite the larger side.]

I-20 In any triangle two sides taken together in any manner are greater than the remaining one. [Triangle Inequality]

I-21 If on one of the sides of a triangle, from its extremities, there be constructed two straight lines meeting within the triangle, the straight lines so constructed will be less than the remaining two sides of the triangle, but will contain a greater angle. [Given triangle ABC, if we construct triangle DBC with D inside ABC, then $DB + DC < AB + AC$, and the angle at D will be greater than the angle at A.]

I-22 Out of three straight lines, which are equal to three given straight lines, to construct a triangle: thus it is necessary that two of the straight lines taken together in any manner should be greater than the remaining one. [To construct a triangle from three lengths, it is necessary that when you add any pair of lengths, the sum is greater than the other length.]

I-23 On a given straight line and at a point on it, to construct a rectilinear angle equal to a given rectilinear angle. [Angles can be copied.]

I-24 If two triangles have the two sides equal to two sides, respectively, but have one of the angles contained by the equal sides greater than the other, they will also have the base greater than the base.

I-25 If two triangles have the two sides equal to two sides, respectively, but have the base greater than the base, they will also have one of the angles contained by the equal straight lines greater than the other.

I-26 If two triangles have the two angles equal to two angles, respectively, and one side equal to one side, namely, either the side adjoining the equal angles, or that subtending one of the equal angles, they will also have the remaining sides equal to the remaining sides and the remaining angle to the remaining angle. [AAS and ASA Congruence]

I-27 If a straight line falling on two straight lines makes the alternate angles equal to one another, the straight lines will be parallel to one another. [Given two lines cut by a third, if the alternate interior angles are congruent then the lines are parallel.]

I-28 If a straight line falling on two straight lines makes the exterior angle equal to the interior and opposite angle on the same side, or the interior angles on the same side equal to two right angles, the straight lines will be parallel to one another. [Given two lines cut by a third, if the exterior and opposite interior angles on the same side of the cutting line are congruent, the lines are parallel. Or, if the interior angles on the same side add to 180, the lines are parallel.]

> **The first 28 propositions listed are independent of Euclid's fifth postulate. They are called *neutral* propositions. Proposition 29 is the first proposition where Euclid explicitly requires the fifth postulate to carry out the proof.**

I-29 A straight line falling on parallel lines makes the alternate angles equal to one another, the exterior angle equal to the interior and opposite angle, and the interior angles on the same side equal to two right angles.

I-30 Straight lines parallel to the same straight line are parallel to one another.

I-31 Through a given point, to draw a straight line parallel to a given straight line. [Given a line and a point not on the line, we can construct a parallel to the line through the point.]

I-32 In any triangle, if one of the sides be produced, the exterior angle is equal to the two opposite and interior angles, and the three interior angles of the triangle are equal to two right angles. [The sum of the angles of a triangle is 180 degrees. The exterior angle is equal to the sum of the opposite interior angles.]

I-33 The straight lines joining equal and parallel straight lines (at the extremities that are) in the same directions (respectively) are themselves equal and parallel. [Given a quadrilateral $ABCD$ with $\overline{AB}$ congruent and parallel to $\overline{CD}$, then $\overline{AD}$ must be congruent and parallel to $\overline{BC}$.]

I-34 In parallelogrammic areas the opposite sides and angles are equal to one another, and the diameter bisects the areas. [Given parallelogram $ABCD$, both pairs of opposite sides are congruent, and both pairs of opposite angles are congruent. Also, the diagonals split the parallelogram into two equal parts.]

I-35 Parallelograms that are on the same base and in the same parallels are equal to one another. [Given two parallelograms $ABCD$ and $ABEF$ with C, D, E, F collinear, then the parallelograms have the same area.]

I-36 Parallelograms that are on equal bases and in the same parallels are equal to one another. [Given two parallelograms $ABCD$ and $EFGH$, with $\overline{AB}$ congruent to $\overline{EF}$, A, B, E, F collinear, and C, D, G, H collinear, then the parallelograms have the same area.]

I-37 Triangles that are on the same base and in the same parallels are equal to one another. [Given two triangles ABC and ABD, with $\overline{CD}$ parallel to $\overline{AB}$, then the triangles have the same area.]

I-38 Triangles that are on equal bases and in the same parallels are equal to one another. [Given two triangles ABC and DEF, with $\overline{AB}$ congruent to $\overline{DE}$, A, B, D, E collinear, and $\overline{CF}$ parallel to $\overline{AB}$, then the triangles have the same area.]

I-39 Equal triangles that are on the same base and on the same side are also in the same parallels. [Given two triangles ABC and ABD

having the same area and on the same side of $\overline{AB}$, then $\overline{CD}$ must be parallel to $\overline{AB}$.]

I-40 Equal triangles that are on equal bases and on the same side are also in the same parallels. [Given two triangles ABC and DEF having the same area, with $\overline{AB}$ congruent to $\overline{DE}$, A, B, D, E collinear, and the two triangles being on the same side of $\overline{AB}$, then $\overline{CF}$ must be parallel to $\overline{AB}$.]

I-41 If a parallelogram has the same base with a triangle and is in the same parallels, the parallelogram is double that of the triangle. [The area of a triangle is half that of a parallelogram with the same base and height.]

I-42 To construct, in a given rectilinear angle, a parallelogram equal to a given triangle. [It is possible to construct a parallelogram with a given angle having the same area as a given triangle.]

I-43 In any parallelogram the complements of the parallelograms about the diameter are equal to one another.

[Given parallelogram $ABCD$ and its diameter $\overline{AC}$, let K be a point on the diameter. Draw parallels $\overline{EF}$ to $\overline{BC}$ and $\overline{GH}$ to $\overline{AB}$, both through K. Then, complements $EKBG$ and $HKFD$ have the same area.]

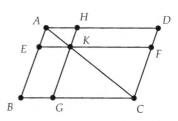

I-44 To a given straight line, to apply in a given rectilinear angle a parallelogram equal to a given triangle. [It is possible to construct a parallelogram on a given segment with a given angle having the same area as a given triangle.]

I-45 To construct in a given rectilinear angle a parallelogram equal to a given rectilinear figure. [It is possible to construct a parallelogram with a given angle having the same area as a given polygon.]

I-46 On a given straight line to describe a square. [It is possible to construct a square on a given segment.]

I-47 In right-angled triangles the square on the side subtending the right angle is equal to the squares on the sides containing the right angle. [Pythagorean Theorem]

I-48 If in a triangle the square on one of the sides be equal to the squares on the remaining two sides of the triangle, the angle contained by the remaining two sides of the triangle is right. [Converse to the Pythagorean Theorem]

Birkhoff's Axioms for Euclidean Geometry

Undefined terms are *point* and *line*, as well as two real-valued functions—a *distance* function, $d(A, B)$, which takes two points and returns a non-negative real number, and an *angle* function, $m(\angle A, O, B)$, which takes an ordered triple of points ($\{A, O, B\}$ with $A \neq O$ and $B \neq O$) and returns a number between 0 and 2π. The point O is called the *vertex* of the angle.

C.1 AXIOMS

The Ruler Postulate The points of any line can be put into one-to-one correspondence with the real numbers x so that if x_A corresponds to A and x_B corresponds to B, then $|x_A - x_B| = d(A, B)$ for all points A, B on the line.

The Euclidean Postulate One and only one line contains any two distinct points.

The Protractor Postulate Given any point O, the rays emanating from O can be put into one-to-one correspondence with the set of real numbers (mod 2π) so that if a_m corresponds to ray m and a_n corresponds to ray n, and if A, B are points (other than O) on m, n, respectively, then $m(\angle NOM) = a_m - a_n$ (mod 2π). Furthermore, if the point B varies continuously along a line not containing O, then a_n varies continuously also.

> **The SAS Similarity Postulate**   If in two triangles ABC and $A'B'C'$ and for some real number $k > 0$, we have $d(A', B') = k \, d(A, B)$, $d(A', C') = k \, d(A, C)$, and $m(\angle BAC) = m(\angle B'A'C')$, then the remaining angles are pair-wise equal and $d(B', C') = k \, d(B, C)$.

Other definitions:

A point B is said to be *between* points A and C $(A \neq C)$ if $d(A, B) + d(B, C) = d(A, C)$. A *segment* $\overline{AB}$ consists of the points A and B along with all points between A and B. The *ray* with endpoint O, defined by two points O and A in line l, is the set of all points B on l such that O is not between A and B. Two distinct lines having no point in common are called *parallel*.

Two rays m, n through O form a *straight angle* if $m(\angle MON) = \pi$, where M and N are points on m, n, respectively. The rays form a *right angle* if $m(\angle MON) = \pm\frac{\pi}{2}$. If the rays form a right angle, we say the lines defining the rays are *perpendicular*.

Hilbert's Axioms for Euclidean Geometry

Undefined terms are *point, line, between, on* (or *incident*), and *congruent*.

D.1   INCIDENCE AXIOMS

I-1 Through any two distinct points A and B there is always a line m.

I-2 Through any two distinct points A and B, there is not more than one line m.

I-3 On every line there exist at least two distinct points. There exist at least three points not all on the same line.

I-4 Through any three points not on the same line, there is one and only one plane.

D.2   BETWEENESS AXIOMS

II-1 If B is a point between A and C (denoted $A * B * C$) then A, B, and C are distinct points on the same line and $C * B * A$.

II-2 For any distinct points A and C, there is at least one point B on the line through A and C such that $A * C * B$.

II-3 If A, B, and C are three points on the same line, then exactly one is between the other two.

II-4 (Pasch's Axiom) Let A, B, and C be three non-collinear points and let m be a line in the plane that does not contain any of these points. If m contains a point of segment $\overline{AB}$, then it must also contain a point of either $\overline{AC}$ or $\overline{BC}$.

D.3 CONGRUENCE AXIOMS

III-1 If A and B are distinct points and A' is any other point, then for each ray r from A' there is a unique point B' on r such that $B' \neq A'$ and $\overline{AB} \cong \overline{A'B'}$.

III-2 If $\overline{AB} \cong \overline{CD}$ and $\overline{AB} \cong \overline{EF}$ then $\overline{CD} \cong \overline{EF}$. Also, every segment is congruent to itself.

III-3 If $A * B * C$, $A' * B' * C'$, $\overline{AB} \cong \overline{A'B'}$, and $\overline{BC} \cong \overline{B'C'}$, then $\overline{AC} \cong \overline{A'C'}$.

III-4 Given $\angle ABC$ and given any ray $\overrightarrow{A'B'}$, there is a unique ray $\overrightarrow{A'C'}$ on a given side of $\overleftrightarrow{A'B'}$ such that $\angle ABC \cong \angle A'B'C'$.

III-5 If $\angle ABC \cong \angle A'B'C'$ and $\angle ABC \cong \angle A''B''C''$ then $\angle A'B'C' \cong \angle A''B''C''$. Also, every angle is congruent to itself.

III-6 Given two triangles ABC and $A'B'C'$, if $\overline{AB} \cong \overline{A'B'}$, $\overline{AC} \cong \overline{A'C'}$, and $\angle BAC \cong \angle B'A'C'$, then the two triangles are congruent.

D.4 PARALLELISM AXIOM

IV-1 Given a line l and a point P not on l, it is possible to construct one and only one line through P parallel to l.

D.5 CONTINUITY AXIOM

V-1 (Dedekind's Axiom) If the points on a line l are partitioned into two nonempty subsets Σ_1 and Σ_2 (i.e., $l = \Sigma_1 \cup \Sigma_2$) such that no point of Σ_1 is between two points of Σ_2 and vice versa, then there is a unique point O lying on l such that $P_1 * O * P_2$ if and only if one of P_1 or P_2 is in Σ_1, the other is in Σ_2, and $O \neq P_1$ or P_2.

Note: The Continuity Axiom may be logically replaced by the following two axioms:

V-1a (Archimedean Axiom) Given $\overline{AB}$ and $\overline{CD}$, there exists a number n such that if the segment $\overline{CD}$ is laid off n times on the ray $\overrightarrow{AB}$, starting from A, then a point E is reached, where $A * B * E$.

V-2b (Axiom of Completeness) The system of points on a line, with its betweeness and congruence relations, cannot be extended in such a way that the relations holding among its elements as well as the basic properties arising from axiom groups I - III and V-1a remain valid.

The 17 Wallpaper Groups

E.1 PATTERNS

Here is a listing of the wallpaper patterns for the Euclidean plane. The seventeen groups have traditionally been listed with a special notation consisting of the symbols p, c, m, g, and the integers 1, 2, 3, 4, 6. This is the crystallographic notation adopted by the International Union of Crystallography (IUC) in 1952.

In the IUC system the letter "p" stands for *primitive*. A lattice is generated from a cell that is translated to form the complete lattice. In the case of oblique, rectangular, square, and hexagonal lattices, the cell is precisely the original parallelogram formed by the lattice vectors v and w and, thus, is primitive.

In the case of the centered-rectangle lattice, the cell is a rectangle, together with an *interior* point that is on the lattice. The rectangular cell is larger than the original parallelogram and not primitive. Thus, lattice types can be divided into two classes: primitive ones designated by the letter "p," and non-primitive ones designated by the letter "c."

A reflection is symbolized by the letter "m" and a glide reflection by the letter "g."

The numbers 1, 2, 3, 4, and 6 are used to represent rotations of those orders. For example, 1 would represent a rotation of 0 degrees, while 3 would represent a rotation of 120 degrees.

The symmetries of the wallpaper group are illustrated in the second and third columns by lines and polygons. Rotations are symbolized by diamonds ($\Diamond$) for 180-degree rotations, triangles ($\triangle$) for 120-degree rotations, squares ($\square$) for 90-degree rotations, and hexagons ($\bigcirc$) for 60-degree rotations. Double lines show lines of reflection and dashed lines show the lines for glide reflections.

The symbolization used here was developed by Xah Lee. The images below are from his web page [28] and are used with permission.

Pattern and Symmetries	Basic Cell Symmetries	Generating Region and Symmetries (Non-Translates)

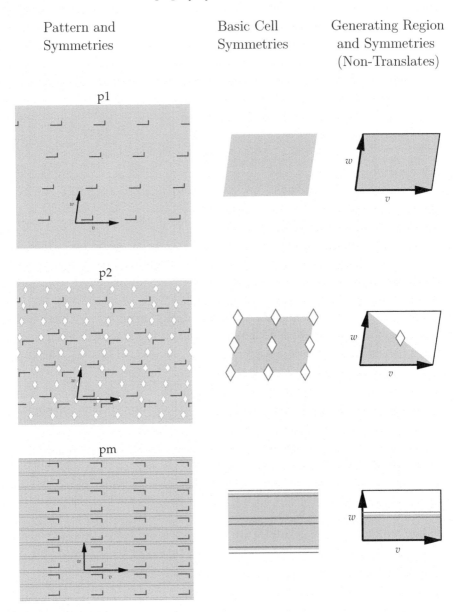

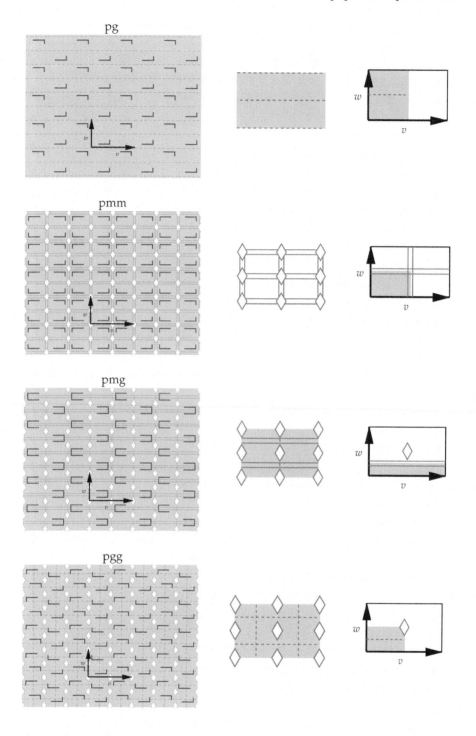

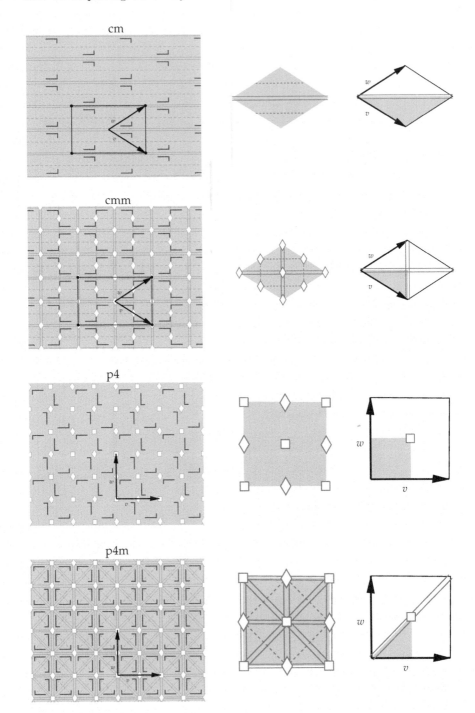

p4g

p3

p3m1

p31m

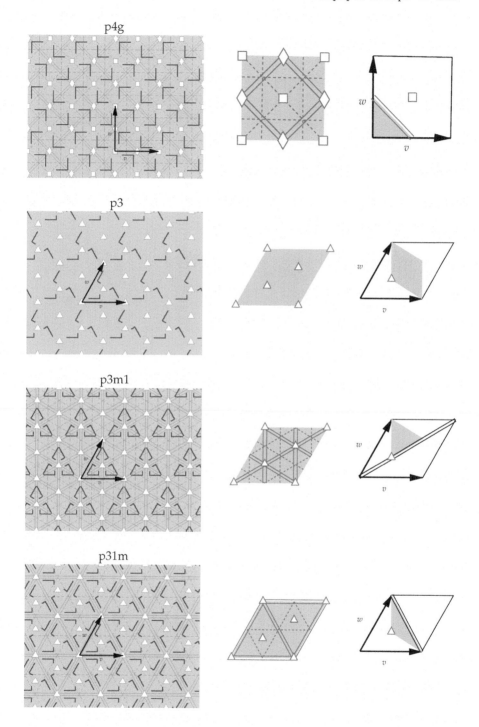

p6

p6m

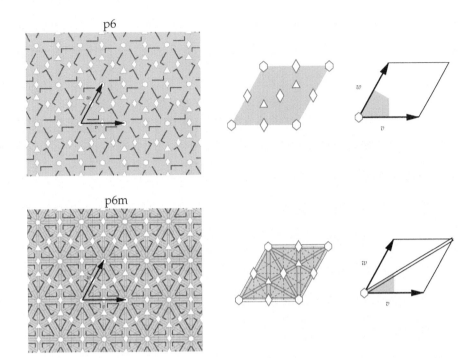

References

[1] Leon Baptista Alberti. *On Painting (Della Pittura)*. Penguin Classics, New York, New York, 1991.

[2] Roger C. Alperin. A mathematical theory of origami constructions and numbers. *The New York Journal of Mathematics*, 6:119–133, 2000.

[3] M. A. Armstrong. *Groups and Symmetry*. Springer-Verlag, New York, 1988.

[4] Michael Barnsley. *Fractals Everywhere*. Academic Press, San Diego, 1988.

[5] Tom Bassarear. *Mathematics for Elementary School Teachers: Explorations*. Houghton Mifflin Company, Boston, second edition, 2001.

[6] G. D. Birkhoff. A set of postulates for plane geometry, based on scale and protractor. *The Annals of Mathematics*, 33:329–345, April 1932.

[7] Carl B. Boyer and Uta C. Merzbach. *A History of Mathematics*. John Wiley and Sons, New York, second edition, 1968.

[8] Arthur Cayley. A sixth memoir on quantics. *Philosophical Transactions of the Royal Society of London*, 149:61–90, 1859.

[9] H. S. M. Coxeter. *Introduction to Geometry*. John Wiley and Sons, New York, second edition, 1961.

[10] H. S. M. Coxeter and S. L. Greitzer. *Geometry Revisited*. The Mathematical Association of America, Washington, D. C., 1967.

[11] Robert Dixon. *Mathographics*. Dover, New York, 1987.

[12] Howard Eves. *Great Moments in Mathematics (Before 1650)*. Mathematical Association of America, Washington, D. C., 1980.

[13] W. T. Fishback. *Projective and Euclidean Geometry*. John Wiley and Sons, Inc., New York, 1969.

[14] J. D. Foley, A. Van Dam, S. K. Feiner, and J. F. Hughes. *Computer Graphics, Principles and Practice*. Addison-Wesley, Reading, Massachusetts, 1990.

[15] David Gans. *Transformations and Geometries*. Appleton-Century-Crofts, New York, New York, 1969.

[16] Marvin Jay Greenberg. *Euclidean and Non-Euclidean Geometries*. W. H. Freeman and Co., New York, second edition, 1980.

[17] Ernst Haeckel. *Art Forms in Nature*. Prestel-Verlag, Munich, Germany, 1998.

[18] Leon Harkleroad. *The Math Behind the Music*. Cambridge University Press, New York, New York, 2006.

[19] Robin Hartshorne. *Geometry: Euclid and Beyond*. Springer-Verlag, New York, 2000.

[20] Robin Hartshorne. *Foundations of Projective Geometry*. Ishi Press International, Bronx, NY, 2009.

[21] J. L. Heilbron. *Geometry Civilized*. Clarendon Press, Oxford, UK, 1998.

[22] David Hilbert. *Foundations of Geometry*. Open Court Press, LaSalle, Illinois, 1971.

[23] David Hilbert and S. Cohn-Vossen. *Geometry and the Imagination*. Chelsea Publishing Co., New York, 1952.

[24] F. S. Hill. *Computer Graphics Using OpenGL*. Prentice-Hall, Upper Saddle River, New Jersey, second edition, 1990.

[25] H. E. Huntley. *The Divine Proportion: A Study in Mathematical Beauty*. Dover Books, New York, 1970.

[26] Humiaki Huzita. Understanding Geometry through Origami Axioms. In J. Smith, editor, *Proceedings of the First International Conference on Origami in Education and Therapy (COET91)*, pages 37–70. British Origami Society, 1992.

[27] Felix Klein. Über die sogenannte nicht-euklidische geometrie. *Mathematische Annalen*, 43:63–100, 1893.

[28] Xah Lee. The discontinuous groups of rotation and translation in the plane. Web page, 1997. http://www.xahlee.org/Wallpaper_dir/c5_17WallpaperGroups.html.

[29] Astrid Lindenmayer and Przemyslaw Prusinkiewicz. *The Algorithmic Beauty of Plants*. Springer-Verlag, New York, 1990.

[30] Benoit Mandelbrot. *The Fractal Geometry of Nature*. W. H. Freeman and Co., New York, 1977.

[31] George E. Martin. *The Foundations of Geometry and the Non-Euclidean Plane*. Springer-Verlag, New York, 1975.

[32] Edwin E. Moise. *Elementary Geometry from an Advanced Standpoint*. Addison-Wesley, Reading, Massachusetts, second edition, 1974.

[33] Jackie Nieder, Tom Davis, and Mason Woo. *OpenGL Programming Guide*. Addison-Wesley, Reading, Massachusetts, 1993.

[34] Seymour Papert. *Mindstorms: Children, Computers, and Powerful Ideas*. Basic Books Inc., New York, 1980.

[35] G. Polya. *How to Solve It: A New Aspect of Mathematical Method*. Princeton University Press, Princeton, New Jersey, second edition, 1957.

[36] Helen Whitson Rose. *Quick-and-Easy Strip Quilting*. Dover Publications, Inc., New York, 1989.

[37] Doris Schattschneider. *M. C. Escher, Visions of Symmetry*. W. H. Freeman and Co., New York, 1990.

[38] Thomas Q. Sibley. *The Geometric Viewpoint*. Addison-Wesley, Reading, Massachusetts, 1998.

[39] University of St. Andrews School of Mathematics. The MacTutor History of Mathematics Archive. Web page, 2002. `http://www-gap.dcs.st-and.ac.uk/~history/`.

[40] Eric W. Weisstein. Desargues, Girard (1593–1662). Web page, 2007. `http://scienceworld.wolfram.com/biography/Desargues.html`.

[41] Hermann Weyl. *Symmetry*. Princeton University Press, Princeton, New Jersey, 1952.

[42] C. R. Wiley. *Introduction to Projective Geometry*. Dover Publications, Inc., Mineola, New York, 1970.

[43] Harold E. Wolfe. *Non-Euclidean Geometry*. Henry Holt and Co., New York, 1945.

Index